Databricks Data Intelligence Platform

Powering the Agentic Era

Second Edition

Jason Yip
Nikhil Gupta
Marcin Wojtyczka

Databricks Data Intelligence Platform: Powering the Agentic Era, Second Edition

Jason Yip
Redmond, WA, USA

Nikhil Gupta
Livingston, NJ, USA

Marcin Wojtyczka
Kappeln, Germany

ISBN-13 (pbk): 979-8-8688-2523-1   ISBN-13 (electronic): 979-8-8688-2524-8
https://doi.org/10.1007/979-8-8688-2524-8

2nd edition of book previously published by Springer Nature

This book is an open access publication.

Managing Director, Apress Media LLC: Welmoed Spahr
Acquisitions Editor: Shaul Elson
Development Editor: Laura Berendson
Editorial Assistant: Gryffin Winkller

Cover designed by eStudioCalamar

Distributed to the book trade worldwide by Springer Science+Business Media New York, 1 New York Plaza, New York, NY 10004. Phone 1-800-SPRINGER, fax (201) 348-4505, e-mail orders-ny@springer-sbm.com, or visit www.springeronline.com. Apress Media, LLC is a Delaware LLC and the sole member (owner) is Springer Science + Business Media Finance Inc (SSBM Finance Inc). SSBM Finance Inc is a **Delaware** corporation.

For information on translations, please e-mail booktranslations@springernature.com; for reprint, paperback, or audio rights, please e-mail bookpermissions@springernature.com.

Apress titles may be purchased in bulk for academic, corporate, or promotional use. eBook versions and licenses are also available for most titles. For more information, reference our Print and eBook Bulk Sales web page at http://www.apress.com/bulk-sales.

Any source code or other supplementary material referenced by the author in this book is available to readers on GitHub. For more detailed information, please visit https://www.apress.com/gp/services/source-code.

If disposing of this product, please recycle the paper

To the Yip family: Toby, Jasmine, Jayne, and Cherry

Table of Contents

About the Authors

Jason Yip is a Databricks Most Valued Professional, recognized by Databricks for his exceptional technical expertise and commitment to the data and AI community. He serves on multiple advisory boards at Databricks, including the Partner Product Advisory Board and the Solution Architect Champion Advisory Board. He currently serves as Director of Data and AI at Tredence, a leading data science and analytics company, and is a Databricks Gold Partner. He advises Fortune 500 companies on implementing data and Generative AI strategies in the cloud. He is a top voice on Databricks and a former Microsoft employee who successfully led the Microsoft Corporate Finance big data transformation using Databricks.

Nikhil Gupta is a seasoned data professional with over 20 years of experience in big data technologies, driving innovation and strategic growth in the field. Nikhil serves as a Solution Architect at Stripe and held the same role at Databricks. Nikhil leverages his expertise to help customers across various industries, including retail, CPG, financial services, banking, and manufacturing, modernize their data and AI implementations on the Databricks platform. His expertise spans a range of big data technologies, including data warehousing, data lakes, and real-time data processing, making him a trusted advisor for Fortune 500 companies.

Marcin Wojtyczka is a Practice Lead Resident Solutions Architect at Databricks and the creator of the open-source **DQX** data quality framework. A former software engineer and data architect, he has spent years building large-scale data platforms and ML/AI systems. Today, he builds reusable libraries and frameworks that enhance and complement the Databricks platform, powering the next generation of data intelligence. Beyond the code, Marcin is a frequent speaker, an active open-source contributor, a product builder, and an expedition sailor.

About the Technical Reviewer

Unmesh Kulkarni is SVP (CTO of AI) at Tredence, a specialist AI and data services provider, where he leads AI technology strategy, ecosystem partnerships, and the solutions architecture team for Agentic and Generative AI.

He brings deep AI expertise for bridging the gap between AI research and enterprise deployments and spearheads the design and delivery of Agentic AI solutions for Tredence's global enterprise clients. Prior to Tredence, Unmesh worked as VP of Product, GTM, and Solutions Architecture at multiple companies in the AI and Data space. With over 15 years of experience across software engineering, product strategy, and AI solutions design, Unmesh focuses on the last mile of AI adoption to unlock enterprise value.

In his current role, he collaborates closely with Databricks AI teams as well as other Hyperscaler platforms for building next-generation Agentic AI and ML systems.

Unmesh holds an MBA from Wharton and an MS in Computer Science from IIT Kanpur.

Praise for *Databricks Data Intelligence Platform*

In this book, Yip provides a great (and practical) technical blueprint for bringing Lakehouse foundations to the GenAI and Agentic era. A lot has happened since his last book, from Agent Bricks to the new Lakebase/OLTP architecture, and he provides a comprehensive resource with the practical focus of someone who works on the platform daily, doing the real, hard work.

—David Meyer, SVP of Products, Databricks

This book is super-exciting! It is an impressively comprehensive resource for both beginners looking to learn the platform and practitioners wanting to understand everything under the covers of Databricks. Jason Yip does a great job going through all the key features: Lakebase, Agent Bricks, composable agents, UC and governance, Lakeflow, and Data Warehousing with DBSQL. I especially found it helpful because of the abundance of inside-the-product screenshots, architecture diagrams, and code examples. This is a must-read for those looking to begin or deepen their Databricks journey!

—Ari Kaplan, Head of Technical Evangelism, Databricks

Jason has done an excellent job of walking the readers through the breadth and depth of the Databricks Data Intelligence platform, including Agent Bricks, Lakebase, Databricks Apps, Unity Catalog, Genie, Lakeflow, Databricks SQL, and MLflow 3. Building on his first book, this is a powerful resource for data and AI practitioners driving transformation and innovation with AI agents built on enterprise data.

—Amit Singh, Global Head of Partner GTM - AI, Lakebase and Genie, Databricks

The Lakehouse paradigm championed by Databricks is changing the way the industry thinks about data. The industry is moving to an Extract Transform Catalog pattern where the data never leaves the lake and is accessible to all who need it. This book gives you insight into how and why a governance-first approach is key to the new data landscape many enterprises are building.

—Robert Thompson, MTS Solutions Architecture, T-Mobile; Databricks MVP

This book clearly captures Databricks evolution from a Lakehouse to a true Data Intelligence Platform. What stands out is its deep, practical treatment of agentic AI, showing not just what agents are but how to operationalize them with evaluation, governance, cost controls, and real-time intelligence. It is a pragmatic blueprint for architects and engineers looking to move AI from experimentation to production at enterprise scale.

—Srivathsan RL, Global Databricks Solution Lead, Cognizant; Databricks AI MVP

This is the definitive guide for organizations looking to operationalize Generative AI. It offers an excellent balance of hands-on implementation and strategic governance, demonstrating exactly how to leverage Databricks data pipelines to transform GenAI concepts into secure, scalable, and high-value real-world solutions.

—Nilton Ueda, Executive Senior Manager, Deloitte; Databricks MVP

The book is an essential read for both newcomers and experienced professionals. It captures nuanced insights that are often missing from standard documentation, enabling readers to develop a broader perspective on its usage and the value it can bring to Databricks projects. Each chapter is thoughtfully structured and clearly articulated, reflecting a high level of expertise. Additionally, the book comprehensively covers the latest advancements and best practices available today.

—Maulik Dixit, VP of Data Engineering, Tredence; Databricks MVP

Databricks Data Intelligence Platform has redefined the modern data stack by unifying Generative AI with the reliable foundation of Delta Lake and the governance of Unity Catalog. The authors expertly deconstruct complex patterns, from serverless GPU orchestration to real-time stateful AI with Lakebase, providing the practical examples needed to build secure, scalable applications. It is the master manual for the next generation of AI engineering.

—Scott Davis, Head of Data and AI, Lumenalta; Databricks MVP

This book is a solid and comprehensive overview of key Databricks components, providing a good foundation of knowledge for both well-known and lesser-known (but equally important) capabilities.

—Josue Bogran, VP of Data + AI Architecture, zeb; Databricks MVP

Moving far beyond the AI hype, this book transforms beginners into seasoned AI practitioners on Databricks. It provides the rare bridge between how to build and why it works, offering clarity that sticks.

—Bartosz Konieczny, Databricks MVP

This is a wonderful book! It felt like someone thoughtfully connecting the dots across the Databricks platform. What stood out to me was how naturally it tied together the platform foundations, governance, SQL, and the newer AI direction, all without losing the practical angle. My personal favorite part was the Agent Bricks coverage, because it made a fast-moving space feel much more understandable and actionable. Highly recommended!!

—Shekhar Shukla, CEO and Founder, EZ; Databricks MVP; MLflow Ambassador

A really strong, engineering-focused book for building modern AI systems on the Databricks Data Intelligence Platform. What I really liked is that it goes beyond explaining how things work and shows how to design, evaluate, and improve them in production.

—Jaco van Gelder, Co-founder, Dtyped; Databricks MVP

Excellent book if you need to prepare for passing an interview about Databricks or before taking Databricks exams.

—Hubert Dudek, Databricks MVP

Finding a resource that covers the full breadth of the Databricks platform while making it visual and accessible is no easy task. This book nails it. A valuable read for anyone looking to get a solid, practical understanding of the platform.

—Maria Vechtomova, Co-founder at Cauchy, Databricks MVP, MLflow Ambassador

This book captured the Databricks Data intelligence. Platform concepts in great details, a good book for beginners and advanced practitioners of Databricks.

—Rajaniesh Kaushikk, Director Technology, Virtusa Corporation,
Databricks MVP, Microsoft MVP

Tredence and Databricks: Advancing Enterprise Data and AI

As intelligent systems move from experimentation to enterprise-scale deployment, modern data platforms have become the backbone of business operations. The partnership between Tredence and Databricks focuses on helping organizations build the architectures required to scale data intelligence. By combining Tredence's domain expertise in applied analytics with the unified architecture of the Databricks Data Intelligence Platform, the collaboration enables organizations to modernize legacy data environments and operationalize advanced analytics and AI.

1. Platform Development and Capabilities

As a Databricks Gold Partner and member of the Databricks Partner Product Advisory Board, Tredence provides practitioner insights drawn from large-scale enterprise transformations. This allows them to contribute strategic perspectives on platform capabilities and architectural evolution.

Tredence's Databricks practice is supported by substantial technical scale, including

- 4,200+ data scientists and engineers

- 950+ Databricks-certified practitioners

- 20+ Databricks Champions and a select group of Databricks MVPs

- 35+ proprietary Databricks AI/ML accelerators designed to rapidly operationalize use cases across industries

2. Industry Application and Recognition

Across sectors like retail, consumer goods, and financial services, enterprises are shifting to real-time, intelligence-driven operations. Tredence helps architect enterprise data platforms that unify machine learning, generative AI, and analytics within a single environment.

The company holds Databricks Brickbuilder status across multiple industries, recognizing its development of repeatable, high-impact solutions. Ecosystem honors include the Databricks Retail and Consumer Goods Partner of the Year for four consecutive years and the 2025 Growth Partner of the Year (Americas).

3. Foundations for the Agentic Enterprise

The next phase of enterprise AI involves agentic systems—architectures capable of autonomous reasoning, real-time adaptation, and continuous optimization. Realizing this requires a unified platform that integrates large-scale data engineering, machine learning lifecycle management, and real-time analytics.

Through their collaboration, Tredence and Databricks provide the technological infrastructure necessary to transition organizations from AI experimentation to operationalizing autonomous, self-improving decision systems at scale.

Introduction

1. Navigating the Data Intelligence Architecture

The transition from passive data storage to active, reasoning AI systems requires a fundamental shift in architecture. The 12 canonical components of an end-to-end AI system diagram presented in Databricks' AI Security Framework outline this modern blueprint—a framework connecting raw data directly to real-world inference.

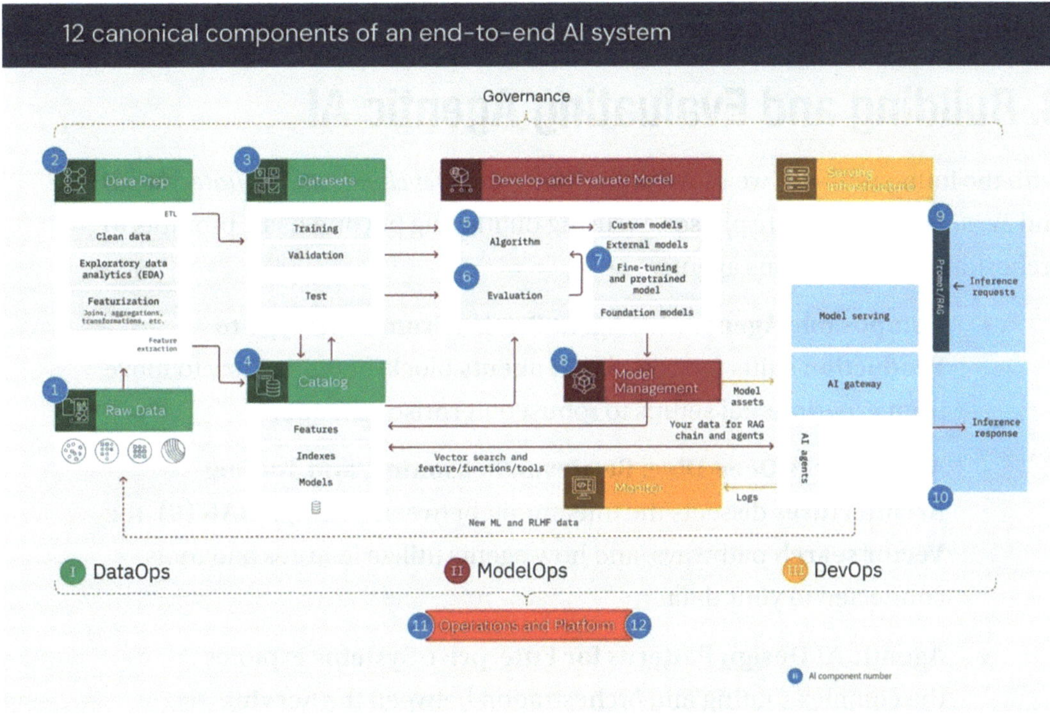

This book is structured as a masterclass curriculum, designed to teach you not just the theory but the practical execution of this architecture. It serves as your step-by-step guide to mastering the Databricks platform, giving you the blueprints to build intelligent, governed, and highly scalable products.

Here is how our journey unfolds across the platform.

2. The Foundation and the Future

We begin by setting the stage for the entire ecosystem, exploring the shift toward a unified, AI-driven architecture.

- **Databricks Platform: From Lakehouse to Data Intelligence** frames the entire architectural blueprint, mapping the journey across all three foundational pillars: **DataOps (I), ModelOps (II)**, and **DevOps (III).**

- **Generative AI on Databricks: Foundations, Capabilities, and Use Cases** introduces the core technologies bridging the Develop and Evaluate Model and Serving Infrastructure domains.

3. Building and Evaluating Agentic AI

With the foundation set, we move directly into the *Develop and Evaluate Model* and *Serving Infrastructure* phases, focusing on moving beyond basic prompts to orchestrating autonomous agents.

- **Composable Agents with Agent Bricks: From Prototype to Production** shifts focus to the **AI agents** block, detailing how to move from experimental setups to robust enterprise execution.

- **Agent Bricks Deep Dive: Retrieval, Reasoning, and Tooling Architectures** dissects the integration between **Prompt/RAG (9)**, the **Vector search** pathways, and how agents utilize features and tools connected to your data.

- **Agentic AI Design Patterns for Enterprise Systems** explores the complex routing and orchestration between the **Serving Infrastructure** and your enterprise **Catalog (4)**.

- **Databricks As an Agentic Platform: Orchestration, Context, and Control** maps to the overarching **Governance** umbrella and the **Operations and Platform (IV)** layers that keep these autonomous systems compliant and well-managed.

- **Quality Tuning and Evaluation with Agent Bricks** corresponds directly to the **Evaluation (6)** and **Fine-tuning (7)** steps, highlighting the critical **New ML and RLHF data** feedback loop to improve model performance.

4. Secure, Scalable Infrastructure

Advanced agents require bulletproof compute, state management, and operational security to survive in production environments.

- **Lakebase: The OLTP Engine for Intelligent Applications** covers the high-concurrency storage underpinning the operational **Datasets (3)** and the indexes within the **Catalog (4)**.

- **Building Security-First, Production-Grade Databricks Apps** formalizes your defense posture and data protection strategies across the entire lifecycle, ensuring your AI systems are production-ready.

- **AI Runtime: Elastic Compute for AI Workloads** details the unseen engine powering both **Algorithm (5)** training and real-time **Model serving**.

- **MLflow 3 and the GenAI Agents** maps directly to **Model Management (8)**, ensuring structured tracking, versioning, and deployment of your model assets.

5. Data Engineering and Governance

Intelligence is only as reliable as the data feeding it. Here, we cover the robust engineering required to manage *Raw Data*, *Data Prep*, and *Datasets*.

- **Real-Time Intelligence with Spark Structured Streaming** covers the low-latency pipelines feeding the **Raw Data (1)** layer from live sources.

- **Lakeflow Connect: Data Ingestion for the Lakehouse** maps to the initial batch and streaming ingestion phase, bringing external sources reliably into **Raw Data (1)**.

- **Open Data Governance with Unity Catalog** explores the **Catalog (4)** component, the central nervous system for managing features, indexes, and models under strict enterprise governance.

- **Delta Lake: The Foundation of Reliable Data and AI** forms the resilient storage bedrock for components 1 through 3 (**Raw Data, Data Prep, Datasets**).

- **Lakeflow Declarative Pipelines: Managing Data and AI Workflows** corresponds to the **ETL** processing block within **Data Prep (2)**, covering data cleaning, exploratory data analytics (EDA), and featurization.

- **Data Warehousing with Databricks SQL** highlights how to serve high-performance analytics directly from your curated **Datasets (3)**.

6. Operations and Real-World Execution

Closing out the lifecycle, continuous oversight is mandatory for live AI systems to track inference, optimize performance, and manage costs.

- **Databricks Pricing and Observability Using System Tables** maps explicitly to the **Monitor** block, utilizing continuous **Logs** to ensure platform health and cost control (**Components 11 and 12**).

- **From Ideation to Creation: Building Intelligent Products with GenAI** brings the entire diagram together, tracing a unified, practical path from the initial **Raw Data (1)** to a final, production-ready **Inference response (10)**.

## CHAPTER 1

Databricks Platform: From Lakehouse to Data Intelligence

The intensifying pace of digital transformation has led companies to amass increasing volumes of diverse data from various sources. This data explosion carries enormous potential for organizations to uncover transformative insights to guide innovation and decision-making through advanced analytics.

In this chapter, we will examine the evolution of data platforms over the last decade or so. Then, we will discuss why today's ideal data platform is a lakehouse and how Databricks established the lakehouse category. We will then go in depth to understand the various facets of the lakehouse platform as it is implemented on Databricks.

Finally, we will discuss how generative AI (GenAI) and large language models (LLMs) have revolutionized the entire artificial intelligence (AI) landscape and how Databricks has embraced this technology to create the Databricks Data Intelligence Platform.

Data Platforms: Historical Perspective

The data landscape has undergone rapid evolution in recent years, necessitated by the exponential growth in information from an ever-expanding variety and volume of data. As organizations deal with this surge in big data, existing infrastructure has struggled to harness its potential effectively. This has led architects and technology leaders to start conceptualizing new integrated systems that can adeptly consolidate the strengths of current data platforms.

© Jason Yip, Nikhil Gupta and Marcin Wojtyczka 2026

J. Yip et al., *Databricks Data Intelligence Platform*, https://doi.org/10.1007/979-8-8688-2524-8_1

Let's start with data warehouses. They provided immense value for decades in descriptive analytics and business intelligence use cases that rely on predefined structured data. However, as the focus and needs expanded to include predictive analytics and the latest machine learning advancements, the nature of workloads moved beyond what traditional warehouses could support effectively. Descriptive analytics for business intelligence based on predefined datasets are no longer enough. Further varied data types, such as unstructured, semi-structured, and streaming use cases, require more extensive and agile processing than data warehouse infrastructures are designed for.

The data lake concept, therefore, gained interest as an alternative to data warehouses, given its natural ability to ingest raw multi-structured data quickly. One of the more popular technologies that was at the forefront of this was Hadoop and its ecosystem. However, lack of transnationality, data quality, and mixed processing modes inhibited unlocking the benefits promised by data lakes. The flexibility came at the cost of governance, reliability, and vital enterprise capabilities. Consequently, the data lakes quickly turned into "data swamps."

Despite all these drawbacks, organizations with no better alternatives began using both these technologies in their data architecture: data warehouses for descriptive and business intelligence (BI) use cases and data lakes for AI/machine learning (ML) use cases, with a variety of processing tools thrown in the mix (sometimes even a single tool for one use case). However, with two completely different systems, solving for two critical types of workloads became problematic. First, it created data silos, which necessitated moving data across the platforms and thus maintaining multiple copies of the same data. Second, the governance models of these disparate platforms were incompatible, requiring separate governance for each system. Finally, organizations started using different tools for BI and ML workloads, reducing operational efficiency and increasing costs. Over time, the complexity of maintaining different systems increased. This was becoming not only costly but also a hindrance to innovation.

More than ever, enterprises needed a unified data infrastructure capable of managing diverse information seamlessly through its entire lifecycle to serve exponentially expanding analytical use cases.

Emergence of the Lakehouse

Let's understand how organizations look at their modern data platforms. First, the platform should be able to store all sorts of data in a single storage location, preferably cost-effectively. Then, that data should have a single governance and access model and, last, a technology that helps them solve all their use cases without moving any data or code.

However, as discussed earlier, organizations using both a data warehouse and a data lake in their architecture are essentially looking at two different piecemeal systems, leading to disconnected data silos, complex integrations, and fragmented governance, severely hampering the building of enterprise-grade analytic solutions that could positively impact the business.

This reality has catalyzed the emergence of a new evolutionary paradigm pioneered by Databricks: the *lakehouse*. The lakehouse architecture aims to bring together the most impactful capabilities of data warehouses and data lakes into an integrated whole on the cloud. Reliable support for varied workloads using consistent data, managed securely under standard governance policies, holds the promise to finally harness big data comprehensively.

With its seismic potential to reshape the analytics landscape, the lakehouse undoubtedly constitutes one of the most pivotal recent data platform innovations.

What Is a Lakehouse?

Let's dig a bit deeper and understand what a Lakehouse is. A *Lakehouse* is a data architecture paradigm that aims to bridge the gap between data lakes and data warehouses. The goal is to provide the flexibility and scalability of a data lake, as well as the performance, reliability, and governance typically associated with a data warehouse. A Lakehouse seeks to implement some of the managed data capabilities seen in warehouses directly on top of object stores. Figure 1-1 compares the three.

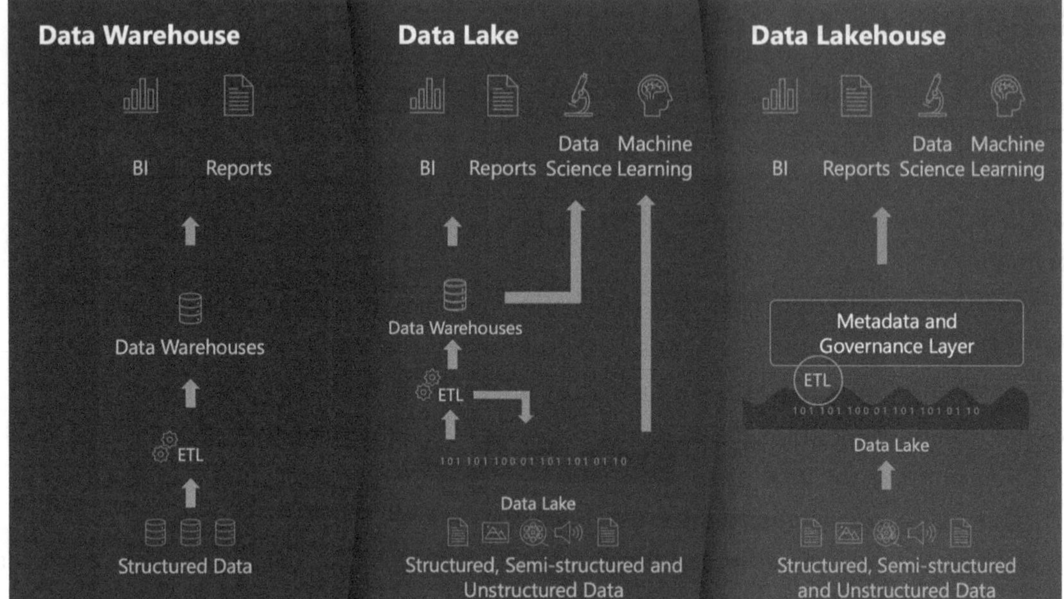

Figure 1-1. *Data warehouse versus data lake versus Data Lakehouse*

The Data Lakehouse construct addresses these gaps by consolidating the capabilities of data warehouses and data lakes:

- Natively manages both structured and varied unstructured data

- Leverages cloud-scale object storage as the foundational data repository

- Provides reliability, security, and governance across storage and processing

- Provides high performance through technologies such as caching, indexing, and partitioning

- Supports real-time and batch workloads via unified streaming architecture

- Provides open extensibility to accommodate rapidly evolving analytics needs

The Lakehouse breaks down data silos and simplifies management by consolidating workloads on a single platform under standard governance policies. This makes it possible to get a single view of information at scale to power advanced analytics. With cloud infrastructure adding unlimited elasticity, lakehouses finally make it feasible to ask bigger questions of data than ever before.

For example, by combining sales, marketing, and finance data, organizations can perform predictive analysis on their business, like which customer segment to target would give the most return for the business. That's just the beginning. Figure 1-2 presents a sample architecture of the use cases that we can achieve with this consolidation.

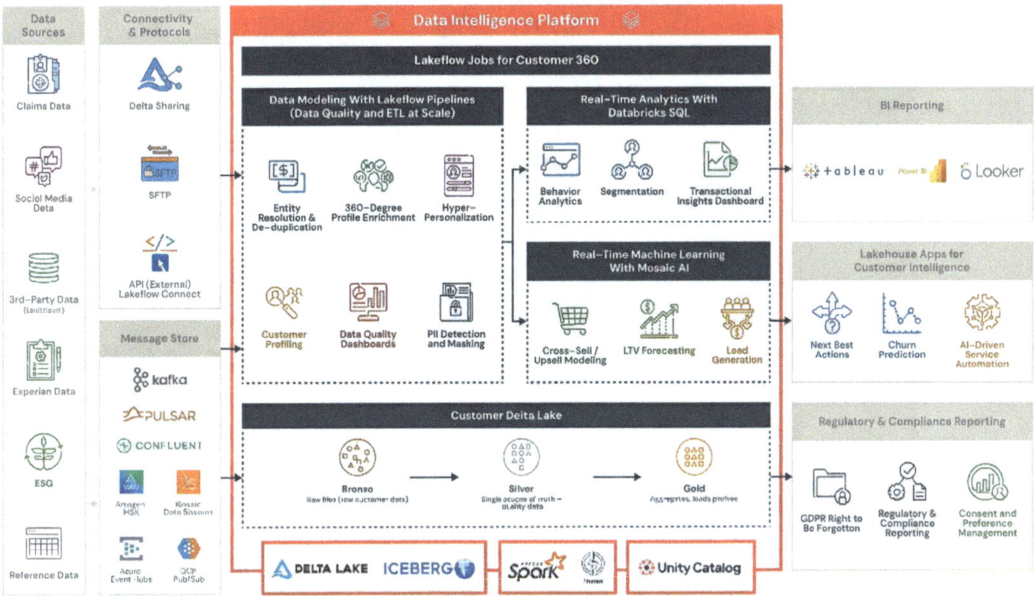

Figure 1-2. *Sample architecture for Customer 360*

If you were to design a new-generation analytical data management system using cheap distributed storage as a foundation, you would end up with something resembling a Lakehouse: flexible schemas but faster queries. The goal is real-time insights without compromising governance. Throughout this book, we will learn how to achieve the business outcomes that were previously not possible without the lakehouse.

What Is the Databricks Lakehouse?

Now that you understand the Lakehouse paradigm, let's move on to see how a
Lakehouse is implemented on Databricks. Published in the Conference on Innovative
Data Systems Research (CIDR) in 2021, Databricks researchers Michael Armbrust et al.
wrote "Lakehouse: A New Generation of Open Platforms that Unify Data Warehousing
and Advanced Analytics" (`https://www.cidrdb.org/cidr2021/papers/cidr2021_`
`paper17.pdf`).

Figure 1-3 shows the Databricks lakehouse platform.

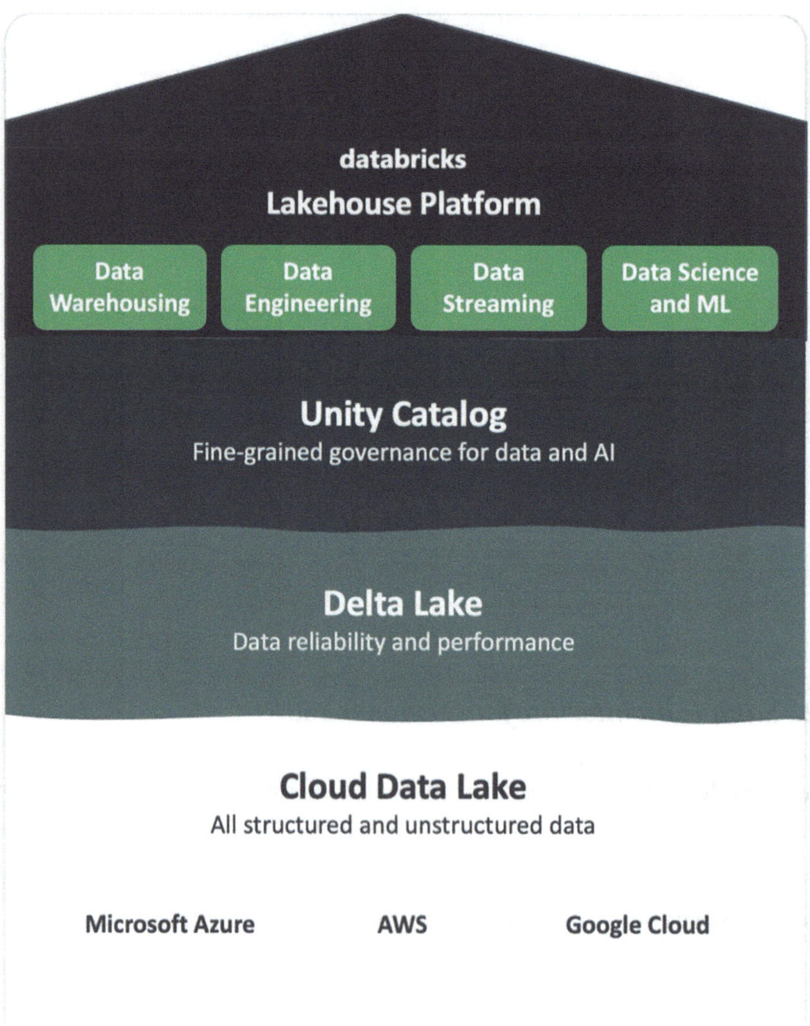

Figure 1-3. *Databricks lakehouse platform*

Databricks, with its Lakehouse architecture, offers a potential solution to consolidate disparate data sources into a single location while avoiding some limitations of existing architectures. Databricks provides one of the most mature enterprise-scale implementations of a Lakehouse architecture through its integrated data and AI platform. Built on open source and open standards, the Databricks Lakehouse architecture simplifies your data estate by eliminating the silos that historically complicate data and AI.

Let's decode this a bit and do a deep dive into a Databricks Lakehouse. Databricks leverages cloud object storage (AWS S3, Azure ADLS, and Google Cloud Storage [GCP]) as its foundational data store. This enables enormous volumes of structured, unstructured, and semi-structured data to be housed in native formats in one of the cheapest storage options available on the cloud. This is what constitutes the "lake" in the lakehouse. Once the data lands in the cloud in raw format, it is moved to Delta or Iceberg format. Please note that data is still in your cloud storage but in Delta or Iceberg format. Delta Lake is an open-source storage layer that brings performance, reliability, and governance to data lakes. Delta Lake applies ACID transactions, schema enforcement, and time travel to make large-scale storage reliable and performant for mission-critical workloads. In short, Delta Lake brings the reliability, performance, and management features of a data warehouse to data stored in your cloud storage. This constitutes the "house" in the Lakehouse architecture.

As shown in Figure 1-2, Unity Catalog provides unified data governance for all data within the Lakehouse. It centrally manages all data assets, including tables, schemas, views, and AI models.

Finally, the Databricks platform provides features that enable all data personas within your organization to build a variety of use cases, including data engineering, data science, streaming, and data warehousing.

To conclude, Databricks provides a unified lakehouse platform built on open-source technologies that supports all three major clouds, aka AWS, Azure, and GCP, and is able to handle diverse use cases at any scale. This platform makes data available for multiple analytics use cases, from business intelligence to machine learning.

Key Features of the Databricks Lakehouse Platform

The Databricks Data Intelligence Platform provides a leading implementation of the Lakehouse paradigm. The following are some core concepts and capabilities:

- **Delta Lake**: This open-source storage layer optimizes the storage of massive volumes of structured, semi-structured, and unstructured data for reliability, performance, and governance.

- **Unified Batch and Streaming**: Databricks processes batch and real-time data on the same platform using Spark Structured Streaming. This enables new ways to combine historical data with streaming data.

- **Unity Catalog**: Unity Catalog captures metadata and usage information across diverse data types and storage systems for unified discovery and governance.

- **Multilanguage Support**: The platform natively integrates languages like SQL, Python, R, Java, and Scala to support various analytics use cases on the same data.

- **Cloud-native Architecture**: By leveraging managed cloud infrastructure, Databricks automates resource management and scaling to meet the needs of the most demanding workloads.

- **Secure, Governed Access**: Comprehensive access controls, encryption, and data masking ensure strict oversight and granular auditing.

- **Autoscaling and Collaboration**: Data teams can quickly scale their work to production while collaborating closely with business users by sharing dashboards, reports, and applications.

Introducing the Databricks Data Intelligence Platform

Looking back at the technology world, 2023 was a groundbreaking year. It was when the world saw the power of GenAI LLMs and their potential. Almost instantly, organizations could envision future use cases that could be built by leveraging them. GenAI became the talk of every boardroom, and everyone was looking to use the technology to take the lead over their competitors.

Databricks, with its Lakehouse platform, is uniquely positioned to leverage this technology not only to enhance its platform but also to help enterprises build their GenAI use cases. Let's talk about these two in detail.

First, Databricks enhanced its lakehouse platform by integrating GenAI capabilities, creating the Databricks data intelligence platform. Databricks uses LLMs in almost every part of its platform, from helping developers troubleshoot coding errors to automatically generating insights from your data. We will discuss each of these features in detail in Chapter 12. The overall platform became more and more intelligent and thus enhanced the user experience.

Second, Databricks has built capabilities and features within the platform that enable organizations to develop their own GenAI use cases. Features like vector search, the fine-tuning API, and Agent Bricks enable organizations to productionalize their GenAI use cases from RAG applications to even building their own model from scratch. We will discuss these features in detail throughout this book.

Thus, Databricks enhanced its platform with LLMs, enabling users to build GenAI applications on it.

To understand the Databricks data intelligence platform, let's use this analogy. Figure 1-4 shows the most powerful spaceship ever built—SpaceX's Starship. It is important to note that at the core, sitting underneath, is the Super Heavy booster, which can withstand 2.8 million pounds of weight while standing and, when in flight, propels the second stage to space with its Raptor engine.

Figure 1-4. *SpaceX spaceship. Used under CC BY-SA 4.0 (`https:// creativecommons.org/licenses/by-sa/4.0/`). `https://commons.wikimedia. org/wiki/File:Full_Stack_starship.jpg` (This image is not included in the book's Creative Commons license.)*

With this concept in mind, it is not hard to understand the Databricks data intelligence platform (see Figure 1-5). It also comprises two major components. At its core, it is powered by the Lakehouse platform and GenAI, making it much more intelligent in meeting user needs and requirements.

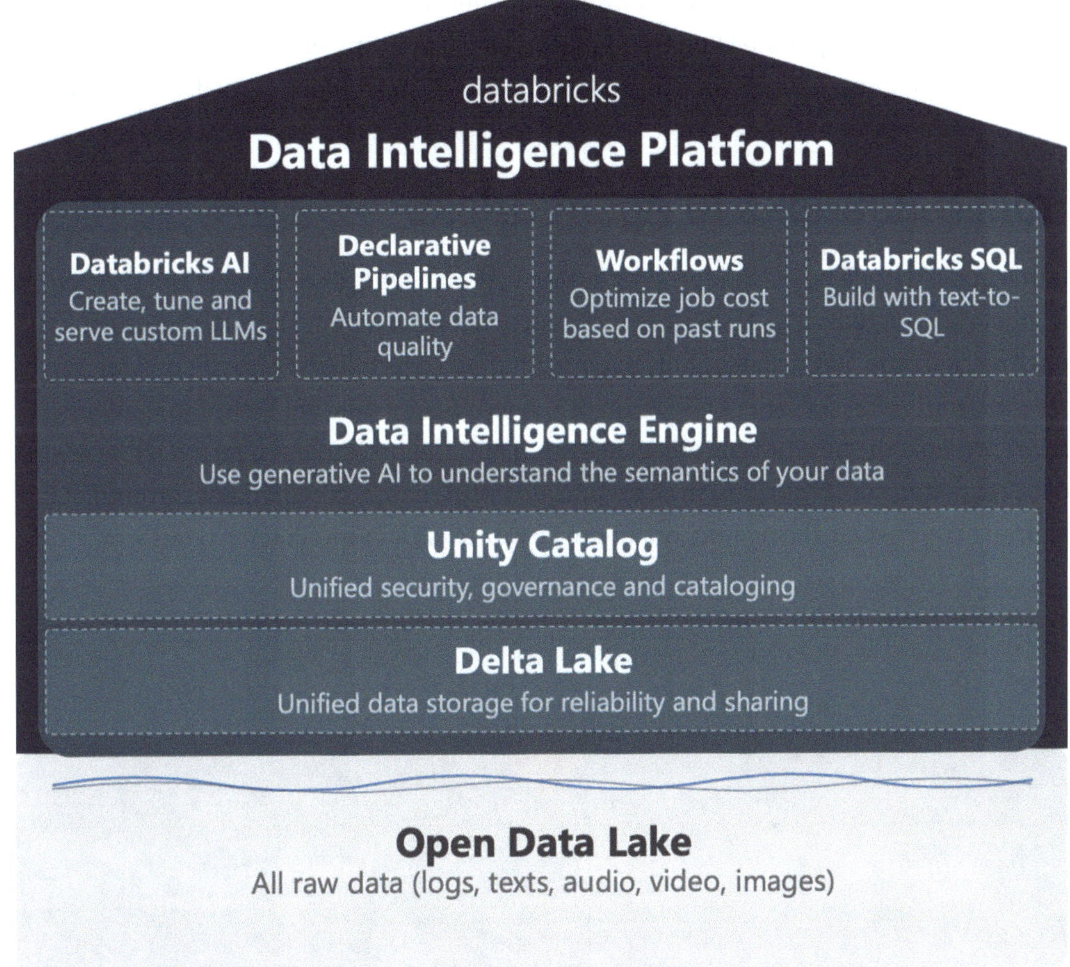

Figure 1-5. *Databricks data intelligence platform*

To conclude, fueled by the latest development of generative AI, Databricks has integrated the Data Intelligence Engine into the core of its offering. This is equivalent to SpaceX's Starship, with the Lakehouse being the booster and the GenAI intelligence being the rocket ship. In short, Databricks has leveraged the latest GenAI models and technology to create the Data Intelligence Engine, which fuels all parts of the platform.

With MosaicML and Databricks Data Intelligence, developers can seamlessly create their workloads as if they were working with a data subject-matter expert (SME) like never before. Databricks AI also allows data scientists to leverage large language models as-is, refreshing their domain-specific knowledge with RAG, fine-tuning with more specialized knowledge, or even training a brand-new LLM from scratch. This powerful second stage can propel the Databricks platform to a new era, enabling organizations to create the next generation of data and AI applications with quality, speed, and agility.

From Databricks to Agent Bricks

In this book, we will have in-depth discussions about agents and the new Databricks product called Agent Bricks. Databricks has evolved from a data engineering and machine learning platform to a fully agentic platform that allows us to not only deploy custom agents with serverless GPUs, but the Agent Bricks offering is a unique offering that represents the best of Databricks Research and their relentless effort to allow users to leverage state-of-the-art research techniques without any research background. Figure 1-6 represents the agentic era of the Databricks Data Intelligence Platform.

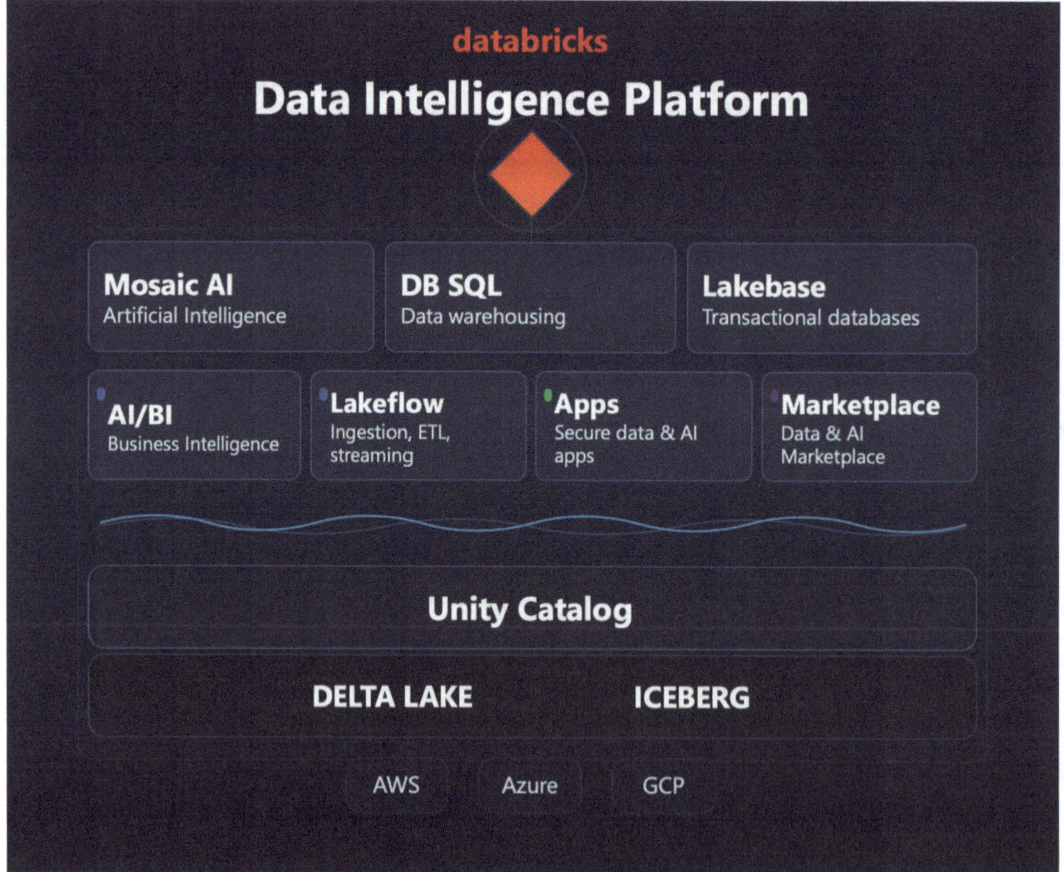

Figure 1-6. *Databricks Data Intelligence Platform in 2025*

Imagine an agent as a software component that can help you accomplish a task. Most of these components are usually small, like a weather gadget—and yes, the term "gadget" used to be all over the place on your desktop, and now they are "agents." When you have a platform like Databricks, it is no longer suitable to rely on a single giant platform to handle all tasks; it needs to be agile enough to work across different scenarios, not just for AI. In other words, by leveraging Databricks SQL as an MCP agentic component, we can make DBSQL part of our agentic workflow for building enterprise data marts.

We will cover every "brick" in Figure 1-6 in depth throughout this book. At a high level, from left to right, top to bottom, we have the following bricks:

- Mosaic AI is the foundation of everything intelligence-related, from reducing the cost of model serving to allowing data scientists to fine-tune an LLM; it is also the platform that powers Agent Bricks

- Databricks SQL is the data warehouse that keeps enhancing its execution, which can now be used in apps for batch AI inference

- Lakebase is the new fully managed and cloud-native transactional database within Databricks for building highly scalable apps with serverless PostgreSQL

- AI/BI is the intelligent dashboard that allows you to build visualizations using natural language right where your data is

- Lakeflow is an AI-enabled workflow suite that allows building a data pipeline in natural language or a declarative manner

- Apps are Python or JavaScript-enabled application platforms for developers to make AI applications live

- Marketplace is the Databricks "app" store where you can find anything from data to ML models

- Everything underneath is governed by Unity Catalog with the Delta and Iceberg formats now unified in one single format

Conclusion

In this chapter, we began by examining the evolution of data platforms. Data warehouses are excellent for BI use cases, while data lakes, with their open storage, are used for ML use cases. However, using both incompatible systems in their architecture created data silos and prevented them from using their full data for business decisions. The Databricks Lakehouse platform enables organizations to store all their data in one place—any data, be it structured, semi-structured, or unstructured data in an open data lake. Unity Catalog provides a single governance layer, and the Databricks platform offers features for use cases from data engineering, data warehousing, streaming, and data science. Finally, we discussed how Databricks built intelligence into its platform by using Mosaic AI to empower GenAI and LLMs to create the Databricks Data Intelligence Platform. Throughout this book, we will deep dive into various parts of the Databricks Platform.

## CHAPTER 2

Generative AI on Databricks: Foundations, Capabilities, and Use Cases

Ever since ChatGPT was released to the public, there has been no shortage of interest in chatbots or generative artificial intelligence (GenAI). But what exactly is GenAI, and how does Databricks come into the picture? And how can it help organizations deploy their own chatbot or develop their own GenAI applications? In this chapter, we will first learn the concepts around GenAI. We will then discuss how Databricks and MosaicML, now an integral part of Databricks, work together to transform the industry once more. This chapter provides background information on the journey of GenAI and introduces the Databricks offering in the GenAI space.

What Is Generative AI?

Generative AI uses several techniques that continue to evolve. Foremost are AI foundation models, which are trained on a broad set of unlabeled data that can be used for different tasks. However, all of them require additional fine-tuning, sophisticated algorithms, and enormous computing power throughout training. Nevertheless, they are, in essence, prediction algorithms.

Today, generative AI most commonly creates content in response to natural language requests—it doesn't require knowledge of, or the need to enter, code—but enterprise use cases are numerous and include innovations in drug and chip design and materials science development.

Figure 2-1 illustrates how generative AI and general artificial intelligence differ. At a high level, generative AI is a technique that generates new content by learning from large amounts of similar content. For example, when generating English-language content, it could have been trained on all Wikipedia text to start with. However, that's just an understatement. The resources required to train these models are not something every household would have access to, despite the number of models increasing by the day. As a result, similar to transfer learning, many data scientists would use so-called foundation models to enhance the AI with internal knowledge.

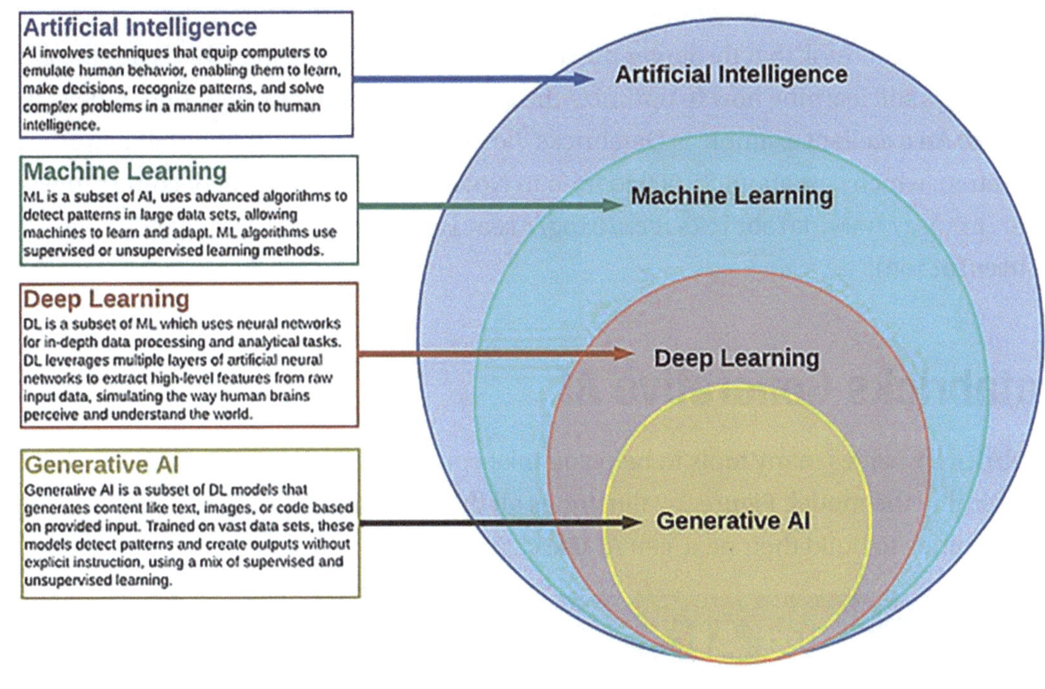

Figure 2-1. Differences between AI models' images by Dr. Lily Popova Zhuhadar, Unraveling AI Complexity—A Comparative View of AI, Machine Learning, Deep Learning, and Generative AI. Wikimedia Commons, `https://commons.wikimedia.org/wiki/File:Unraveling_AI_Complexity_-_A_Comparative_View_of_AI,_Machine_Learning,_Deep_Learning,_and_Generative_AI.png`. Licensed under CC BY-SA 4.0 (`https://creativecommons.org/licenses/by-sa/4.0/`). (This image is not included in the book's Creative Commons license.)

The enhancement process can be performed primarily using retrieval-augmented generation (RAG) or fine-tuning. There is a fundamental difference between the two: RAG optimizes the context provided to the model, whereas fine-tuning optimizes the model's parameters. We will discuss both these in greater detail later in the chapter.

Finally, if resources are available and the goal is to train a fully domain-specific model focused on mitigating bias, MosaicML's training platform will help you do so, albeit at a much lower cost. This process results in a custom large language model (LLM). ChatGPT is an application built on top of LLMs to serve as a chatbot, providing an intuitive interface for the general public. But then, on the other hand, in the case of GenAI, we might think that the larger the model, the better, but in fact, this is not the case. The world is still learning how to optimize the data for optimized throughput of the tasks required. An excellent example is Databricks' bespoke LLM for auto-documentation generation, which cost about $1,000 to train in November 2023 and will become much lower (`https://www.databricks.com/blog/creating-bespoke-llm-ai-generated-documentation`).

Databricks Generative AI

Databricks provides many tools to help you take control of model training, all the way to governing the model. Figure 2-2 illustrates all the capabilities Databricks provides for organizations to build their next-gen AI use cases.

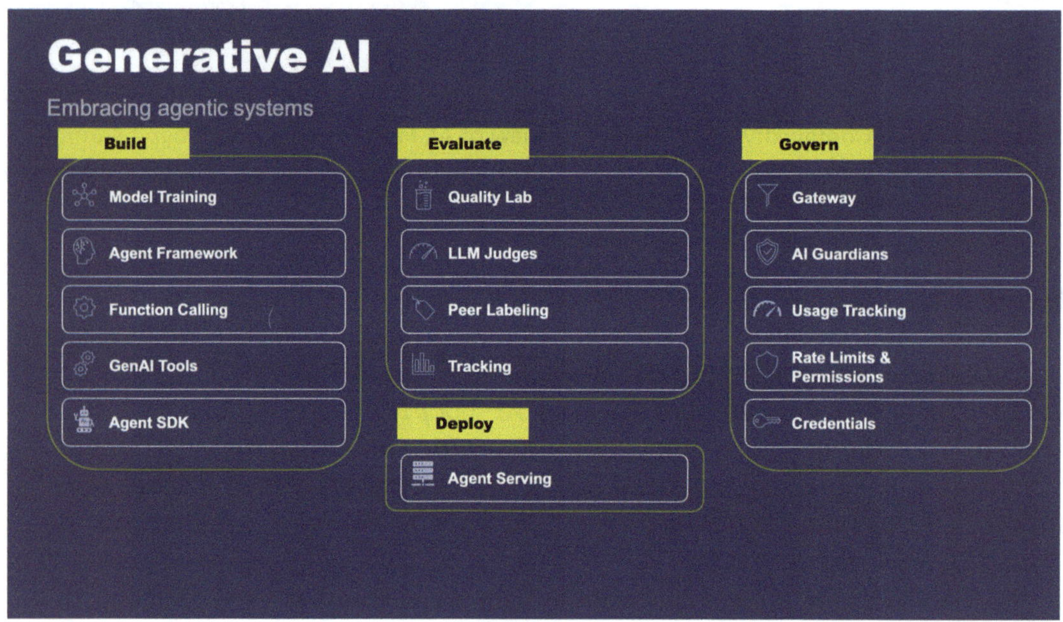

Figure 2-2. *Databricks' generative AI offerings*

In this chapter, we will examine the basic out-of-the-box features Databricks provides, and in subsequent chapters, we will discuss how to leverage advanced tools to enhance your GenAI offering.

While we will discuss the details in later chapters, here are the high-level functionalities of each stack:

- **Build**: This includes the Mosaic AI stack to refine an LLM either through RAG, fine-tuning, or pre-training.

- **Evaluate**: This is part of the AI Agent framework to allow evaluation with metrics as well as getting peer feedback.

- **Deploy**: There is a one-line command to deploy a non-production app for a user acceptance test.

- **Govern**: AI Gateway is an extension of MLflow's core tracking capability along with Unity Catalog to govern internal and external LLM APIs.

The GenAI Journey

With GenAI dominating the world now, many organizations start by allowing their employees to experiment with these models. To combine data within the organization, some with more budget want to train a model of their own from scratch. While others keep the model for use within their organization, some also decide to open it up to the world, making it a foundation model.

Figures 2-3 and 2-4 present two different views of this journey. Figure 2-3 represents the journey or maturity an organization can achieve with GenAI. From left to right: prompt engineering, retrieval-augmented generation, fine-tuning, and, lastly, pre-training.

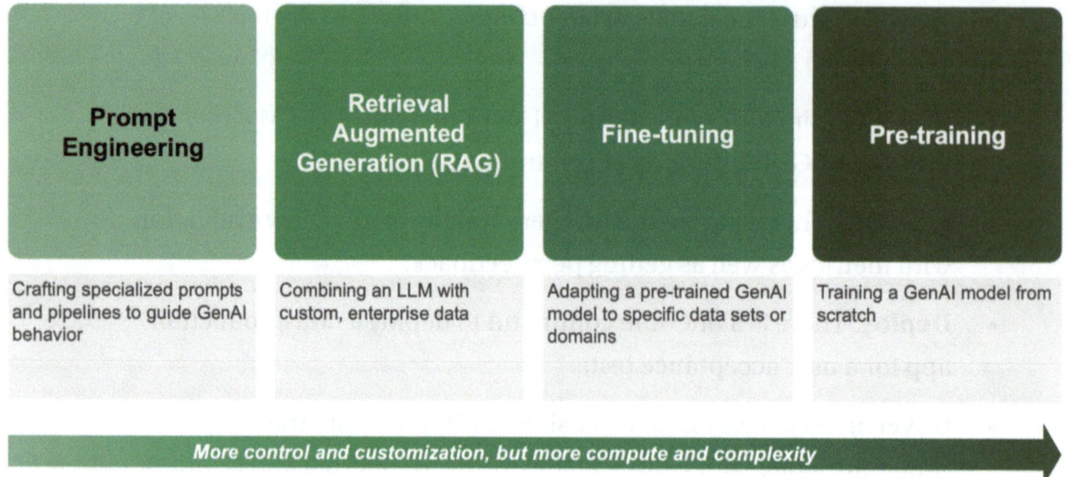

Figure 2-3. *The GenAI journey*

While these building blocks are available, not every organization will need to reach the last step of pre-training due to the computation, cost, and complexity, as well as the expert knowledge required to reach the next level. Figure 2-4 illustrates this idea in another way. Note that pre-training is the most time-consuming and complex process. We will walk you through the process, but before you move to the next step, it is best to consider the trade-off between time and cost, as well as whether experts are available to validate the results.

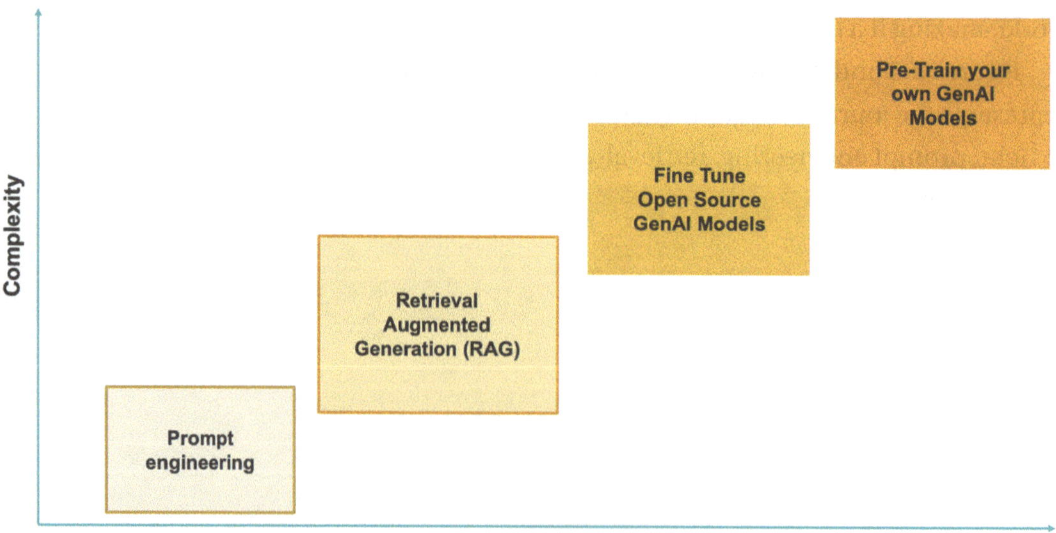

Figure 2-4. *An alternate aspect of the GenAI journey, with time and complexity involved*

Prompt Engineering

Prompt engineering is the art of asking the right questions to get the best output from a **large language model (LLM)** using plain-language prompts. It enables direct interaction with the LLM, allowing you to communicate with it using only natural language instructions. In the past, building these kinds of models within an organization required machine learning models that were typically trained with deep knowledge of datasets, statistics, and modeling techniques. However, today, LLMs, trained by frontier labs, can be "programmed" in English and other languages, making them more accessible to a broader audience.

Here are some key points about prompt engineering:

Best Practices for Prompting:

- **Clear Communication**: Clearly communicate what content or information is most important.

- **Structured Prompts**: Structure your prompts by starting with the role or context, followed by input data and then the instruction.

- **Varied Examples**: Use specific and varied examples to help the model narrow its focus and generate accurate results.

- **Constraints**: Use constraints to limit the scope of the model's output and prevent factual inaccuracies.

- **Break Down Complex Tasks**: Divide complex tasks into a sequence of simpler prompts.

- **Self-Evaluation**: Instruct the model to evaluate or check and revise its own responses before producing them.

- **Creativity**: Be creative! The more open-minded and creative you are, the better your results will be.

Quick Start on Prompts:

- **Direct Prompting (Zero-Shot)**: The simplest type of prompt that provides only an instruction without examples.

- **Example**: "Can you give me a list of ideas for blog posts for tourists visiting New York City for the first time?"

- **Role Prompting**: Assign a role to the model and ask it to understand your goals and objectives before designing a prompt.

- **Example**: "You are a mighty and powerful prompt-generating robot. Design a prompt for the best outcome based on the context and data provided."

- **Chain-of-Thought Prompting**: Break down complex tasks into a sequence of simpler prompts.

Remember, being a great prompt engineer doesn't require coding experience. Creativity and persistence will benefit you greatly on your journey in this evolving field of LLMs and prompt engineering.

AI Playground

Riding on the wave of generative AI, many corporations have released foundational large language models. However, without a background in programming, it is difficult to use these models. Databricks has optimized a few models and curated them in the "Playground" section for experimentation, providing a standard interface for interacting with them. It is also an interactive environment where users can simultaneously experiment and "chat" with various large language models and compare results.

Databricks has partnered with industry-leading partners and provided quick access through Playground. While the list will continue to evolve, Databricks will continue to host leading models on the platform.

The curated models include the following:

1.   OpenAI premium models, including the GPT 5 series

2.   OpenAI open-weight models: GPT OSS 120B and 20B

3.   Anthropic Claude series, up to the latest version

4.   Google Gemini series, including Gemini 3 and beyond

5.   Meta Llama series

Use Cases

Playground is not just for testing foundation models, but it also supports custom endpoints like Agent Bricks endpoint. From the next chapter onward, we will have deep-dive chapters on Agent Bricks. While it is convenient for testing, we can also apply tools and generate code from Playground, making the coding experience smoother.

Sentiment Analysis

You will be provided with a tweet, and your task is to classify its sentiment as positive, neutral, or negative.

Example:

Input: The new Spiderman movie was meh

Output: Sentiment: Neutral

Unstructured Text Parsing

You will be provided with unstructured data, and your task is to parse it into JSON format.

Example:

Input: Advanced Micro Devices, Inc. is an American multinational semiconductor company based in Santa Clara, California, that develops computer processors and related technologies for business and consumer markets. The company was founded in 1969 by Jerry Sanders and a group of other technology professionals.

Output: { "name": "Advanced Micro Devices, Inc.", "location": "Santa Clara, California", "founders": ["Jerry Sanders"], "founding_year": 1969, "products": ["computer processors", "related technologies"], "markets": ["business", "consumer"] }

Summarization

You will be provided with a document and asked to summarize it.

Example:

Input: Copy activity supports the Azure Databricks Delta Lake connector to copy data from any supported source data store to an Azure Databricks Delta Lake table and from the Delta Lake table to any supported sink data store. It leverages your Databricks cluster to perform the data movement; see details in the Prerequisites section.

Output:

1. The copy activity in a data integration tool supports the use of the Azure Databricks Delta Lake connector.

2. This feature enables data copying from any supported source data store to an Azure Databricks Delta Lake table and from the Delta Lake table to any supported sink data store.

3. The copy activity utilizes a user's Databricks cluster to perform data movement.

4. Users should refer to the Prerequisites section for further details and requirements.

Document Q&A

You will be provided with a document and asked a question about it.

Example:

Document:

Farm Utility Exemption Notice

The full text can be found on this website due to its length:

```
https://www.dfa.arkansas.gov/images/uploads/exciseTaxOffice/et1441.pdf
```

Question: Is a greenhouse a qualifying agricultural structure?

Answer: Yes, a greenhouse is a qualifying agricultural structure according to the document. It is listed under C of the definition of qualifying agricultural structures as a greenhouse used for commercial production.

While the playground allows developers to experiment with model use cases, behind the scenes, it uses Databricks' serverless technology to power the LLM API. You can use the UI, the Foundation Models APIs, the Python SDK, the MLflow Deployments SDK, or the REST API to query supported models. The APIs are compatible with OpenAI, so you can even use the OpenAI client for querying.

Tools

"Tools" has evolved into something powerful that enables us to connect many agentic components within and outside Databricks. In the next few chapters, we will discuss Agent Bricks and how to use Databricks as an agentic component in your GenAI applications. Below are the tools that are currently supported within Playground:

1. **MCP Servers**: Model Context Protocol (MCP) is a standard for agent communication. We can host custom logic in Databricks Apps or use Unity Catalog functions. This is the preferred way that Databricks recommends.

2. **UC Functions**: We can create user-defined functions and store them in Unity Catalog (UC). However, these functions execute in a secure, isolated environment and do not have access to file systems or internal services. For external services, we can leverage the requests package.

MCP Servers vs. Unity Catalog functions

Databricks recommends using MCP servers or defining the logic directly in agent code for faster execution, per-user authentication, and greater flexibility.

For structured data retrieval tools when the query is known ahead of time and the agent provides the parameters, Databricks recommends using Unity Catalog functions as agent tools specifically.

1. **Vector Search**: Vector search indexes are defined in Databricks and are built primarily for retrieval-augmented generation or RAG, which we will discuss in the next section.

2. **MCP Servers**: A universal interface for connecting LLMs to external tools and data sources, also known as the USB for LLMs. We will discuss MCPs in depth in Chapter 6. Databricks supports built-in and external MCP servers, so there is no limit to what we can build within Databricks.

Retrieval-Augmented Generation (RAG)

In Chapter 4, we will have a deep dive into RAG using Agent Bricks' Knowledge Assistant. However, understanding the basics of creating an index will allow us to troubleshoot any RAG applications, including Knowledge Assistant. First, the vector search endpoints are located in the compute tab, where we manage our clusters, as shown in Figure 2-5.

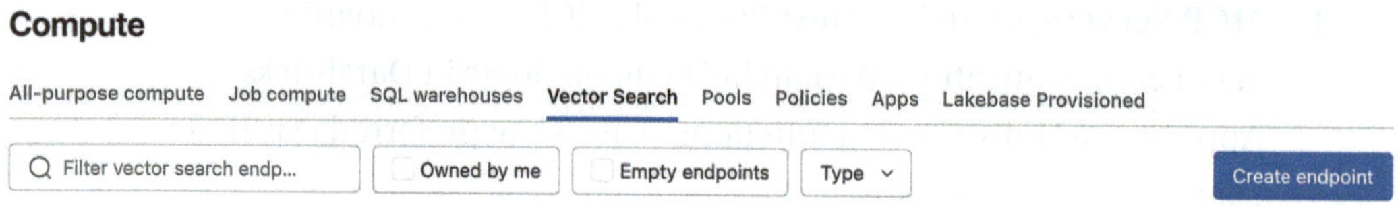
Compute
All-purpose compute
Job compute
SQL warehouses
Vector Search
Pools
Policies
Apps
Lakebase Provisioned
Filter vector search endp...
Owned by me
Empty endpoints
Type
Create endpoint

Figure 2-5. *Creating a vector search endpoint*

For beginners, the interface for creating a vector search index can be overwhelming, as shown in Figure 2-6. For illustrative purposes, we will explain the options. This will allow greater flexibility in operations, even when working with the fully automated Agent Bricks.

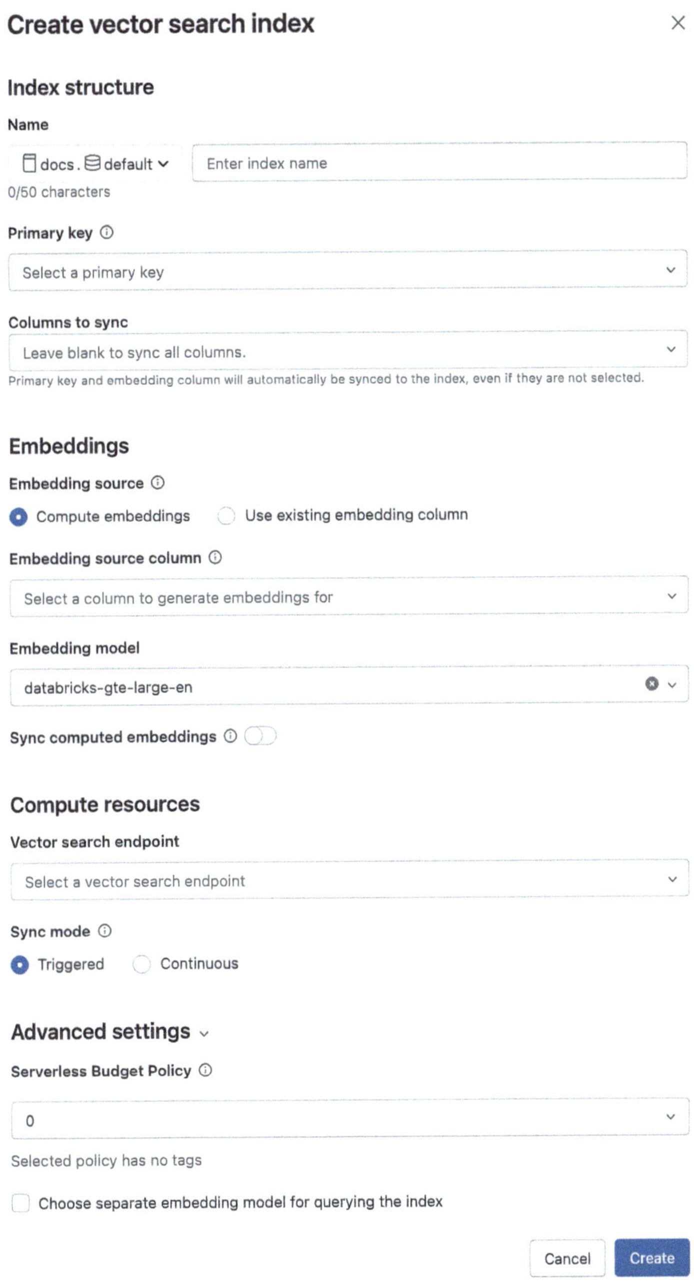

Figure 2-6. *Creating a vector search index*

As shown in Figure 2-6, we can either let Databricks calculate the embedding or calculate it ourselves. Unless there is a specific embedding model that works well for the team, we can let Databricks calculate it for us. The confusion is with "Columns to sync," which does not refer to the embedding column; it refers to columns other than the embedding column.

While you can choose to store only the primary key and the embedding, Mosaic AI Vector Search allows you to sync additional columns for two critical reasons: filtering and retrieval efficiency.

Metadata Filtering

The most common reason to sync additional columns is to enable metadata filtering, which lets you narrow results based on specific column values. Hybrid search, which combines vector similarity with keyword/full-text search, is a related but distinct feature.

1. **Example**: If you are searching through documentation, you might want to find the most relevant paragraphs but only from the "Version 2.0" manual.

2. If the version column isn't synced to the index, you cannot apply that filter during the search, and the engine might return irrelevant "Version 1.0" results.

Result Enrichment Without Lookups

If you store only the primary key and the vector, your search will return only an ID and a similarity score.

1. To show the user the actual text or a title, your application would have to perform a separate table lookup (a join) back to the Delta table for every single result.

2. Syncing these as "metadata columns" allows the search index to return the text and relevant context immediately in a single response.

Testing the Vector Search Index

Playground allows us to connect Vector Search indexes as tools, and we can then export the code by clicking "Get code" ➤ "Create agent notebook." The code will use Databricks' Agent SDK with the VectorSearchRetrieverTool. Listing 2-1 shows where you can find the usage of VectorSearchRetrieverTool.

The package's source code can be found at the following repo:

```
https://github.com/databricks/databricks-ai-bridge
```

Listing 2-1. Code excerpt from get code in playground

```
VECTOR_SEARCH_TOOLS.append(
        VectorSearchRetrieverTool(
            index_name="unitygo.rag_chatbot.databricks_documentation_
            vs_index",
        )
    )
for vs_tool in VECTOR_SEARCH_TOOLS:
    TOOL_INFOS.append(create_tool_info(vs_tool.tool, vs_tool.execute))
```

Mosaic AI Fine-Tuning API

Having explored retrieval-augmented generation (RAG) and Vector Search, we now move to the next key stage in the GenAI journey: model fine-tuning.

In the world of LLMs, training costs and hardware requirements increase as the stage progresses from prompt engineering to pre-training. Not only that, but the technical knowledge required also increases. Table 2-1 illustrates the skill and hardware requirements for each stage.

Table 2-1. *Role and hardware requirements for each step of the GenAI journey*

	Prompting	**RAG**	**Fine-Tuning**	**Pre-Training**
Role	English	Data Engineers	Data Scientists	Research Scientists
Hardware	CPU	CPU	GPU	GPU clusters

Fine-tuning lets you get more out of the models available by providing:

- Higher-quality results than prompting

- Ability to train on more examples than can fit in a prompt

- Token savings due to shorter prompts

- Lower-latency requests

Before committing to fine-tuning, evaluate the requirements in Table 2-1 and attempt to refine your prompts first. Increasing the volume of ground truth data is a necessary step to improve the model's precision on specific topics during the fine-tuning phase.

As the name suggests, fine-tuning is a process for gaining specific knowledge faster, but it comes with a cost that must be justified. For example, the model cannot answer some specialized medical questions that often involve many nuances; only trained medical professionals can answer them. Then it can be a good use case for fine-tuning.

OpenAI has published a detailed guide on prompt engineering, and we can try these with Mosaic ML Playground:

```
https://platform.openai.com/docs/guides/prompt-engineering
```

Fine-Tuning Example

Fine-tuning is re-teaching the large language something very specific, allowing it to refocus the knowledge. For example, in a telecom scenario, we want customer representatives to respond in accordance with internal training, which requires specific language. We'd then provide a list of questions and answers (i.e., prompts and responses) to the LLM and tune it to behave very specifically. Databricks has integrated Mosaic ML's fine-tuning API into the platform. The details of the fine-tuning API can be found at the MosaicML website:

```
https://docs.mosaicml.com/projects/mcli/en/latest/finetuning/
finetuning.html
```

We can also find it in the Experiment tab, in the "Foundation Model Fine-tuning" section, as shown in Figure 2-7. However, with the release of Agent Bricks and serverless GPUs, these experiments are there for someone who has a very good understanding of the process; the fine-tuned model can degrade or will not give desired results.

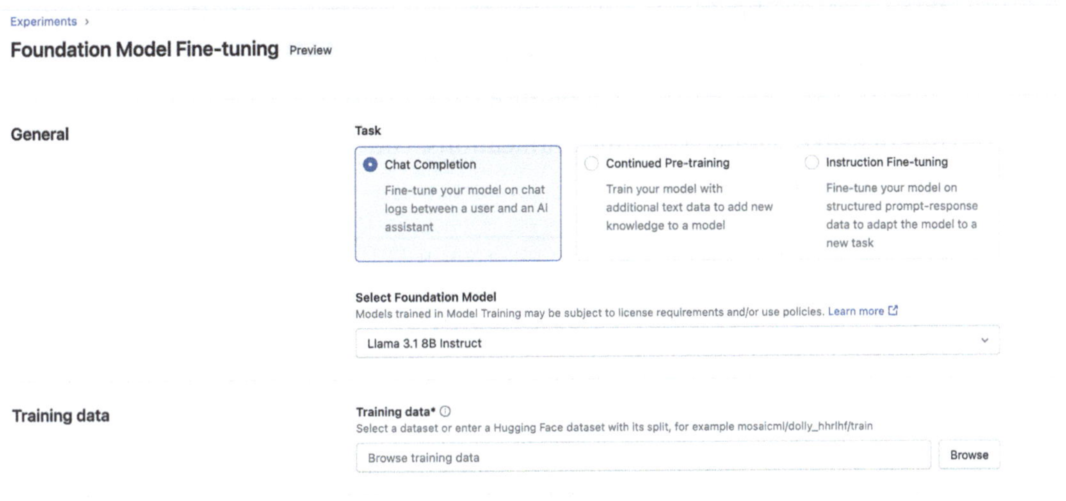

Figure 2-7. Foundation model fine-tuning in Databricks

The advantage of integrating MosaicML with Databricks is that the fine-tuned model will now be supported by the Databricks platform through Model Serving and Model Registry, and it can also take advantage of the managed MLflow feature. Everything is integrated into a single environment.

Pre-Training

Pre-training is the most costly and would require the most effort to accomplish (`https://www.databricks.com/blog/ai2-olmo-is-here`). Because everything will be created from scratch, one must build a model as in traditional deep learning, but it will likely require thousands to millions of times more data and much more commodity hardware, which is not something a small-to-medium enterprise would want to do.

A Case Study of AI2's OLMo, a Truly Open-Source Large Language Model

The Open Language Model (OLMo) is a collaboration between Databricks and the Allen Institute for AI, and we will examine the requirements to re-create this model (`https://arxiv.org/pdf/2402.00838.pdf`). See Figure 2-8.

Dataset: In traditional deep learning, the sample size required per category is about a few thousand. By comparison, the Dolma dataset is an open dataset of **3 trillion** tokens from a diverse mix of web content, academic publications, code, books, and encyclopedic materials.

Source	Type	UTF-8 bytes (GB)	Docs (millions)	Tokens (billions)
Common Crawl	web pages	9,812	3,734	2,180
GitHub	code	1,043	210	342
Reddit	social media	339	377	80
Semantic Scholar	papers	268	38.8	57
Project Gutenberg	books	20.4	0.056	5.2
Wikipedia	encyclopedic	16.2	6.2	3.7
Total		**11,519**	**4,367**	**2,668**

Figure 2-8. *Composition of the data used in the model training*

Model Training: Because the data volume is so large, it no longer fits on a single GPU. It is required to distribute across multiple GPUs. Section 3.1 of the AI2 paper discusses the distributed framework in detail.

Model Architecture: A proper architecture must be implemented. Unlike using a pre-existing foundation model, pre-training requires designing or choosing an architecture from scratch. Section 2.1 of the AI2 paper discusses such an architecture for 1B, 7B, and 65B parameters.

Hardware: This might be the most expensive and most difficult part to achieve. Not to mention, there is currently a very limited availability of high-end GPUs on the market. They are reserved for researchers who would deliver ultimate value to the company.

The OLMo model uses MosaicML with 27 nodes on the cluster, each with 8x NVIDIA A100 GPUs, 40GB of memory, and an 800Gbps interconnect. In total, 216 GPUs will be required to pretrain this model. Unless an organization has someone who really understands the ins and outs of LLMs and the projects have a high ROI, organizations usually stop their GenAI journey at fine-tuning.

GenAI Pricing

While the pricing of the GenAI infra is usually billed per hour and can be found at the following Databricks websites. For information about DBU hours, please refer to Chapter 18.

Model Serving:

```
https://www.databricks.com/product/pricing/model-serving
https://www.databricks.com/product/pricing/foundation-model-serving
```

Vector Search:

```
https://www.databricks.com/product/pricing/vector-search
```

Model Training:

```
https://www.databricks.com/product/pricing/mosaic-training
```

One concept not fully explained in the documentation is pay per token. The price is per 1 million tokens. Please note that one token does not directly correspond to a single English word or a fixed number of bytes. For example, ASCII is 1 byte, and Unicode ranges from 1 byte to 4 bytes. The concept in LLMs is similar, but it is not so straightforward.

What Are Tokens and Tokenizers?

The very short version is to split text into smaller chunks for the model to consume, as every model has a limit on the amount of text (its context window) it can process at one time, known as tokens. Tokenizers split text into subwords, aka tokens. To learn more about tokenizers, please refer to this blog post from Huggingface: `https://huggingface.co/docs/transformers/main/tokenizer_summary`

More importantly, the real question is, how do we estimate how many tokens my input text will generate? To answer this question, we first need to understand which tokenizer each model uses.

Byte-Pair Encoding (BPE)

Byte-Pair Encoding (BPE) was initially developed as an algorithm for text compression and later used by OpenAI for tokenization during pretraining of the GPT model. It's used by a lot of Transformer models, including the GPT series, Qwen series, and Claude, which uses a variant of BPE.

Understanding the tokenizer is just the first step. While you can easily load a tokenizer with one line of code, as shown in Listing 2-2, we also need a way to estimate the number of tokens required for our task to support cost estimation.

Listing 2-2. Getting the tokenizer from the model

```
from transformers import AutoTokenizer, AutoModelForCausalLM
tokenizer = AutoTokenizer.from_pretrained("name-of-tokenizer")
```

The good news is that there are some Python or JavaScript tools that we can utilize to estimate the number of tokens:

- **Open AI**: `https://cookbook.openai.com/examples/how_to_count_tokens_with_tiktoken`

- **Anthropic (Claude)**: `https://platform.claude.com/docs/en/build-with-claude/token-counting`

- **Qwen (Chinese)**: `https://github.com/imoneoi/mistral-tokenizer/`

With these tools, we can easily calculate the number of tokens in input text to estimate the cost of using the GenAI services. Of course, the cost can also be found in system tables.

Token Usage from Databricks Foundation Model APIs (FMAPI)

Databricks hosts frontier models such as the OpenAI GPT series and Anthropic's Claude series on its platform. The added benefits are beyond governance and privacy because Databricks also provides cost tracking out of the box via the `system.serving` schema. The full schema can be found here:

`https://docs.databricks.com/aws/en/ai-gateway/configure-ai-gateway-endpoints#usage-tracking-table-schemas`

For FMAPI calls, we can also track the usage from the response directly as shown in Listing 2-3.

Listing 2-3. Tracking token usage from FMAPI

```
# Example: Track token usage
response = model.generate(prompt)
input_tokens = response.usage.prompt_tokens
output_tokens = response.usage.completion_tokens
total_tokens = response.usage.total_tokens
```

Conclusion

Navigating the LLM world is challenging, and addressing those blockers with the right solution requires careful consideration, especially from a model lifecycle perspective. The Databricks GenAI stack provides a powerful solution for accelerating machine learning and AI capabilities at competitive pricing points. Databricks provides flexibility and customization options that traditional ML platforms lack or provide at a higher price. With the GenAI capabilities in Databricks, organizations can focus on creating value, whether it is managing data, tracking experiments, packaging code, or deploying models into Unity Catalog, thereby streamlining the entire LLM life cycle with governance.

## CHAPTER 3

Composable Agents with Agent Bricks: From Prototype to Production

In the rapidly evolving world of Generative AI (GenAI), there is an elephant in the room (Figure 3-1). While the potential of GenAI is immense, according to a 2024 Forbes report, 90% of GenAI projects never reach production.[1] Further, in August 2025, an MIT survey states that 95% of GenAI projects don't see a good return on investment.[2] This predicament still holds true today in 2026. The challenges often boil down to two core issues: difficulty building confidence in the agent's quality and a lack of a clear Return on Investment (ROI) due to high operational costs.

---

[1] Peter Bendor-Samuel, "Reasons Why Generative AI Pilots Fail To Move Into Production," *Forbes*, January 8, 2024, `https://www.forbes.com/sites/peterbendorsamuel/2024/01/08/reasons-why-generative-ai-pilots-fail-to-move-into-production/`.

[2] Jaime Catmull, "MIT Says 95% of Enterprise AI Fails—Here's What the 5% Are Doing Right," *Forbes*, August 22, 2025, `https://www.forbes.com/sites/jaimecatmull/2025/08/22/mit-says-95-of-enterprise-ai-failsheres-what-the-5-are-doing-right/`.

Figure 3-1.   *The elephant in the room (Image by Sam Hood. Public Domain)*

To understand this, consider the real-world dilemma of all GenAI projects. The path to production is blocked by a series of hurdles:

- **Choosing the Right Toolkit:** AI developers are overwhelmed by the sheer number of model choices, from open source to proprietary, each with its own complexities.

- **Teaching the Domain:** Developers are relying on brittle and manual prompt engineering. This leads to unwieldy prompts with thousands of tokens and creates unpredictable regressions, where fixing one issue breaks another.

- **Evaluating Quality:** The evaluation process depends on manual "vibe checks" and ad hoc feedback from experts. This makes improvement slow and difficult to measure systematically.

- **Productionizing at Scale:** The cost to process millions of documents or unstructured data is prohibitive, blocking the project from ever being deployed.

These struggles are common. Many companies are sitting on a goldmine of unstructured data—patient notes, customer reviews, contracts—but they struggle with too many technical "knobs" to turn, unreliable evaluation methods, and a difficult trade-off between cost and quality.

Agent Bricks: Databricks' No-Code Solution for Production Agents

From the user-facing perspective, Agent Bricks is packed with easy-to-use interfaces and evaluation toolsets. Agent Bricks' core promise is to deliver production-ready AI agents that are optimized on your data and for your specific domain. It is designed to solve the production problem by abstracting away complexity, automating evaluation, and optimizing costs without sacrificing quality.

Agent Bricks provides four main paths, broken down into two categories, for its use cases, as seen in Figure 3-2:

- **Information Extraction:** This is for turning unstructured data into structured insights. For example, you could use it to summarize contracts or classify customer reviews by product and sentiment.

- **Custom LLM:** This path focuses on specializing a Large Language Model (LLM) endpoint to adhere to domain-specific guidelines on your enterprise data, such as tailoring a chat model to follow company policies.

- **Knowledge Assistant:** Retrieval Augmented Generation use case. This is probably the most common use case because most of the AI interaction is still happening on the chat interface.

- **Multi-Agent Supervisor:** It can route questions to three different types of agents. They are Genie Spaces, Agent Bricks agents, and MCP servers.

- **Custom Agents:** For more advanced use cases that are beyond the provided templates, Databricks Agents SDK is available for integration with popular libraries like LangGraph and LlamaIndex. For more information, please refer to the Databricks documentation:

  https://docs.databricks.com/aws/en/generative-ai/agent-framework/author-agent

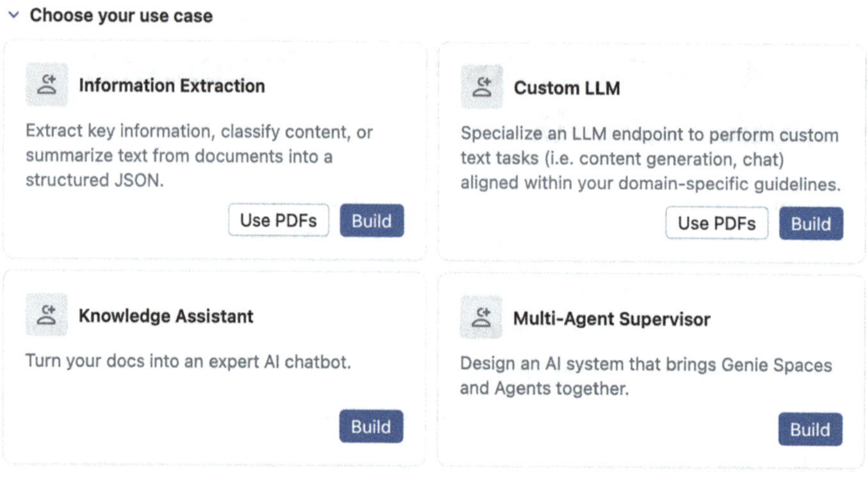

Figure 3-2. *Agent Bricks use cases*

An Agent Bricks Walkthrough

After learning about the promise of Agent Bricks, we need to put it in action. Agent Bricks has two main categories, which are worth explaining because their purpose is not immediately obvious from the names alone.

> **Generative Bricks:** Grounding on the truth, generating from the source is the essence of Generative Bricks, and they are information extraction and custom LLM. The reason why we call them "Generative Bricks" is that these bricks don't necessarily need to "extract" information from the source. While summarization and extractions are definitely very common use cases, in the above section, we mentioned classification,

similar to an ML classifier, and sentiment analysis (a task traditionally handled by models like BERT). Or we can have an agent to "summarize" medical notes at the same time, expanding abbreviations into their full forms. These pieces of information don't exist in the source, but they can be generated—completely grounded by the source documents provided.

Non-Generative Bricks: Knowledge Assistant and Multi-Agent Supervisor don't generate anything. The Knowledge Assistant, aka No-code RAG, is based on similarity search in a vector index. Everything is the same as building a RAG yourself; the only difference now is that there is no code involved and there is no need for LangChain or LangGraph, but of course, you are still welcome to build one on top of the index that Agent Bricks created. Multi-Agent Supervisor is even less generative because the whole purpose is to route the questions to appropriate agents.

In this chapter, we will first do a deep dive into the two Generative Bricks, and in the next chapter, we will walk through the non-Generative Bricks in detail.

Information Extraction

We begin in the Databricks workspace and navigate to the "Agents" tab to build an information extraction agent. The first step is to point the agent to our data, in this case, a Delta table in Unity Catalog containing the call transcripts. We will be using the `talkmap/telecom-conversation-corpus` dataset from Hugging Face. We will roll up all the turns in the conversations into one conversation id in each row as shown in Figure 3-3.

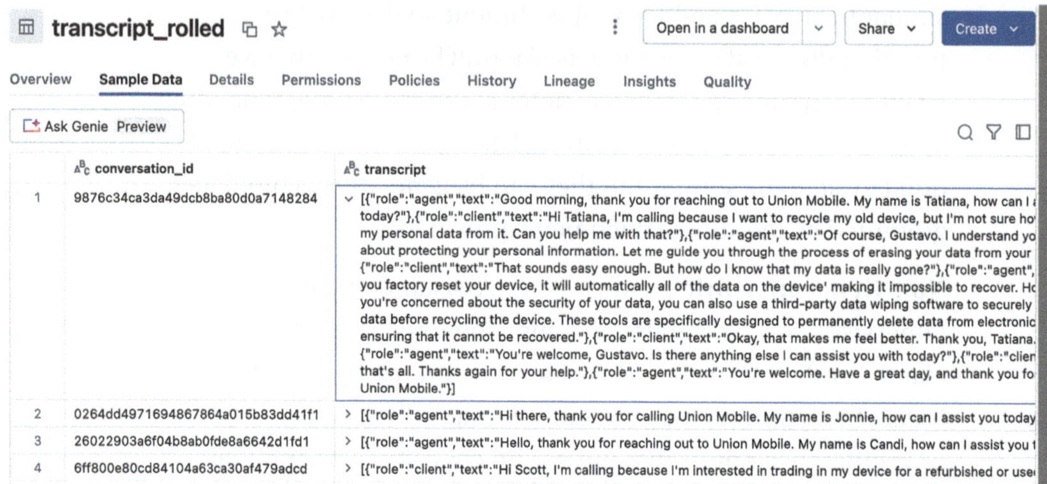

Figure 3-3. Transcripts are being rolled up at the conversation id level

Agent Bricks then autogenerates a schema based on the data it infers. But we can also provide our own schema. For our example, this could include fields like customer sentiment, issue categories, upsell opportunities, churn risk, whether the issue was resolved, the call date, and a summary. As illustrated in Figure 3-4, we are providing a sample schema, where we will be able to provide more guidance in the next step on what to expect from the agent's output.

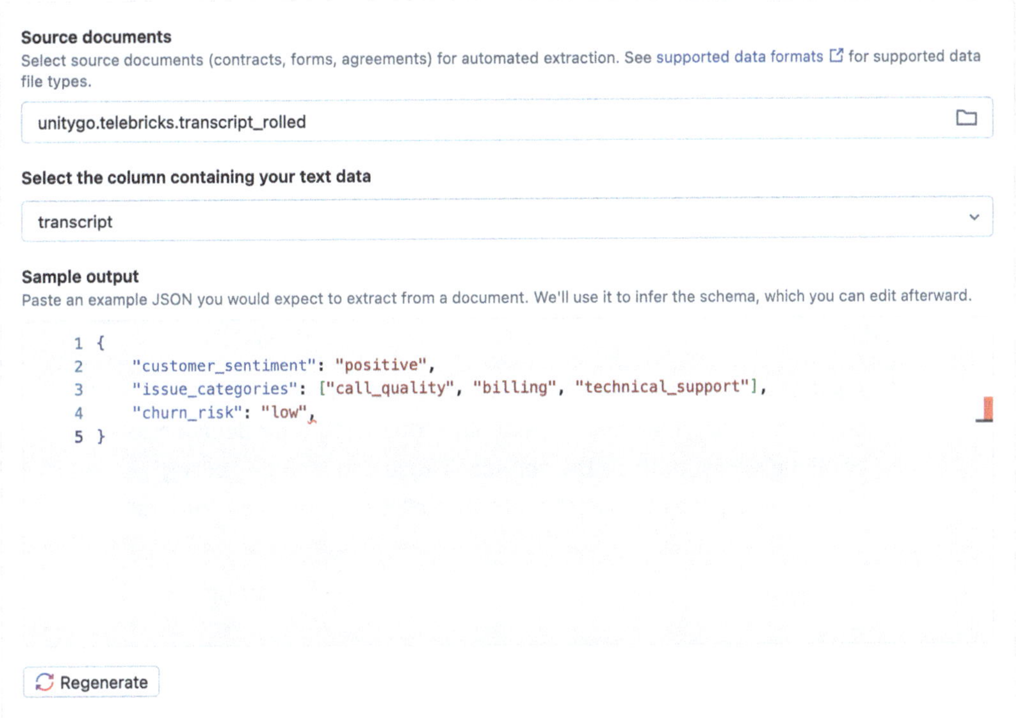

Source documents

Select source documents (contracts, forms, agreements) for automated extraction. See supported data formats ☐ for supported data file types.

unitygo.telebricks.transcript_rolled

Select the column containing your text data

transcript

Sample output

Paste an example JSON you would expect to extract from a document. We'll use it to infer the schema, which you can edit afterward.

```
1  {
2      "customer_sentiment": "positive",
3      "issue_categories": ["call_quality", "billing", "technical_support"],
4      "churn_risk": "low",
5  }
```

↻ Regenerate

Figure 3-4. *Sample output expected for the agent in information extraction*

Iterating on Quality

With the initial agent created, the next step is to improve its quality. Agent Bricks provides autogenerated descriptions for each schema field and offers recommendations for refinement. For instance, if the "issue categories" field is too open-ended, the system might suggest providing a specific list of possible categories like "network connectivity" or "billing issues" to ensure consistency. In Figure 3-5, it illustrates the open text fields pre-populated with Agent Bricks' suggestion according to the agent's understanding of the data and the field definitions provided. Because Agent Bricks is 100% natural language-driven, you are encouraged to give a very precise description of what each field really means. A more advanced usage is to specify the "categories" we want the agent to classify, very much like a multi-class machine learning classifier, but this time around AI is doing it for us. Controlling the agents requires a mix of programming and natural language mindsets. Figure 3-5 illustrates the description based on the JSON schema specified. You can edit both the JSON and the field description.

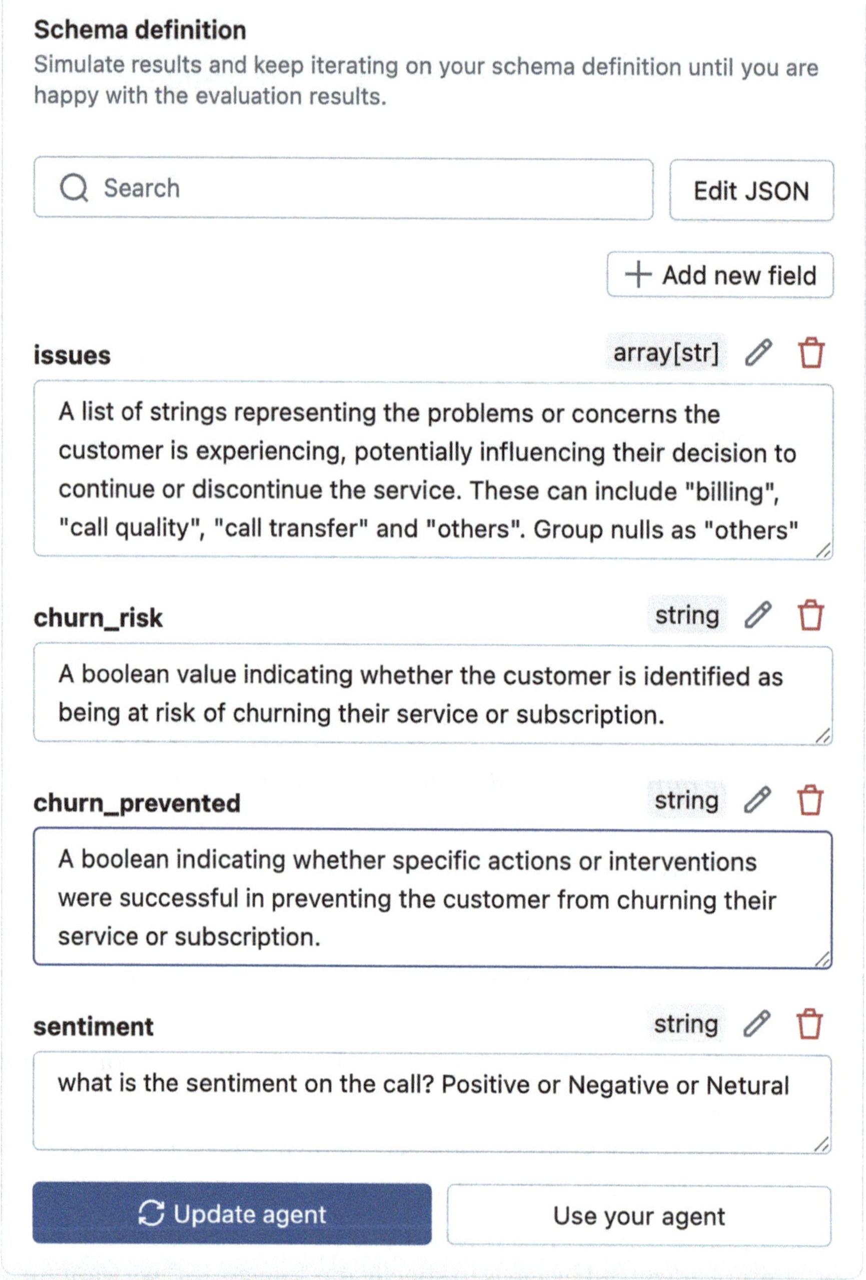

Schema definition

Simulate results and keep iterating on your schema definition until you are happy with the evaluation results.

Search

Edit JSON

+ Add new field

issues — array[str]

A list of strings representing the problems or concerns the customer is experiencing, potentially influencing their decision to continue or discontinue the service. These can include "billing", "call quality", "call transfer" and "others". Group nulls as "others"

churn_risk — string

A boolean value indicating whether the customer is identified as being at risk of churning their service or subscription.

churn_prevented — string

A boolean indicating whether specific actions or interventions were successful in preventing the customer from churning their service or subscription.

sentiment — string

what is the sentiment on the call? Positive or Negative or Netural

Update agent

Use your agent

Figure 3-5. JSON schema descriptions and their respective data types

The interface allows you to immediately review sample outputs. This allows you to validate the agent's performance, identify inaccuracies, and refine the schema descriptions accordingly. Remember, the process should be iterative and continue to improve via "Update Agent" and human feedback. After updating the agent, a new endpoint will be deployed, and the system provides a few examples to confirm that the changes produced the desired outcome, as shown in Figure 3-6.

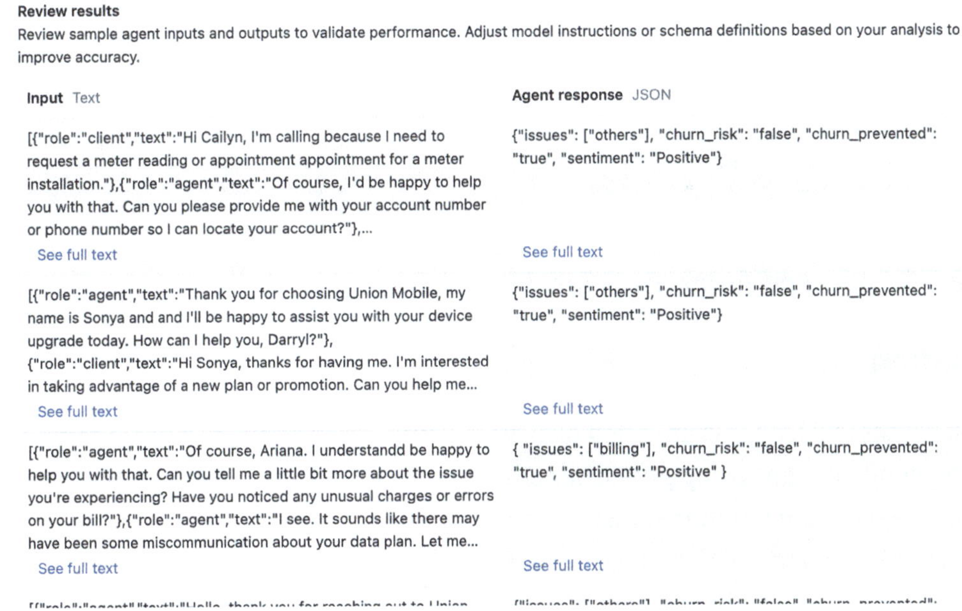

Figure 3-6. *Sample agent responses from information extraction*

Optimization and Production

Fine-tuning is the process of reinforcing a large language model (LLM) to prioritize a very specific task. In our case, this is the customer-agent interaction and the specific fields we want to extract from it. However, it is easier said than done because it requires specialized knowledge to understand the architecture of an LLM; otherwise, the LLM would not be able to generalize the knowledge. Databricks provides fine-tuning APIs out of the box, but they are intended for advanced users who are already familiar with the process.

Alternatively, Agent Bricks can do the heavy lifting for you. Once you are satisfied with the agent's quality, the "Optimize" function creates a smaller, faster, and cheaper endpoint. The platform provides visual comparison graphs showing the performance of the optimized model versus the active one, breaking down the overall score, quality versus cost, and throughput, as shown in Figure 3-7.

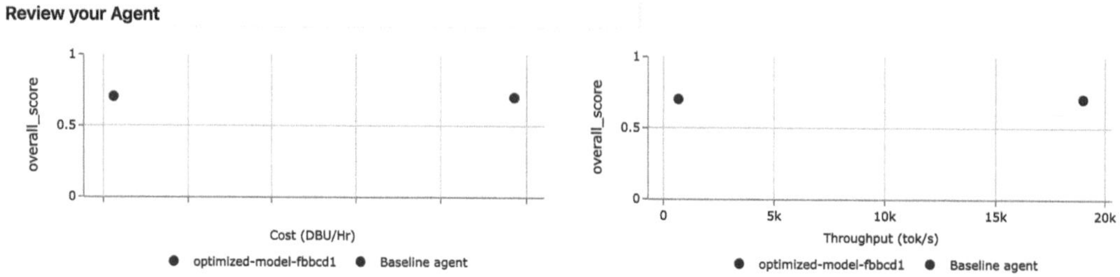

Figure 3-7. *Agent performance before and after optimization*

Evaluation

When it comes to AI, people are often okay with being "good enough," but we can't quantify "good enough" using public benchmarks because the fact that an LLM is good at math doesn't mean it understands customer sentiment in a call transcript. Databricks auto-evaluation reports (Figure 3-8) give a detailed comparison of the input data, showing where the optimized model matches or even improves on quality. For example, it might show that the "issue resolved" category is 7% more accurate with the optimized model. With this observability, you can gain confidence—in this example, achieving 90%+ accuracy. However, remember this is an iterative experiment. It is through clear expectations and criteria that we can achieve high accuracy.

	call_date	call_summary	churn_preven...	churn_risk	customer_sen...	is_schema_m...
	TRUE **100%**	TRUE **91%** -4%	TRUE **100%**	TRUE **93%** -2%	TRUE **98%**	TRUE **100%**
	True +0%	True -4%	True +0%	True -2%	True +0%	True +0%
	null +0	False +4%	null +0	False +2%	False +0%	null +0
		null +0		null +0	null +0	
?1b75375083", "customer... ●	⊘ True	⊘ True	⊘ True	⊘ True	⊘ True	⊘ True
○	⊘ True	⊘ True	⊘ True	⊘ True	⊘ True	⊘ True
?7d6cafa9509", "custome... ●	⊘ True	⊘ True	⊘ True	⊘ True	⊘ True	⊘ True
○	⊘ True	⊘ True	⊘ True	⊘ True	⊘ True	⊘ True
?c0f05f29830", "custome... ●	⊘ True	⊘ True	⊘ True	⊘ True	⊘ True	⊘ True
○	⊘ True	⊘ True	⊘ True	⊘ True	⊘ True	⊘ True
9e73bb546b3", "custom... ●	⊘ True	⊘ True ↗	⊘ True	⊘ True	⊘ True	⊘ True
○	⊘ True	⊗ False	⊘ True	⊘ True	⊘ True	⊘ True

Figure 3-8. *Agent Bricks evaluation*

For final productionization, the optimized endpoint can be called using the `ai_query SQL function` (Figure 3-9) for batch processing. This enables autoscaling and task parallelization, making it time-effective to process millions of data points and finally visualize the insights in a dashboard.

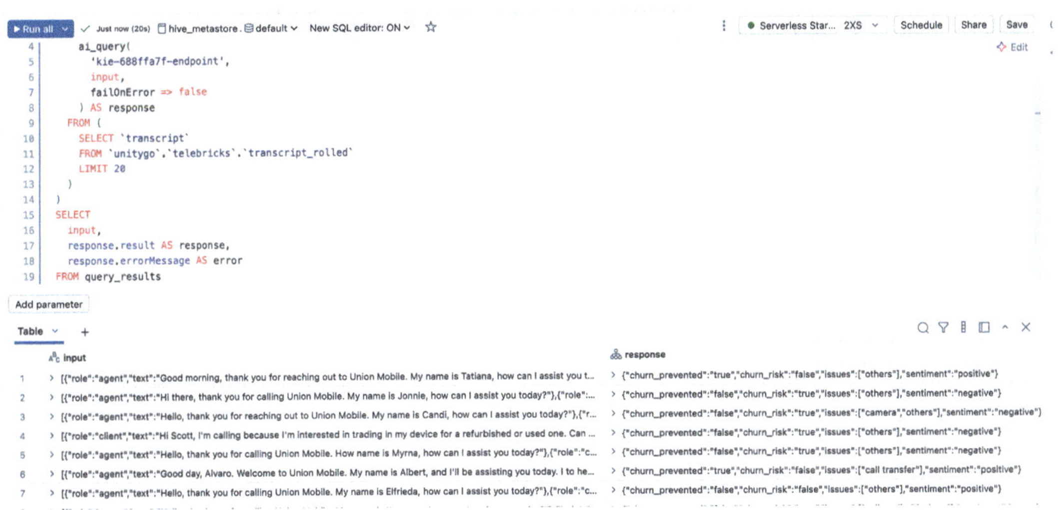

Figure 3-9. `ai_query` *batch inference results*

Behind the Scenes: The Research and Engineering Sauce

Agent Bricks achieves these results by integrating state-of-the-art research and engineering behind the scenes. These include:

- **Abstraction of "Knobs":** The platform handles model selection, prompting techniques like Chain-of-Thought, and other complexities for you. There is no need to tune parameters or compare model performance.

- **Automated Evaluation:** It uses the Databricks Mosaic AI Agent Eval platform, which features custom LLM judges trained on your guidelines to provide quantitative performance views and track regressions.

- **Model Distillation:** To create the optimized endpoint, Agent Bricks trains a smaller, cost-efficient "student" model from the larger, more intelligent "teacher" model, allowing it to retain performance at a fraction of the cost. This technique helps eliminate the trade-off between quality and cost. For example, Agent Bricks can deliver performance comparable to state-of-the-art models at 1/50th the cost.

- **Agent Learning and Stability:** The system is designed to iterate on your feedback, using natural-language guidelines to inform a prompt-optimization loop and a Reinforcement Learning from Human Feedback (RLHF) process for further quality enhancement.

This end-to-end solution also supports various data formats, including Delta tables, text files, PDFs (using the *ai_parse_document* function), and images, with more on the way.

Custom LLM

In the last section, we discussed information extraction, which is the first variation of Generative Bricks. The second variation is called Custom LLM. Behind the scenes, they are very similar and share a similar set of tools like automated prompting, iterating on quality with "Update Agent," automated evaluation as well as optimized endpoints.

This suite is unique to Agent Bricks because each component would otherwise require a dedicated team to develop and maintain. Databricks has embedded their research team's expertise behind the scenes.

Information Extraction is suitable for scenarios where we want a highly structured output in JSON format. And we can only tune it using English descriptions. A Custom LLM, on the other hand, gives us more flexibility in terms of the output—structured or unstructured. It also allows us to provide expected responses. This allows the LLM to produce a more accurate output. To illustrate its full generative power, we will build a chess commentary model using a Custom LLM. Figure 3-10 illustrates the intuitive interface for getting started with a Custom LLM.

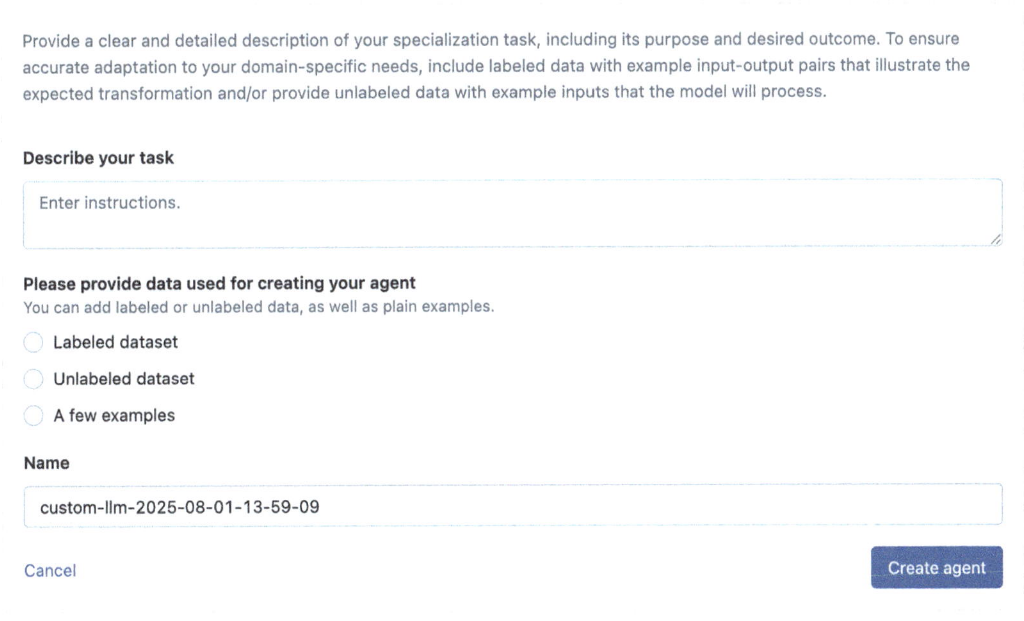

Provide a clear and detailed description of your specialization task, including its purpose and desired outcome. To ensure accurate adaptation to your domain-specific needs, include labeled data with example input-output pairs that illustrate the expected transformation and/or provide unlabeled data with example inputs that the model will process.

Describe your task

Enter instructions.

Please provide data used for creating your agent
You can add labeled or unlabeled data, as well as plain examples.

- Labeled dataset
- Unlabeled dataset
- A few examples

Name

custom-llm-2025-08-01-13-59-09

Cancel   Create agent

Figure 3-10. *Custom LLM interface*

Chess is played around the world, but for people who are not professional chess players, the moves may not be meaningful and often require professional analysis. To tune a Custom LLM, we can put a simple prompt saying, "Summarize the chess games. Tell me how each player did in a simple and easy to understand way." While Custom LLM is capable of taking a "Labeled dataset," we will choose an "Unlabeled dataset," which means providing raw data without expected outputs. In our case, it's an unannotated Portable Game Notation (PGN), which only contains chess moves and is completely meaningless to chess novices. Listing 3-1 is a game between Chess World

Champion Magnus Carlsen and chess grandmaster Alireza Firouzja during the final of ESport Chess World Cup 2025.

Listing 3-1. Magnus Carlsen vs. Alireza Firouzja during ESport Chess World Cup 2025

```
[Event "Finals I Playoffs"]
[Site "Riyadh, Saudi Arabia"]
[Date "2025.08.01"]
[Round "03-01-04"]
[White "Firouzja, Alireza"]
[Black "Carlsen, Magnus"]
[Result "0-1"]
[WhiteElo "2858"]
[BlackElo "2935"]
[TimeControl "600"]

1. e4 e5 2. Nf3 Nc6 3. Bc4 Bc5 4. c3 Nf6 5. d3 h6 6. b4 Bb6 7. a4 a6 8. 0-0
0-0 9. Nbd2 d6 10. h3 Ba7 11. Re1 Ne7 12. Nf1 Ng6 13. Ng3 c6 14. Bb3 d5 15.
exd5 Nxd5 16. Bd2 Re8 17. Qc2 Be6 18. Rad1 Qc7 19. c4 Ndf4 20. Bxf4 Nxf4
21. c5 Bxb3 22. Qxb3 a5 23. d4 exd4 24. Rxe8+ Rxe8 25. Rxd4 Rd8 26. Re4
axb4 27. Rxb4 Rd7 28. Re4 Bxc5 29. Qc4 Bxf2+ $1 30. Kxf2 Nd3+ 31. Ke3 Qxg3
32. Re8+ Kh7 33. Qe4+ g6 34. Re7 Qf2# 0-1
```

We can download a list of chess games from pgnmentor.com and save it in a table, since a Custom LLM requires structured data as input. For unstructured data, we can use the "Use PDF" option to set up a workflow and parse the data into JSON. This function currently supports PDF, DOC(X), PPT(X), JPG, and PNG formats. Figure 3-11 shows the interface for parsing the mentioned unstructured data, which includes image formats. For other formats, we can always leverage the Python package "Unstructured." For more information about the *ai_parse_document* function, please visit:

```
https://docs.databricks.com/aws/en/sql/language-manual/functions/ai_parse_
document
```

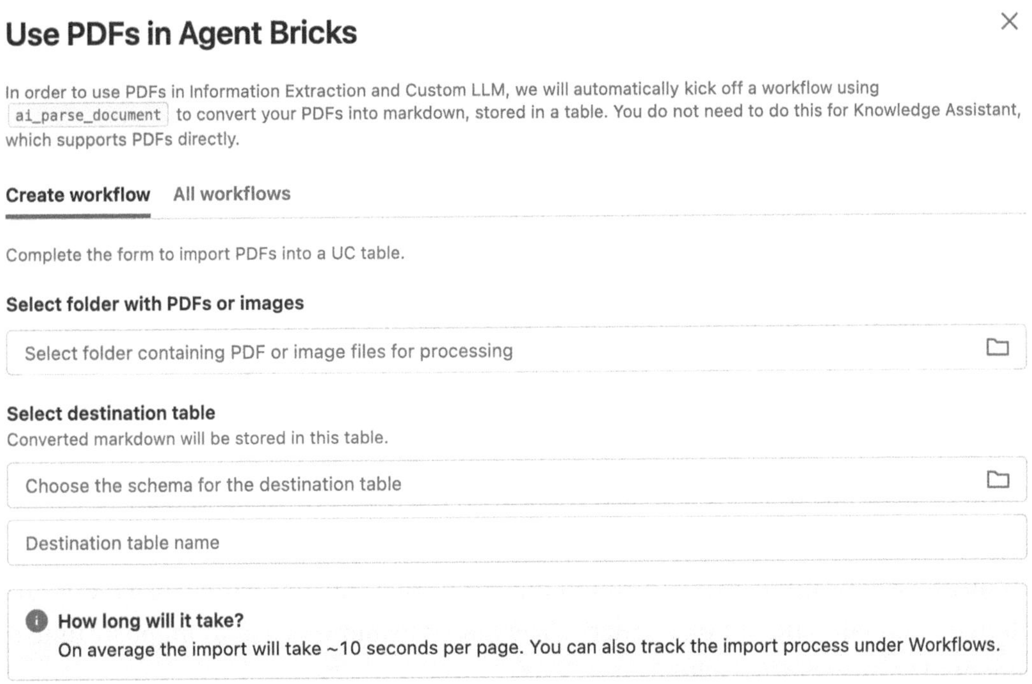

Use PDFs in Agent Bricks

In order to use PDFs in Information Extraction and Custom LLM, we will automatically kick off a workflow using `ai_parse_document` to convert your PDFs into markdown, stored in a table. You do not need to do this for Knowledge Assistant, which supports PDFs directly.

Create workflow   All workflows

Complete the form to import PDFs into a UC table.

Select folder with PDFs or images

Select folder containing PDF or image files for processing

Select destination table
Converted markdown will be stored in this table.

Choose the schema for the destination table

Destination table name

How long will it take?
On average the import will take ~10 seconds per page. You can also track the import process under Workflows.

Figure 3-11. *Use unstructured files in Agent Bricks*

Iterating on Quality in Custom LLM

In the beginning, Agent Bricks will sample a small number of records from the input data. Based on the records, it will automatically generate recommendations, similar to information extraction. But this recommendation applies to the whole output, not just a specific field. Figure 3-12 shows one suggestion from Agent Bricks.

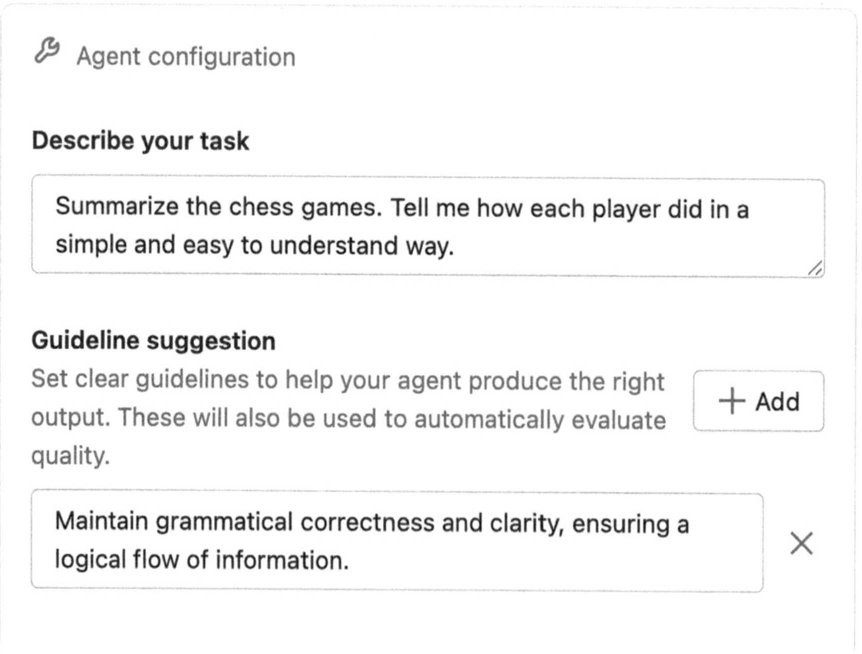

Figure 3-12. *Agent Bricks recommendations*

These recommendations will be displayed on the right side of the screen as
"Guideline suggestion" (Figure 3-13). By updating the agent and rating the sample
outputs, we will continue to get recommendations. Alternatively, we can use the "+Add"
button to manually add our own criteria. The idea is to allow us to continuously tune the
agent until we are satisfied with it.

Figure 3-13. *Agent configuration and guideline suggestion*

The initial output will not be very impressive. As seen in Figure 3-14, the agent response is very basic and doesn't really capture any essence of the game.

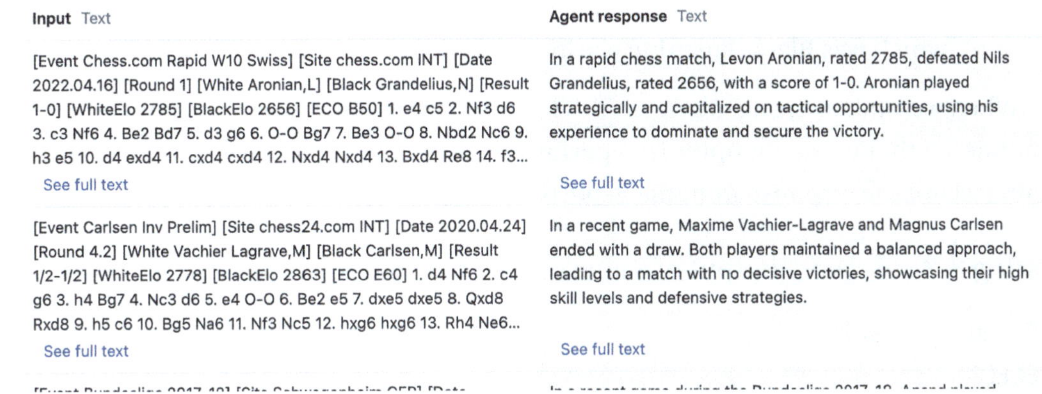

Figure 3-14. *Initial agent responses to the games*

Adding Appropriate Criteria

Chess commentaries can be technical at times, and the game can also be long. For anyone who is not a chess enthusiast but also wants to learn more about the game, we wanted it to be educational and able to capture key moments and explain them a little bit. With a bit of help from the agent and some iterations, we settled on the guidelines below:

- Maintain grammatical correctness and clarity, ensuring a logical flow of information.

- Provide a concise summary of key moments or turning points in the game that contributed to the final result.

- Use simple and easy-to-understand language, avoiding technical chess jargon to ensure readability for non-experts.

- Identify and clearly state the main outcome of the game for each player, such as win, loss, or draw.

- Identify chess tactics used in the game.

- Identify chess openings.

- Mention the decisive move number during the midgame; no need to mention the final move number.

- The summary should accurately identify which player was White and which was Black, ensuring clarity on player roles.

When we update the agent, Agent Bricks would try to optimize the prompt and generate synthetic data for optimized performance behind the scenes. We can then once again test out the response from the Esports chess game as shown in Listing 3-2:

Listing 3-2. ai_query output of a chess game from an Agent Bricks endpoint

```
WITH query_results AS (
 SELECT
   pgn AS input,
   ai_query(
     't2t-ce7c44b6-endpoint',
     input,
     failOnError => false
   ) AS response
 FROM (
   SELECT '[Event "Finals I Playoffs"]
[Site "Riyadh, Saudi Arabia"]
[Date "2025.08.01"]
[Round "03-01-04"]
[White "Firouzja, Alireza"]
[Black "Carlsen, Magnus"]
[Result "0-1"]
[WhiteElo "2858"]
[BlackElo "2935"]
[TimeControl "600"]

1. e4 e5 2. Nf3 Nc6 3. Bc4 Bc5 4. c3 Nf6 5. d3 h6 6. b4 Bb6 7. a4 a6 8. 0-0
0-0 9. Nbd2 d6 10. h3 Ba7 11. Re1 Ne7 12. Nf1 Ng6 13. Ng3 c6 14. Bb3 d5 15.
Exd5 Nxd5 16. Bd2 Re8 17. Qc2 Be6 18. Rad1 Qc7 19. c4 Ndf4 20. Bxf4 Nxf4
21. c5 Bxb3 22. Qxb3 a5 23. d4 exd4 24. Rxe8+ Rxe8 25. Rxd4 Rd8 26. Re4
axb4 27. Rxb4 Rd7 28. Re4 Bxc5 29. Qc4 Bxf2+ $1 30. Kxf2 Nd3+ 31. Ke3 Qxg3
32. Re8+ Kh7 33. Qe4+ g6 34. Re7 Qf2# 0-1' as pgn
```

```
  )
)
SELECT
 input,
 response.result AS response,
 response.errorMessage AS error
FROM query_results
```

This time the agent response is much more interesting, as shown in Listing 3-3. It not only identified the winner and the loser of the game; it also mentioned the chess opening and the decisive move number during the mid-game. The output is also simple and easy to understand for non-experts. It's a very well-done job, as shown in Listing 3-3.

Listing 3-3. Agent Bricks generated commentary on a chess game

```
In the match between Alireza Firouzja and Magnus Carlsen, Carlsen won
decisively with the final score of 0-1. Firouzja started well with a calm,
balanced opening known as the Giuoco Piano. During the critical middle
game, Carlsen took advantage with a series of tactical moves leading to an
overwhelming position. The decisive part of the game was Carlsen's precise
combo starting from 29. Bxf2+! leading to a quick end. Despite Firouzja's
strong start, Carlsen's tactical brilliance secured his victory.
```

How to Do Proper Evaluations?

The advantage of a Custom LLM is that we can provide a label, and then we can build a similarity check between prediction and ground truth on human expert labels. But beyond simple similarity scores, how do we tune the agents so that we can evaluate properly?

The core difficulty in evaluating AI lies in its behavioral specifications. Unlike software where code defines behavior, an LLM's output is often "fuzzy" and open-ended, meaning there can be many correct answers. This makes traditional, verifiable testing insufficient. The key to bridging this measurement gap is to use calibrated LLM judges. These are LLMs that have been carefully trained to emulate human judgment, providing a scalable and automated way to assess subjective quality.

A Structured Evaluation Framework: The 3x3 Approach

To bring order to the evaluation process, a structured framework is essential. The 3x3 approach, developed by Databricks, aligns evaluation strategies with specific stages of the AI lifecycle, from initial prototypes to full production.

The three pillars of expected usage include:

- **Must-Haves:** Core functionalities and unique problems the application is designed to solve (signs of life).

- **Must-Not-Haves:** Undesirable behaviors, often informed by legal/regulatory requirements (e.g., don't insult customers, avoid certain topics).

- **Actual Usage:** What users genuinely do with the application, which may differ from intended use.

The three stages of the GenAI life cycle include:

- **Proof of Concept:** Focus on initial signal, basic functionality. "Pro vibes" are acceptable here to quickly check for promise. Less emphasis on actual usage.

- **Pilot:** Aim for consistent, high-quality results for meaningful use at scale. This is the "Bermuda Triangle of GenAI," where many projects get stuck due to incorrect metrics or directionality.

- **Production:** Focus on monitoring and observability to ensure reliable operation at scale.

This framework ensures that evaluation efforts are appropriately scoped for each stage of development, preventing wasted effort and providing a clear roadmap for achieving a production-ready system.

Real-Life Example

As illustrated in Figure 3-8, Databricks evaluates each criterion against data samples, a process vital for measuring agent quality. Much like regression testing in software engineering, which ensures new changes don't break existing functionality, our goal here was to ensure the agent could identify the game's turning point. Listing 3-4 highlights this 'decisive moment' with the following comment:

Listing 3-4. Chess comments on a decisive move

```
"The decisive part of the game was Carlsen's precise combo starting from
29. Bxf2+! leading to a quick end."
```

We can load the game into a chess game analyzer with engine evaluation. The engine evaluation is represented by the bar on the left, which indicates which side is winning. As seen in Figure 3-15, the engine evaluation illustrates that Black has a slight advantage, but overall, both sides are still balanced in power, especially in a game between masters.

Figure 3-15. *Engine evaluation on move 28*

Moving on to move 29 (Figure 3-16), we can see that the evaluation bar dropped significantly on White's side. This significant drop indicates a tactical blunder by White, confirming the agent's accurate identification of move 29 as the decisive turning point. This aligns with the "must-haves" pillar of the 3x3 framework, demonstrating how the agent's output reflects real-world game analysis.

Figure 3-16. *Engine evaluation on move 29*

Conclusion

Agent Bricks is a powerful tool for taking domain-specific AI from prototype to
production. As noted earlier, a critical industry problem remains: most GenAI projects
never progress past experimentation. Despite initial promise, they often stall due to
brittle pipelines, unclear quality signals, and operational costs that make full-scale
deployment impractical.

From Bespoke Orchestration to Integrated Workflows

Agent Bricks directly addresses these hurdles by automating prompt optimization
and abstracting model-level complexity. When coupled with built-in evaluation and
cost-quality optimization, the result is a dramatically faster time-to-market. Instead of
spending months engineering custom pipelines and manual tuning workflows, teams
can leverage integrated components that are ready to ship.

Maximizing Business Value

This shift allows organizations to move beyond the technical pipeline and focus on delivering strategic value. With features such as model distillation and natural-language-driven configuration, teams can iterate on domain-specific logic rather than on the underlying infrastructure. This enables AI solutions to evolve with market needs far more efficiently than traditional, code-first approaches.

In a landscape where many AI initiatives fail to cross the finish line, Agent Bricks provides a mature, practical, end-to-end path. It empowers enterprises to deliver reliable, high-performing solutions that are cost-predictable and fundamentally aligned with business outcomes.

## CHAPTER 4

Agent Bricks Deep Dive: Retrieval, Reasoning, and Tooling Architectures

In Chapter 3, we discussed the first two use cases for Agent Bricks: Information Extraction and Custom LLM. To recap, Agent Bricks is a UI-based product that packs the best of Databricks research behind the scenes, so we don't need to learn or keep up with the latest techniques to make our GenAI solutions work with our data. There are two different types of bricks: generative bricks and non-generative bricks.

Generative Bricks can act like a machine learning classifier or a live commentary person; the choice is yours. We just need to be creative and think outside of the box. We can also parse unstructured data, such as PDFs and images, and use it in Generative Bricks. There are unlimited options for us to leverage the generative power. The built-in LLM judges allow us to evaluate the output to ensure that the agents are not hallucinating. Databricks aims to help deliver production agents, and these toolsets are available out of the box.

Non-generative bricks are referring to knowledge assistant and multi-agent supervisor. While generative bricks are optimized for batch inference, they are designed to infer from unseen data. Naturally, it'd make sense to manage these bricks with evaluations using an LLM as a judge. However, when evaluating a chatbot like retrieval-augmented generation (RAG), it makes more sense to use a human-in-the-loop approach, as it is designed to interact with humans. Having this expectation in mind, we will take a deep dive into this non-generative brick.

J. Yip et al., *Databricks Data Intelligence Platform*, https://doi.org/10.1007/979-8-8688-2524-8_4

Agent Bricks promised to abstract all the tunings, aka knobs, from users. However, similar to the "Developer Mode" in web browsers, it will still be relevant to some individuals, even as the need for custom web development has become less common. In other words, it is still beneficial to understand what happens under the hood of these agents, as it requires less specialized knowledge than before, and it will also allow us to incorporate agents outside of Databricks into our solutions—whether you choose to host the agents on Databricks or not.

Retrieval Augmented Generation: The Knowledge Assistant

While ChatGPT democratized LLM-based chatbots for consumer use, companies need to deploy personalized models that answer their needs:

- Privacy requirements on sensitive information

- Preventing hallucination

- Specialized content, not available on the Internet

- Specific behavior for customer tasks

- Control over speed and cost

- Deploy models on private infrastructure for security reasons

To accomplish this, organizations often need to provide internal documents to ground the model in truth. This process requires converting some context into something called embeddings. Embeddings are mathematical representations (vectors) of the semantic content of data, typically text or image data. While there are numerous methods for creating embeddings, Generative AI in Databricks typically utilizes large language models, such as BAAI's BGE-Large-EN, to generate them. These embeddings are essential for GenAI applications that rely on similarity-based document or image retrieval. The result of embedding is a series of numbers produced by the algorithm. Listing 4-1 shows how we can generate an embedding based on an input text.

Listing 4-1. Generate embedding with given tex

```python
from databricks.sdk import WorkspaceClient
w = WorkspaceClient()
response = w.serving_endpoints.query(
    name="databricks-bge-large-en",
    input="What is Databricks?"
)
display(response.data[0].embedding)
```

Figure 4-1 is an illustration of embedding representations.

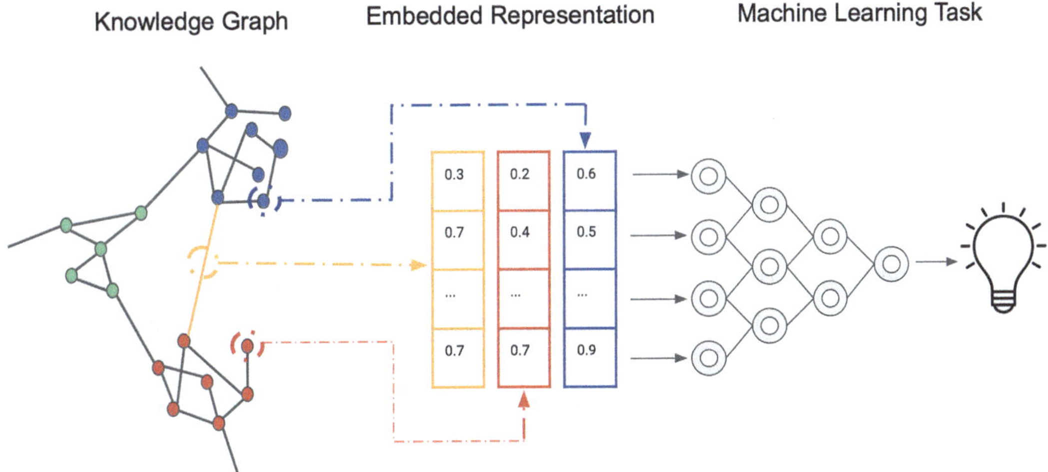

Figure 4-1. *Representation of embedding from a knowledge graph*

From there on, we will need to build a workflow, generally known as a chain, to leverage the embeddings to find relevant documents. Figure 4-2 shows a sample workflow using LangChain to connect document embeddings to Databricks' vector index and sync them with Databricks' vector database.

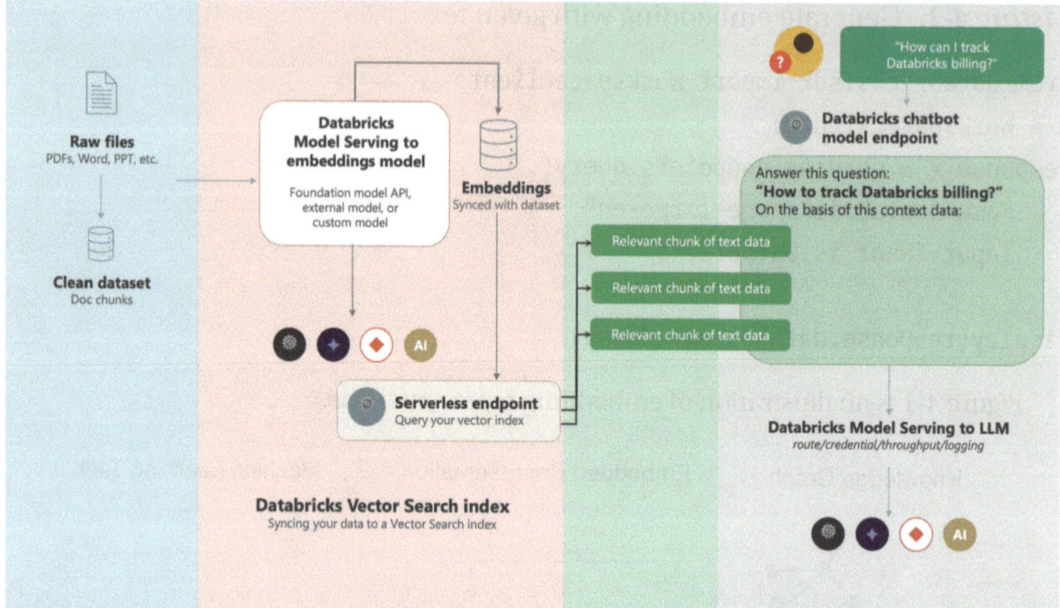

Figure 4-2. *AI Agent workflow with Databricks vector database*

As data volumes increase, it has become increasingly difficult to optimize the performance of large-scale data applications. To address performance issues, Databricks has released a suite of tools that enable developers to focus on building the pipeline to achieve higher quality rather than on performance tuning and infrastructure maintenance. These tools include:

- Fully managed foundation models providing pay-per-token-based LLMs or dedicated compute via provisioned throughput

- The first step in our LLM workflow is to generate embeddings from text or binary data. The Databricks Foundation Model API provides performance guarantees for certain foundation models across different use cases. For embedding, BGE Large (English) and GTE Large (English) are provided with an API interface, allowing developers to compute embeddings at scale. You are encouraged to check the documentation for the latest list of models.

- A vector search service to power semantic search on existing tables in your lakehouse.

- A vector database is a specialized database to store embeddings. To ensure performance, a vector index will be created for a specific column. Mosaic AI Vector Search will either calculate the embeddings for you if it is a text column in a Delta table. Alternatively, it will sync the embeddings to an index when the values are stored in a Delta table. Finally, API can be used to sync the index if no table is provided. In any case, embeddings will need to be available in a highly performant, scalable format for *similarity search*.

In summary, Databricks provides multiple types of vector search indexes, as shown in Figure 4-3:

- **Managed Embeddings:** You provide a text column and endpoint name, and Databricks synchronizes the index with your Delta table

- **Self-Managed Embeddings:** You compute the embeddings and save them as a field of your Delta table. Databricks will then synchronize the index

- **Direct Index:** When you want to use and update the index without having a Delta table

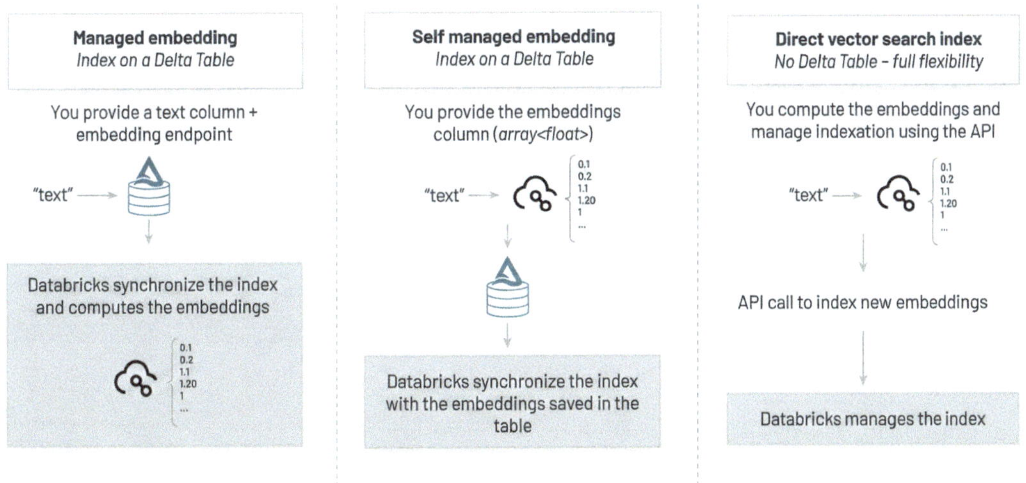

Figure 4-3. *Mosaic AI vector search index types*

Databricks selected **gte-large-en** and **bge-large-en** as its primary hosted embedding models because they offer an optimal balance of size, performance, and cost. Other models (like OpenAI or Qwen) are better suited for specific needs like multilingual support, massive context windows, or extreme accuracy. Table 4-1 outlines when to use each index type.

Table 4-1. *Choosing between index types*

Model/Type	When to use it?
BGE/GTE (Managed embedding)	Managed by Databricks, hassle-free
OpenAI/Qwen or other models (Self-managed embedding)	Specific use cases like maximum accuracy and multi-language support
Direct vector search index	Real-time applications like copilot

Similarity Search: The Magic Behind the Scenes

In the section above, we discussed that vector search is based on a similarity algorithm. The text is first encoded as a vector, and the similarity between the question and the answer is measured to determine the best answer. Cosine similarities (and the related dot product) are algebraic operations that take two equal-length sequences of numbers (usually coordinate vectors) and return a single number. For cosine similarity, when the result is 0, the vectors are orthogonal (unrelated); when the result is 1, they are identical; and when the result is -1, they are diametrically opposed. Vector Search uses this function to find relevant documents that answer a specific question. Figure 4-4 is a two-dimensional representation of cosine similarity.

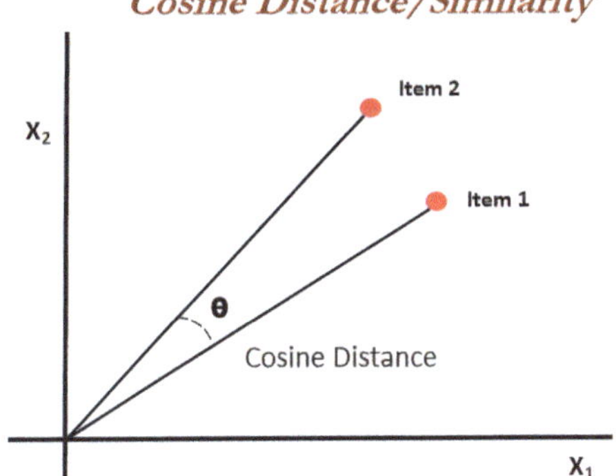

Figure 4-4. *Cosine similarity*

That was an introduction to cosine similarity. However, Mosaic AI Vector Search in Databricks does not use cosine similarity by default. According to Databricks, Mosaic AI Vector Search uses the Hierarchical Navigable Small World (HNSW) algorithm for approximate nearest neighbor searches and the L2 distance metric to measure similarity between embedding vectors (`https://docs.databricks.com/aws/en/vector-search/vector-search`). If you want to use cosine similarity, you need to normalize your embedding vectors to unit length before feeding them into Vector Search.

In other words, HNSW is the indexing algorithm. It organizes your data into a graph structure that enables the system to quickly find the "approximate nearest neighbors" (ANN) of a query, even with millions of records.

Think of it as the "map" used to navigate through your vectors. While HNSW is the navigation tool, the distance metric determines how "closeness" is measured.

- **Databricks Default:** Mosaic AI Vector Search uses the L2 distance (Euclidean distance) metric by default.

- **Cosine Similarity:** To use **Cosine Similarity** within this HNSW framework, you must normalize the embedding vectors before feeding them into the index.

Figure 4-5 is an illustration of HNSW.

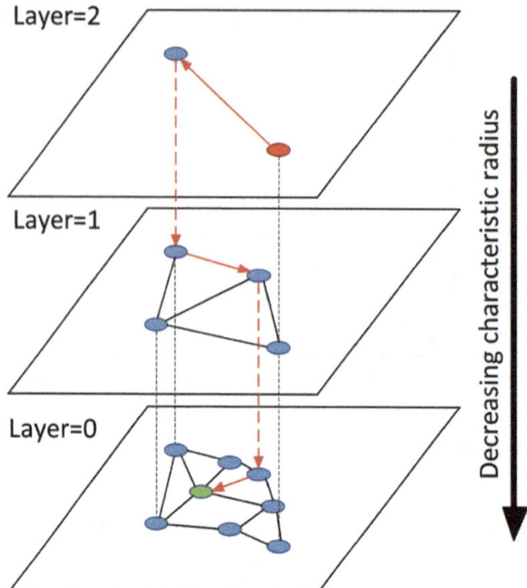

Figure 4-5. *Illustration of the hierarchical NSW idea*

Beyond similarity search, Databricks also combines keyword search to enhance the relevancy of the results. At a high level, keyword search computes a relevance score for candidate results, ranking them based on how well they match the query keywords. All text or string columns are searched, including the source text column and any metadata columns in text or string format. The full calculation is beyond the scope of this book but can be found on the Databricks documentation:

```
https://docs.databricks.com/aws/en/generative-ai/vector-search
```

Figure 4-6 shows a sample workflow of vector search.

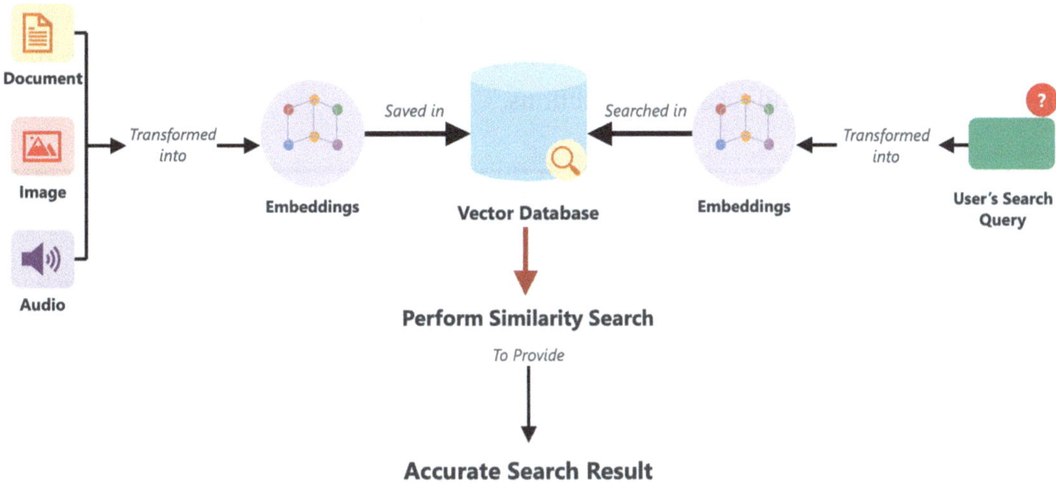

Figure 4-6. *A visual guide of vector search*

Reranking

The core trade-off in vector search is that it prioritizes *speed* over *perfect accuracy*. Because vector search is designed to return results extremely quickly, especially at a large scale, it may retrieve approximate matches rather than the most precise ones.

The reranking step is one way to fix this by reranking the top results using a more accurate (but slower) model. Figure 4-7 is a high-level architecture of these two processes combined.

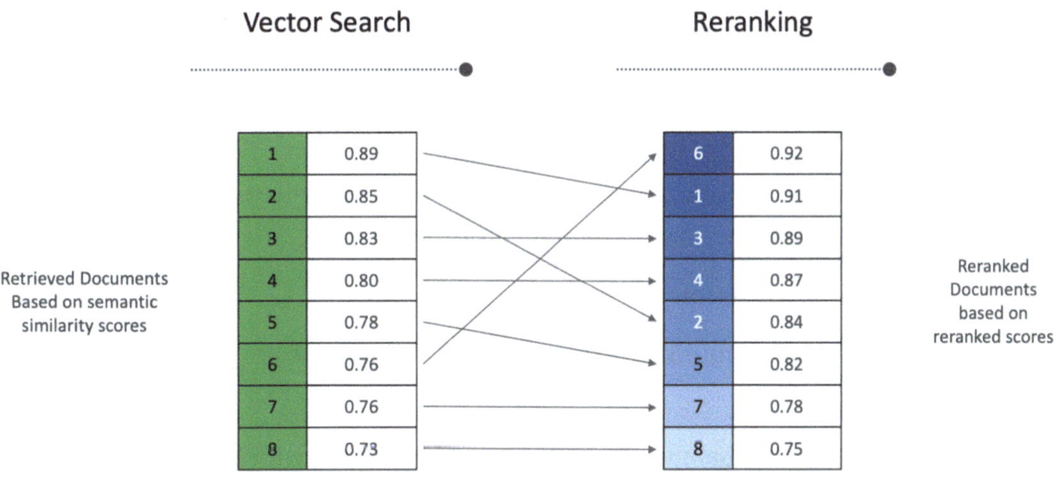

Figure 4-7. *The twin process of vector search and reranking*

In data science terms, vector search prioritizes *Recall,* while reranking prioritizes *Precision.* To balance these two metrics, the F1 score is often used to evaluate the trade-off between precision and recall in predictions.

DATA SCIENCE 101

In the context of Data Science, we often measure precision and recall against a "class," which refers to a specific category or group of data points. For example, if you're building a model to identify relevant documents, "relevant" would be one class, and "not relevant" would be another.

- **Precision** measures how many of the data points the model *identifies as belonging* to a certain class actually *do* belong to that class.

- **Recall** measures how well the model identifies *all* the data points that truly belong to a certain class.

- **F1 Score** combines precision and recall into a single metric, providing a balanced measure of a model's accuracy, and is often used as the ultimate guide for the model's performance.

Reranking improves retrieval quality through a three-stage process:

1. **Initial Vector Retrieval**

   The system retrieves the top K most similar chunks using vector search based on a similarity score (high recall).

2. **Reranking**

   A reranker evaluates each retrieved chunk together with the query to assign a relevance score (high precision).

3. **Sorting and Selection**

   The final top few chunks are passed to the LLM for answer generation.

The above architecture balances speed (fast vector search) with accuracy, optimizing both precision and recall.

Vector Search Infrastructure

Spark is a pioneer of distributed computing with decoupled storage. This architecture has been available since Lakehouse's inception. For this reason, as the RAG solution becomes increasingly popular, Databricks has rebuilt the vector search index using the same architecture, called Mosaic AI Vector Search. In addition to the compute and storage decoupling, when *comparing with the standard vector search*, there are other advantages:

- **Massive Scalability:** Supports multi-billion vector capacity for enterprise-grade applications and large-scale knowledge bases, ensuring seamless growth.

- **Cost Efficiency:** Achieve up to 7x lower operational costs through optimized infrastructure and streamlined data management.

- **Blazing Fast Ingestion:** Experience 20x faster indexing speeds, enabling real-time integration of new data and enhanced AI model responsiveness.

- **Flexible Data Retrieval:** Utilize SQL-style filtering for precise searches, combining vector similarity with structured metadata queries for accurate results.

Figure 4-8 shows the two options when creating standard and storage-optimized endpoints.

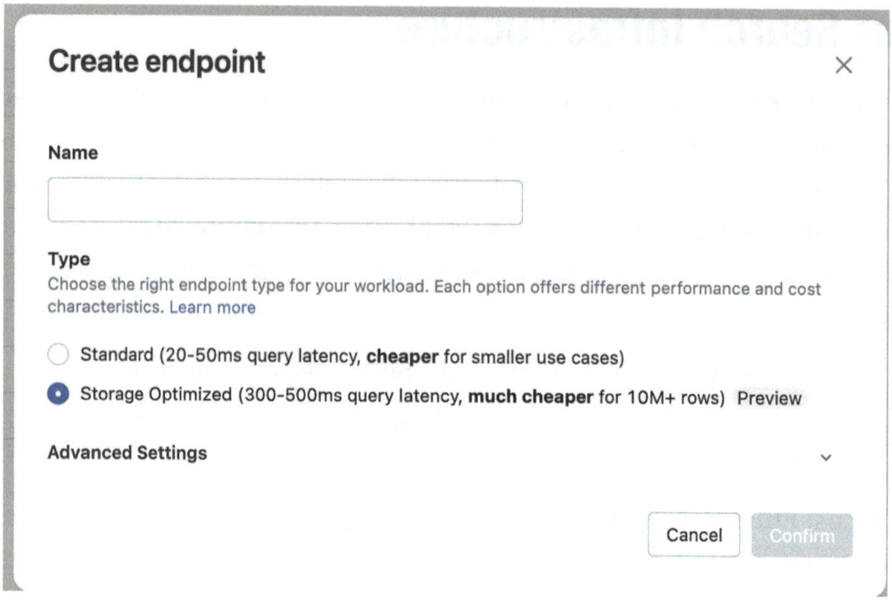

Figure 4-8. *Standard and storage-optimized endpoints*

However, whenever there is a huge advantage, there will always be fine print that comes with it. The storage-optimized endpoint currently does not support continuous sync. In other words, it does not support real-time mode, and the synchronization can only be triggered, and every sync fully rebuilds the vector search index. But as technology continues to evolve, this limitation will eventually disappear. For more information about the limitations and how to choose an index, please refer to Databricks documentation:

```
https://docs.databricks.com/aws/en/generative-ai/vector-search#so-
endpoints-limitations
```

A Crash Course on Chunking

One of the more important knobs for vector search performance is called chunking. Chunking for RAG is the process of breaking large documents into smaller segments, called chunks. Chunking is necessary because large documents exceed embedding model limits, also known as context windows, and often contain multiple unrelated topics. Embedding the entire document as one unit would produce a single vector that dilutes meaning and performs poorly in retrieval. By breaking documents into smaller, semantically meaningful chunks, the system can retrieve only the most relevant portions, improving both precision and recall in RAG systems.

A core goal of chunking is to preserve the document's semantic integrity, ensuring that each chunk represents a coherent idea. When chunks capture natural boundaries such as paragraphs, sections, or topic transitions, retrieval quality improves dramatically because the model retrieves self-contained meaningful units. Below we will visit different types of chucking. Figure 4-9 illustrates the chucking process.

Chunking for RAG

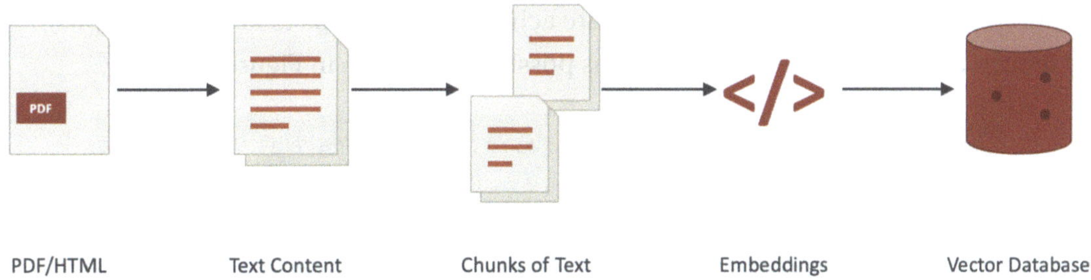

Figure 4-9. *The process of chucking in RAG*

While we don't dive deep into how to chunk the documents, we'll briefly introduce them below to help you understand the internal workings. Table 4-2 also summarizes when to use these methodologies.

Fixed-Size Chunking

This is the most straightforward approach, in which text is divided into chunks of a fixed size (e.g., 400 characters). An optional overlap feature can be introduced to repeat the end of one chunk at the beginning of the next, helping preserve context across boundaries. As seen in Figure 4-10.

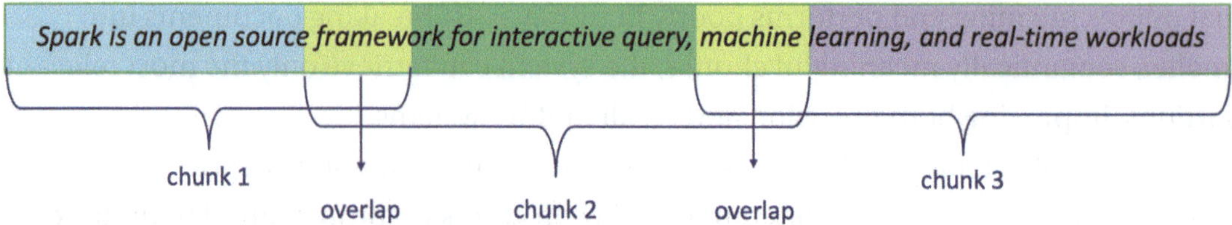

Figure 4-10. Fixed-size chunking

Recursive Chunking

This method uses a set of default separators (like newlines, spaces, or even empty strings) in a hierarchical order. It attempts to split the text based on these separators, iterating until chunks of the desired size are achieved. This helps to keep paragraphs, sentences, and words together as much as possible, as shown in Figure 4-11.

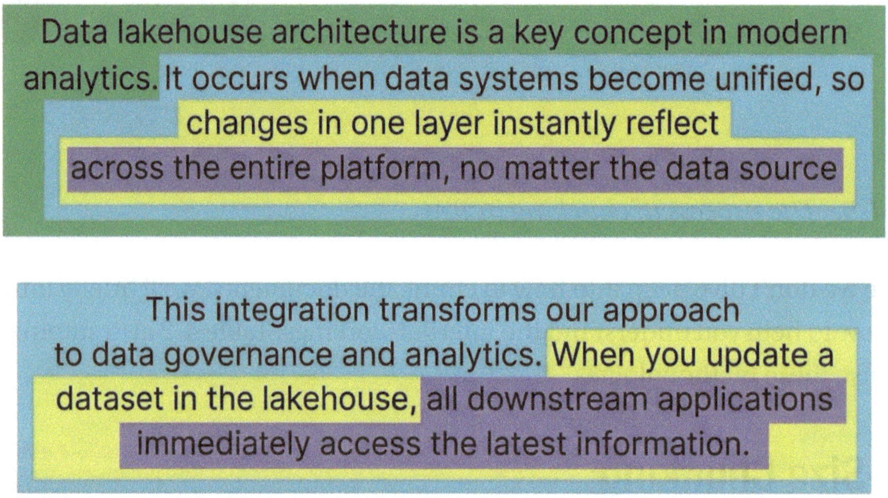

Figure 4-11. Recursive chunking

Semantic Chunking

This strategy splits text based on the semantic similarity of sentence embeddings. Sentences with high semantic similarity are grouped together, resulting in context-aware chunks that are more likely to contain coherent ideas, as shown in Figure 4-12.

Databricks is a unified data analytics platform designed to simplify and accelerate data engineering, data science, and AI initiatives. It is built on an open lakehouse architecture, comiing the best aspects of data lakes and data warehouses for scalable, reliable data storage and processing. Key platform features include collaborative notebooks, integrated Apache Spark, Delta Lake, and powerful machine learning tools for unified workflows. This integration enables faster time-to-insight, improved team collaboration, and a simplified infrastructure for mission-critical AI applications.

Figure 4-12. *Semantic chunking*

Document-Based Chunking

This approach splits text based on the document's inherent structure. It can use elements such as Markdown text, images, tables, or even code classes and functions to determine chunk boundaries, as shown in Figure 4-13.

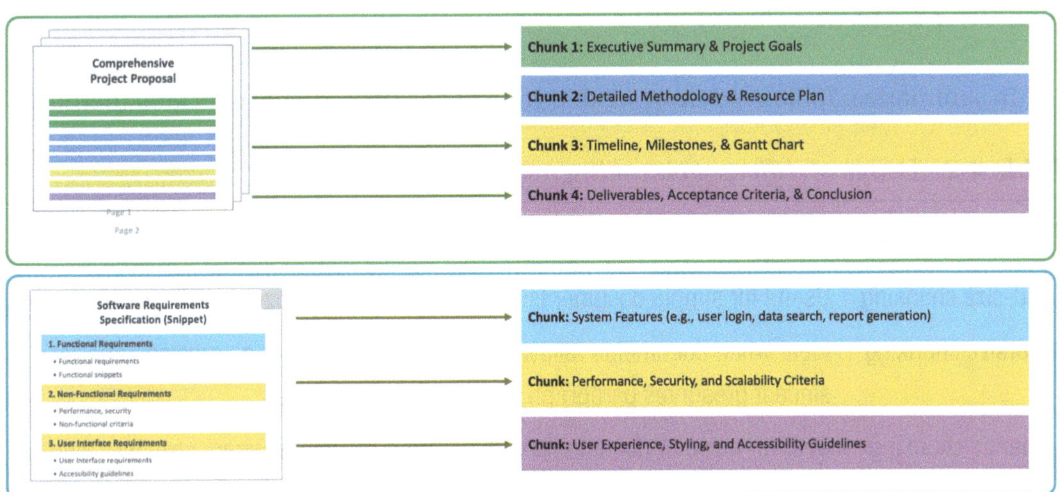

Figure 4-13. *Document-based chunking*

Agentic Chunking

This is an experimental approach that uses an LLM (agentic AI) to determine an appropriate document-splitting strategy. The LLM can consider semantic meaning, content structure, paragraph types, section headings, and more, thereby simulating human reasoning when processing long documents, as shown in Figure 4-14.

Figure 4-14. Agentic chunking

To summarize, Table 4-2 is how we determine when to use each chucking method.

Table 4-2. Usage of different chunking methods

Chucking method	When to use it?
Fixed-size chunking	Useful for simple documents or pipelines that prioritize speed over accuracy.
Recursive chunking	Ideal when documents follow consistent formatting (e.g., articles, reports), since it preserves paragraph and sentence structure.
Semantic chunking	Best for noise or documents with different meanings where structure alone is insufficient. It groups sentences based on meaning.
Document-based chunking	Effective for technical or markdown-style documents with clear hierarchical structure, like research papers.
Agentic chunking	Recommended when documents vary widely in structure and semantics. This method uses an LLM to determine intelligent boundaries.

For a deep dive into the pros and cons and learn how to do them in code, please refer to the blog post below from Databricks:

```
https://community.databricks.com/t5/technical-blog/the-ultimate-guide-to-
chunking-strategies-for-rag-applications/ba-p/113089
```

After learning the basic concepts and building blocks of RAG, we will look at an example of how to create an end-to-end RAG application using Knowledge Assistant. You'll find that all the underlying configuration details are abstracted away for you.

A Practical Example for RAG: Using Unstructured Data

Knowledge Assistant helps us to build a RAG very easily without worrying about all the knobs to tune and the technology behind it. In this section, we will investigate how to build a chatbot based on unstructured data, and then we will discuss how to tune the quality. The quality tunings in these nongenerative bricks are done using examples. We will also understand the reasons behind these examples and what to expect when providing examples to the agent.

First, we need to select the data for ingestion. Currently Knowledge Assistant supports the following formats: txt, pdf, md, ppt/pptx, and doc/docx. You will need to have access to a volume to process these files. However, if you are looking for additional data sources, the open-source package unstructured.io provides comprehensive support for unstructured data. We will walk through how to work around the limitations to use unstructured.io or other packages of your choice, and you can still take advantage of the Knowledge Assistant interface for quality tuning. Figure 4-15 allows you to select a folder for your files.

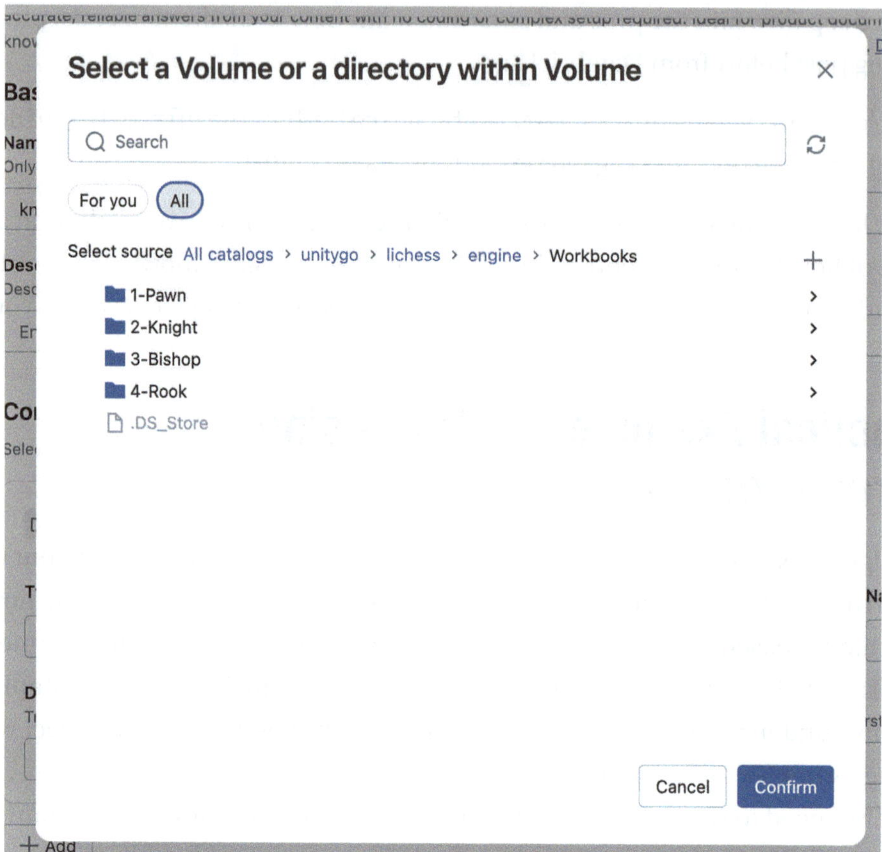

Figure 4-15. *A folder that contains all our PDFs*

Once the files finish syncing, we will be able to start chatting with our files. There really isn't much to do, which greatly simplifies the technical barrier to getting started. Figure 4-16 shows that the sync is successful.

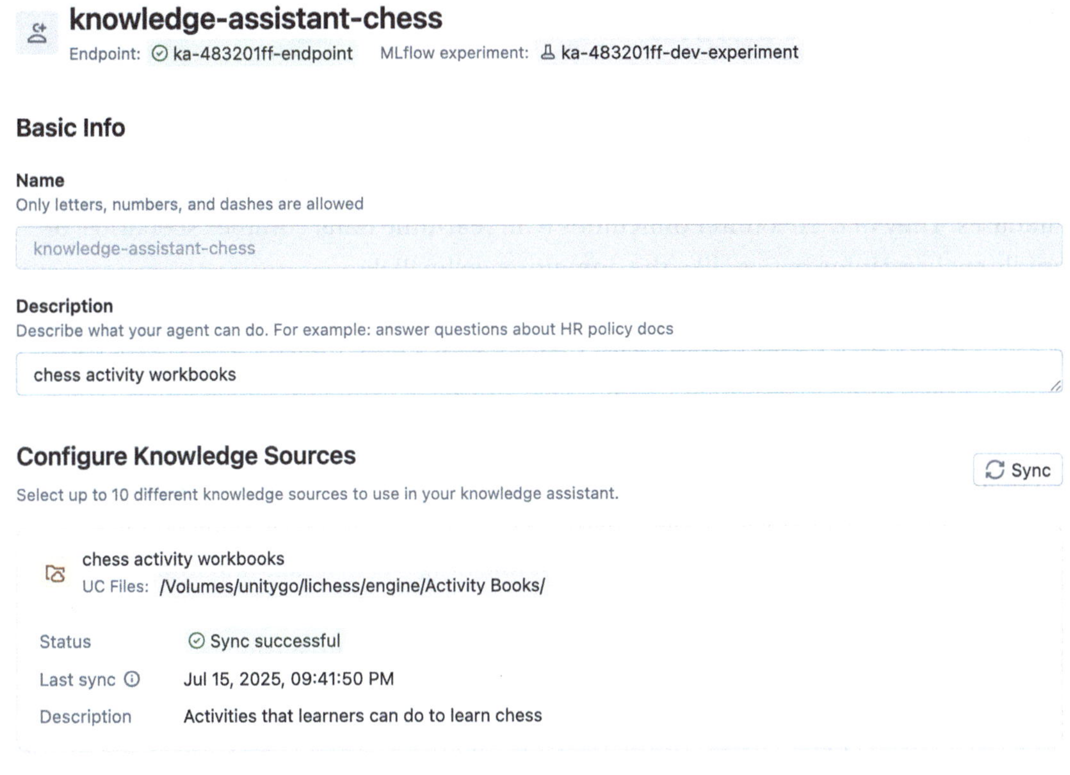

Figure 4-16. *Knowledge assistant sync status*

If you have read the section above, it is not hard to imagine that the sync is building a vector search index behind the scenes. And we can find a corresponding vector index in Databricks to confirm this. All the vector search indexes, whether they're generated by Agent Bricks or created manually, can be found in the Compute plane in the Vector Search tab, as shown in Figure 4-17.

Figure 4-17. *Vector search indexes*

Quality Improvement

Agent Bricks aims to abstract away the technical details from users so we can focus on the outcome rather than the technical know-how. As discussed above, large language models (LLMs) are probabilistic and highly dependent on the similarity search in vector databases. They may encounter difficulties with real-time data, complex scenarios, or rapidly evolving information, like the domain-specific PDFs.

Databricks solves this problem by integrating a feedback mechanism into Agent Bricks. Without that, teams would resort to iterative prompt adjustments or extensive manual intervention to maintain quality. This approach is highly inefficient and time-consuming, resulting in months before production. Consequently, RAG systems should be engineered for continuous learning based on their operational usage, rather than solely relying on initial training, because high-quality data is essential for building effective systems. This can be achieved by establishing clear signals and integrating feedback loops into the product design.

Databricks has included a Review App in Agent Bricks, which can also be used as a standalone component, to make it easy to do reviews. For more information about using the Review App as a standalone component, please refer to Chapter 11. A labeling session can be shared with human experts. This is essentially a question-and-answer session where we can curate a list of questions that we want to answer with this chatbot and have our experts set the tone and expectations. The review app is tightly integrated with Knowledge Assistant, so we don't need to build additional applications to manage these reviews.

As shown in Figures 4-18 and 4-19, human experts can give guidelines, which are similar to the guidelines in a Custom LLM to tune the agent as well as provide feedback on the answer. The differences are that the guidelines are there to teach the agent, while the feedback is more free-form comments.

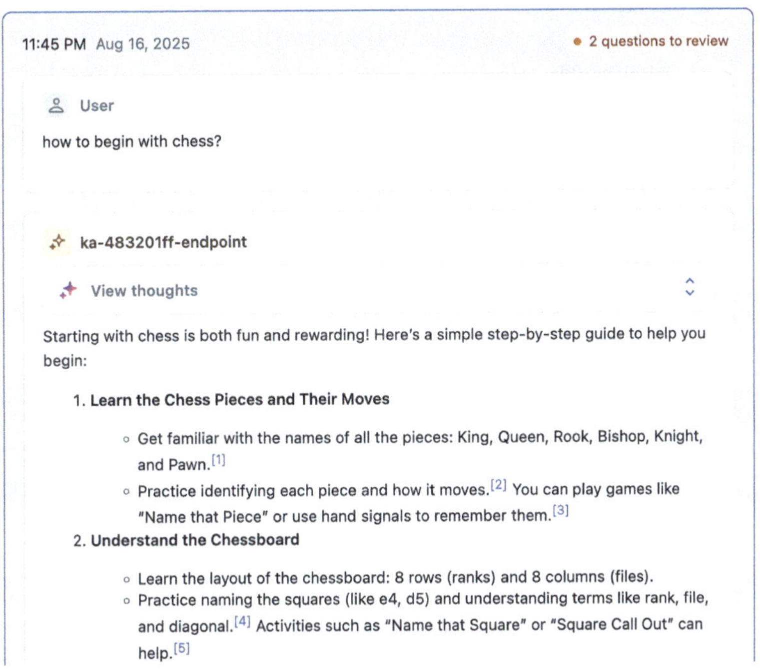

Figure 4-18. *Review session with agent output*

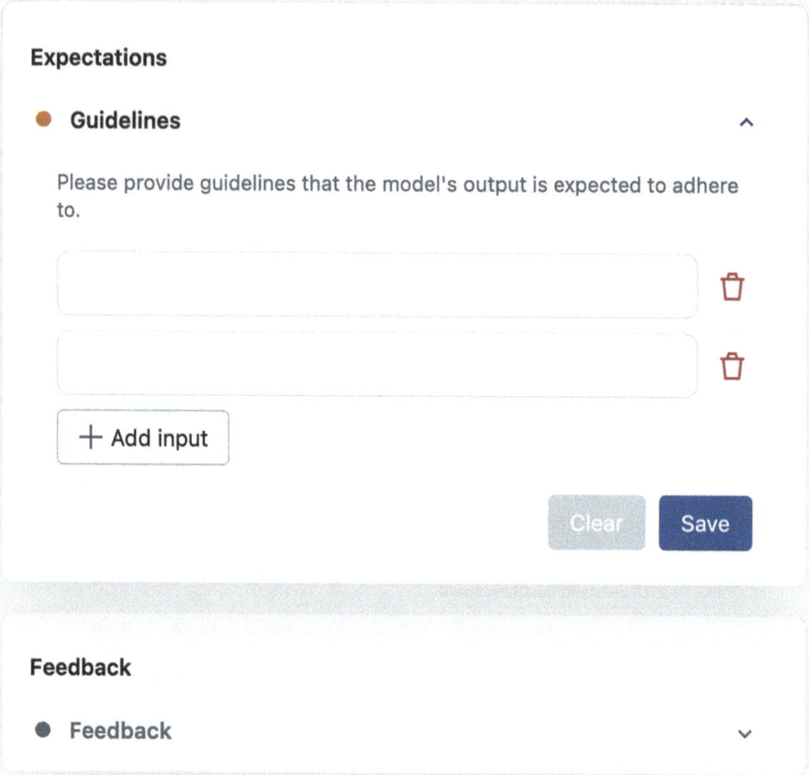

Expectations

● **Guidelines**

Please provide guidelines that the model's output is expected to adhere to.

+ Add input

Clear   Save

Feedback

● **Feedback**

Figure 4-19. *Review session's feedback section*

How Many Iterations Are Needed to Improve Quality?

The question of how many iterations are needed to improve the quality of a knowledge assistant is not simple, as it depends heavily on the learning paradigm used. In traditional machine learning models, where systems learn from vast, static datasets, significant improvements often require hundreds or thousands of labeled examples, and the F1 score must be monitored continuously for various drift scenarios. However, Databricks pioneers Agent Learning from Human Feedback (ALHF), which fundamentally changes this dynamic, as demonstrated by the following Databricks research blog post:

```
https://www.databricks.com/blog/agent-learning-human-feedback-alhf-
databricks-knowledge-assistant-case-study
```

Instead of relying on a large, static dataset, ALHF allows an agent to learn directly from minimal natural-language feedback from human experts. Knowledge Assistant leverages ALHF behind the scenes, "drastically boosts the overall answer quality on Databricks DocsQA with as few as 4 feedback records" (`https://www.databricks.com/blog/agent-learning-human-feedback-alhf-databricks-knowledge-assistant-case-study`). Figure 4-20 shows the answer quality on the DocsQA dataset. The DocsQA evaluation can be found on Databricks Labs' GitHub:

`https://github.com/databrickslabs/doc-qa`

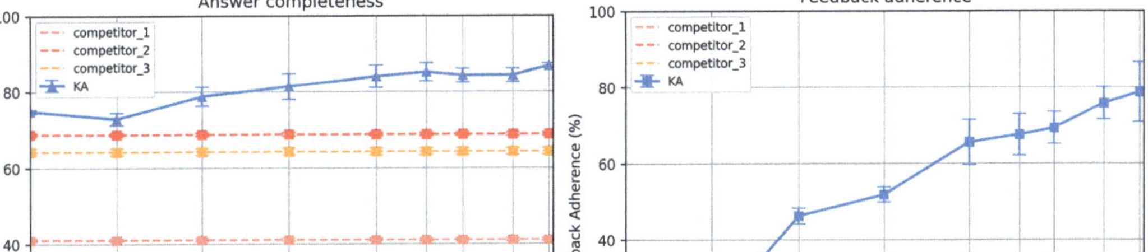

Figure 4-20. *Answer quality on DocsQA. Data from Databricks Labs, DocsQA repository.* `https://github.com/databrickslabs/doc-qa.` *Used under the Databricks License (© 2023 Databricks, Inc.). (This image is not included in the book's Creative Commons license.)*

The results are even more compelling with a slightly larger number of iterations. The research states, "With just 32 feedback records, we more than quadruple the answer quality over the static baselines." This rapid improvement with a small number of feedback rounds is a testament to the "sample-efficient" nature of the ALHF approach.

The document explicitly notes that "you don't need hundreds or thousands" of feedback examples to achieve significant results.

This high efficiency is made possible by the agent's ability to generalize from a single piece of feedback. The Databricks research explains this through two core technical challenges that were successfully addressed:

- **Learning When to Apply Feedback (Scoping):** The agent uses a memory-based approach to record all prior feedback. When a new question is posed, the agent can intelligently determine which feedback is relevant and apply it dynamically. For example, feedback on a PostgreSQL-specific SQL query isn't limited to that one question; it informs all future SQL-related questions.

- **Adapting the Right System Components (Assignment):** The KA is a multi-component system, and ALHF ensures that feedback is routed to the correct part of the pipeline. This allows the agent to refine specific behaviors, such as modifying its search queries to better suit the expert's expectations.

Conclusion

In this chapter, we have discussed the internal workings of Retrieval Augmented Generation (RAG) and the transformative power of Databricks' Knowledge Assistant within the Agent Bricks framework. RAG is essential because LLMs hallucinate when they lack grounded information. Enterprises often require answers based on private, proprietary, or frequently updated content, none of which is available in the model's pretraining data. RAG allows the model to retrieve factual, contextual data from internal sources and generate answers based on real evidence, dramatically improving accuracy, trust, and compliance.

From the foundational concepts of embeddings and the crucial role of vector databases to performance optimization techniques, Databricks offers various chunking strategies and the twin processes of similarity search and reranking. We now understand what is under the hood of Agent Bricks Knowledge Assistant.

## CHAPTER 5

Agentic AI Design Patterns for Enterprise Systems

In the previous chapters, we have discussed different types of agents, including Information Extraction, Custom LLM, and Knowledge Assistant. These are great use cases, but at the same time, they are isolated for a specific use case. This is obviously by design because they are built for domain-specific purposes. What if we can chain all these agents together to create a very powerful system? This is called an Agentic AI workflow. We will first discuss the various Agentic AI design patterns as the foundation and then examine Agent Bricks' multi-agent supervisor.

AI agent workflows will likely drive massive progress in the industry, as we have already seen use cases such as agentic coding and business process automation. Foundation models will fade into the background as the engine, with less attention paid to the version (like GPT-4 or GPT-5). Instead of a single-pass generation, an agentic workflow allows an LLM to iterate on a task. This could involve planning an outline, using tools to gather information, drafting a response, and then reflecting on that draft to spot weaknesses and revise it. This iterative process yields much better results.

A Case Study

The performance gains from this shift are significant. According to DeepLearning.ai's experiment (`https://www.deeplearning.ai/the-batch/how-agents-can-improve-llm-performance/`), even in older versions of GPT-3.5 vs. GPT-4, we can observe differences in benchmarking techniques and the tools developed compared to the base model. There have been many discussions about how people don't care about benchmarking because they want their agents to understand their data. However, would you hire an accountant who is not certified to work on your finances? That's the reason we still need to care about benchmarking, but benchmarking alone is only the accreditation, not the work itself.

Figure 5-1 outlines the performance of GPT-3.5 and GPT-4. The zero-shot (green circle) indicates baseline performance without any tuning or agentic patterns. As we move to the right, we can observe patterns that will be used to improve agent performance. As we can see, GPT-3.5's zero-shot score of 48.1% is significantly lower than GPT-4's 67.0%. However, when GPT-3.5 is wrapped in an agent loop using advanced strategies, its performance can soar as high as 95.1%, far surpassing the base GPT-4 model. On the other hand, the multi-agent coder on GPT-4 achieved a score of 96%, which is nearly perfect. As mentioned before, a straight A student won't be able to solve every coding problem on earth, but it provides good credentials. That's how we should look at these benchmarking experiments.

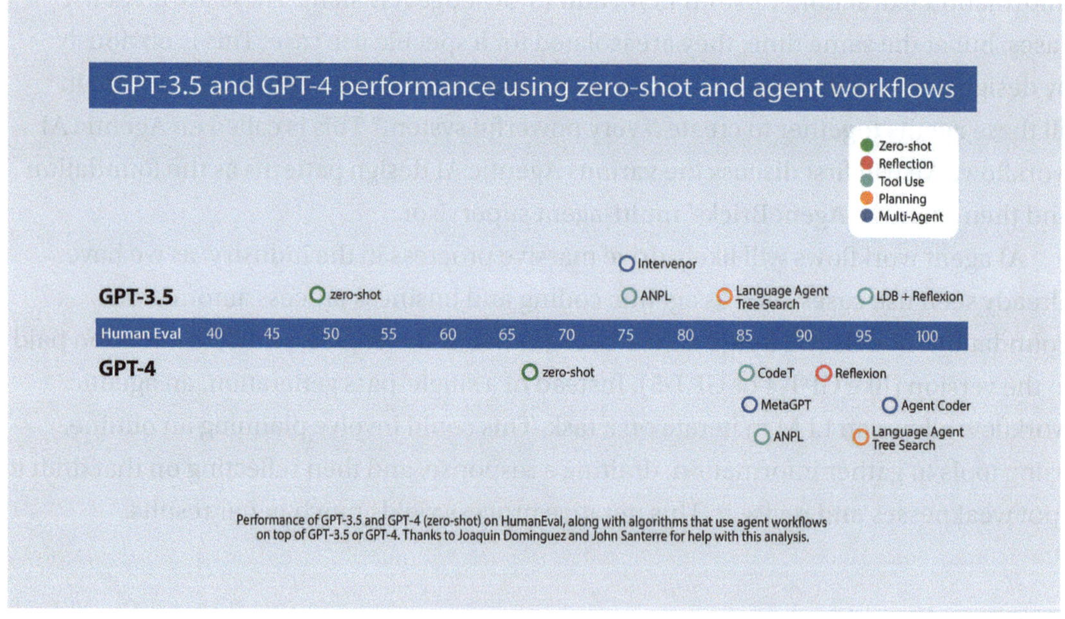

Figure 5-1. *Performance comparison chart of GPT-3.5 vs. GPT-4*

In this chapter, we will examine the key design patterns that enable this leap in performance. These patterns are the foundational building blocks for creating robust, reliable, and intelligent AI agents.

Deep Dive into Agentic Patterns

While the world of AI agents is evolving rapidly, a set of core design patterns has emerged and is successfully used in production applications today. Understanding the fundamentals will allow us to choose the right tool for the right use case. Below, we will explore five of these foundational patterns: Tool Use, Reflection, Planning, Multi-agent Collaboration, and ReAct (Reasoning and Acting).

Visualizing Agents with LangGraph

As we go through the design patterns, the easiest way to learn how they work and troubleshoot any issues we may encounter is through visualization. However, a static diagram can only go so far. In this chapter, we will also provide code in LangGraph that you can visualize and execute using Databricks' Foundation Model API.

Tool Use

To understand the tool use pattern, we must first understand what tools are. The definition of tools changes over time, but it's more than a pre-defined piece of logic. While LLMs can code and figure out many things on their own, organizations often have their own business rules. For example, the fiscal year start and end dates vary by organization. The question of annual fiscal revenue would likely be different if we treated it as a calendar year. Figure 5-2 is an illustration of the Tool Use pattern.

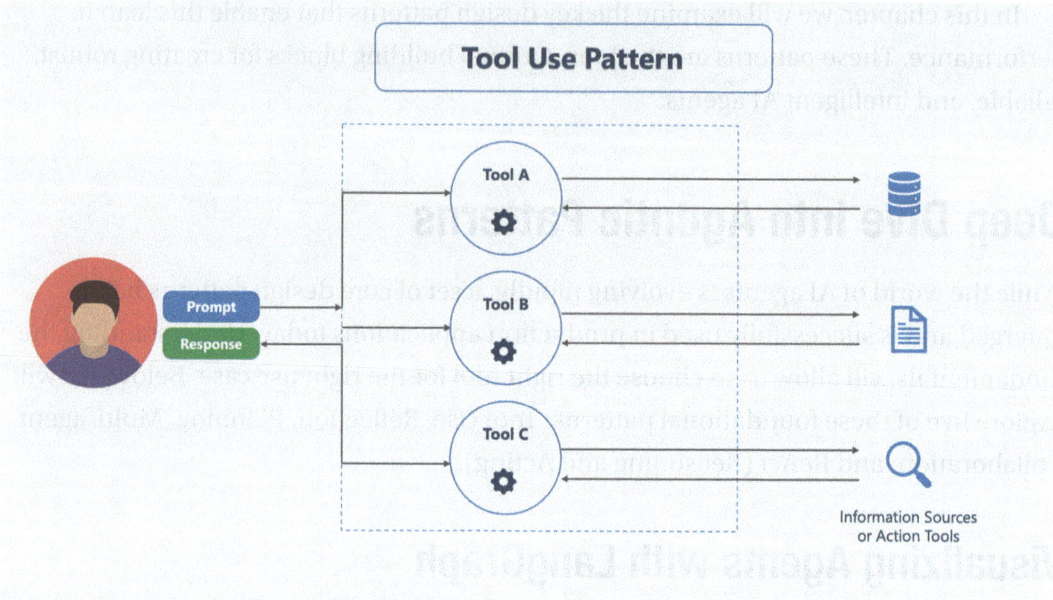

Figure 5-2. *The Tool Use pattern*

The Tool Use pattern moves an agent from a simple advisor to an active operator by calling the functions provided. This is why tool use is often referred to as function calling. It gives an LLM access to external functions it can call to gather information, take action, or process data. Instead of relying solely on its pre-trained knowledge, the agent can interact directly with enterprise systems, APIs, and other information sources.

The main difference between Tool Use and traditional programming is that the LLM will choose which function to call and when. The LLM dynamically decides which tool to invoke and with what arguments at runtime, rather than following a predetermined control flow. For example, when faced with a calculation, it could generate a request to run Python code, such as `principal * (1 + interest_rate) ** years,` to compute compound interest accurately. This bridges the gap between knowledge and action, allowing agents to complete tasks, update records, and orchestrate workflows. Figure 5-3 is in graph form, and the code in Listing 5-1 will display the graph and allow us to execute it.

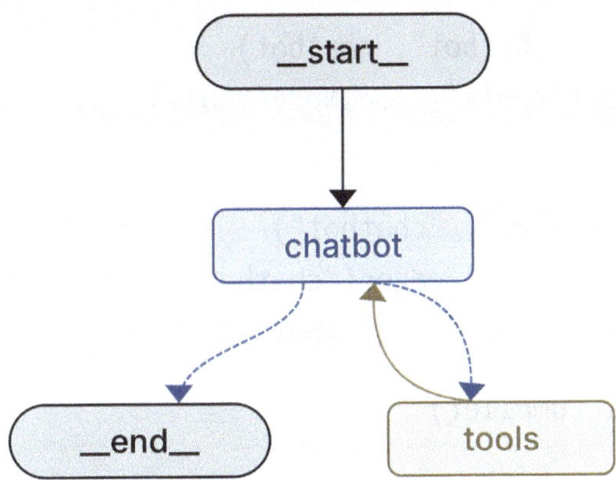

Figure 5-3. *The LangGraph workflow of the Tool Use pattern*

Listing 5-1. Tool Use pattern in LangGraph with ChatDatabricks

```
import os
from dotenv import load_dotenv

from langchain_tavily import TavilySearch
from databricks_langchain import ChatDatabricks
from langgraph.graph import StateGraph, START, END, MessagesState
from langgraph.prebuilt import ToolNode, tools_condition

load_dotenv()

# Create tools
tool = TavilySearch(max_results=2)
tools = [tool]

# Create LLM with tools
llm = ChatDatabricks(endpoint="databricks-claude-sonnet-4")
llm_with_tools = llm.bind_tools(tools)

def chatbot(state: MessagesState) -> dict:
    return {"messages": [llm_with_tools.invoke(state["messages"])]}

# Build the graph
graph_builder = StateGraph(MessagesState)
```

```python
# Add nodes
graph_builder.add_node("chatbot", chatbot)
graph_builder.add_node("tools", ToolNode(tools))

# Add edges
graph_builder.add_edge(START, "chatbot")
graph_builder.add_conditional_edges("chatbot", tools_condition)
graph_builder.add_edge("tools", "chatbot")

graph = graph_builder.compile()
app = graph
```

A real-world example of tool use is chess-playing agents. Although the AI chess problem is not new, LLMs' chess knowledge can be laughable at times. This is because it is a general AI rather than a specialized one. By equipping an LLM with a chess evaluation function, we can elevate it to chess master level in an instant, without teaching it more chess knowledge. The following blog post discusses this in depth:

```
https://medium.com/@jasonyip_77999/agentic-chess-against-
stockfish-1-0-288bf7e6114
```

Reflection

The Reflection pattern focuses on self-improvement to enhance reliability. It enables the agent to assess and improve its own outputs without constant human supervision. At its core, reflection involves automating the process of providing critical feedback. Figure 5-4 outlines this pattern.

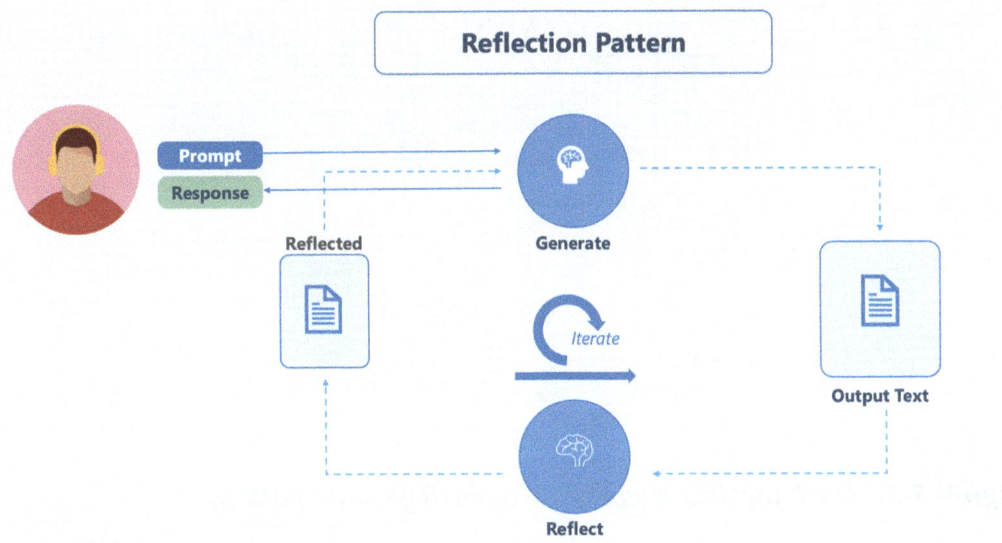

Figure 5-4. *The Reflection pattern*

The process is straightforward but effective. First, the LLM is prompted to complete a task, such as writing code. Then, in a subsequent step, it is prompted to reflect on its own work. For example, a prompt might be: "Check the code carefully for correctness, style, and efficiency, and give constructive criticism for how to improve it." This often causes the LLM to identify its own errors. Finally, the agent is prompted again to rewrite the code, taking its own feedback into account. This loop can be repeated to yield further improvements. This process can be further enhanced by giving the agent tools to help its evaluation, such as running code against unit tests to check for errors. Figure 5-5 is in graph form, and the code in Listing 5-2 will display the graph and allow us to execute it.

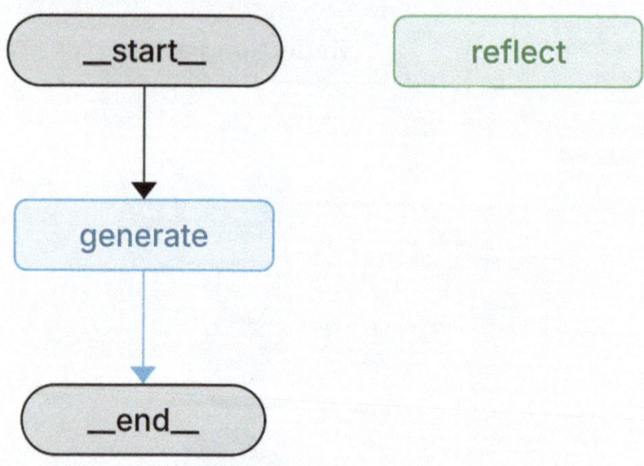

Figure 5-5. *The LangGraph workflow of the Reflection pattern*

Listing 5-2. Reflection pattern in LangGraph with ChatDatabricks

```python
import os
from dotenv import load_dotenv
from langchain_core.messages import AIMessage, BaseMessage, HumanMessage
from langchain_core.prompts import ChatPromptTemplate, MessagesPlaceholder
from databricks_langchain import ChatDatabricks
from langgraph.graph import END, StateGraph, START, MessagesState

load_dotenv()

llm = ChatDatabricks(
    endpoint="databricks-claude-sonnet-4"
)

# Generation prompt and chain
generation_prompt = ChatPromptTemplate.from_messages([
    (
        "system",
        """You are an essay assistant tasked with writing excellent
        5-paragraph essays. Generate the best essay possible for the user's
        request. If the user provides critique, respond with a revised
        version of your previous attempts.""",
    ),
```

```python
    MessagesPlaceholder(variable_name="messages"),
])
generate = generation_prompt | llm

# Reflection prompt and chain
reflection_prompt = ChatPromptTemplate.from_messages([
    (
        "system",
        """You are a teacher grading an essay submission. Generate critique
        and recommendations for the user's submission. Provide detailed
        recommendations, including requests for length, depth, style, etc.""",
    ),
    MessagesPlaceholder(variable_name="messages"),
])
reflect = reflection_prompt | llm

async def generation_node(state: MessagesState) -> MessagesState:
    """Generate or revise the essay based on current messages"""
    return {"messages": [await generate.ainvoke(state["messages"])]}

async def reflection_node(state: MessagesState) -> MessagesState:
    """Reflect on the generated essay and provide critique"""
    # Swap message roles: reflector sees AI-generated essay as human input
      cls_map = {"ai": HumanMessage, "human": AIMessage}
    translated = [state["messages"][0]] + [  # Keep original request as-is
        cls_map[msg.type](content=msg.content) for msg in
        state["messages"][1:]
    ]
    res = await reflect.ainvoke(translated)
    # Return critique as human feedback for the generator
    return {"messages": [HumanMessage(content=res.content)]}

def should_continue(state: MessagesState) -> str:
    """Decide whether to continue reflecting or end the process"""
    ai_messages = [msg for msg in state["messages"] if msg.type == "ai"]
    if len(ai_messages) >= 3:
        return END
    return "reflect"
```

```python
# Build the graph
builder = StateGraph(MessagesState)
builder.add_node("generate", generation_node)
builder.add_node("reflect", reflection_node)

# Add edges
builder.add_edge(START, "generate")
builder.add_conditional_edges("generate", should_continue)
builder.add_edge("reflect", "generate")

graph = builder.compile()
app = graph
```

A real-world example is Anthropic's "think tool," which allows the LLM to stop and think about whether it has all the information it needs to move forward. According to Anthropic, "think tool" can almost double its baseline performance without any tools, as shown in Figure 5-6, which is another powerful confirmation of the performance shown in Figure 5-1 with tool use.

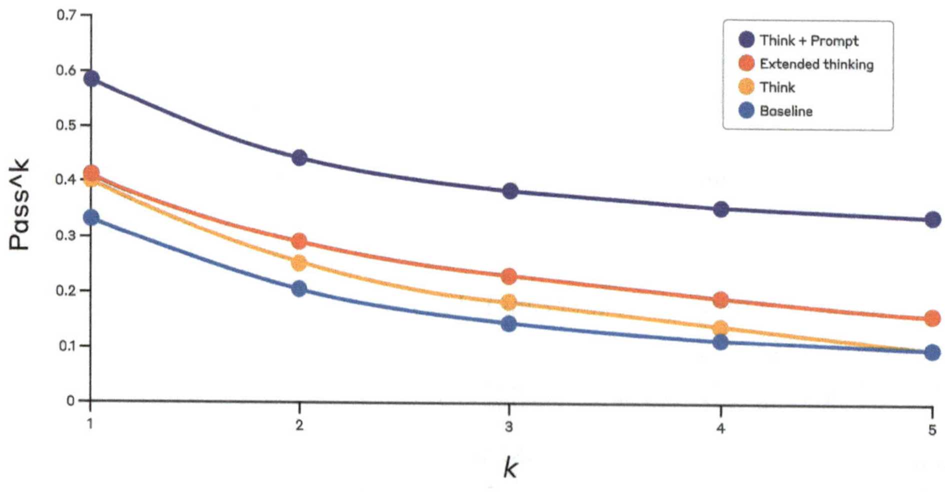

Figure 5-6. *Claude 3.7 Sonnet's performance on the "airline" domain of the Tau-Bench eval*

For more information about the "think tool," please refer to Anthropic's blog:
`https://www.anthropic.com/engineering/claude-think-tool`

Planning

Most real-world business processes involve multiple steps with complex
dependencies. The Planning pattern addresses this by enabling an agent to break down a
high-level goal into a sequence of actionable tasks. The agent can then track its progress
and adapt the plan as requirements change. Figure 5-7 illustrates this pattern.

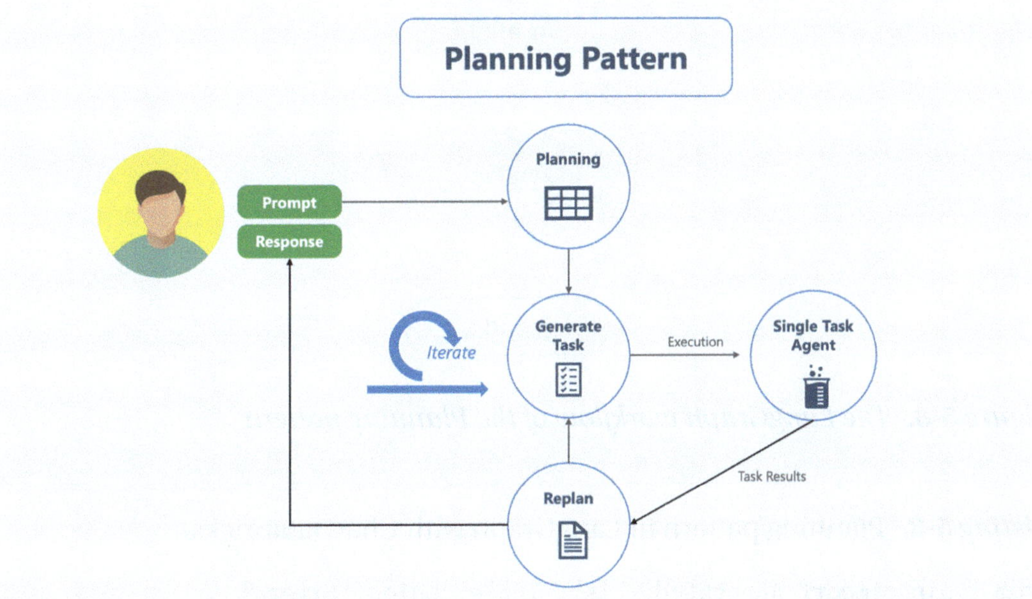

Figure 5-7. *Planning pattern*

For example, if an agent is asked to research a topic, a planning LLM could first
break down the objective into subtasks like "research specific subtopics," "synthesize
findings," and "compile a report." An agent can then execute these steps in order.
This pattern is powerful for complex tasks where the path to a solution isn't known in
advance. However, as seen in Figure 5-8, it is a less mature technology than Reflection;
it is nevertheless an important pattern to guide some tasks that users don't have clarity
to begin with. Figure 5-8 is the graph form, and the code from Listing 5-3 will show the
graph and allow us to execute.

Figure 5-8. *The LangGraph workflow of the Planning pattern*

Listing 5-3. Planning pattern in LangGraph with ChatDatabricks

```python
from typing import Annotated, List, Tuple, Union, Literal
from typing_extensions import TypedDict
from pydantic import BaseModel, Field

from langchain_core.prompts import ChatPromptTemplate
from langchain_community.tools.tavily_search import TavilySearchResults
from databricks_langchain import ChatDatabricks
from langgraph.prebuilt import create_react_agent
from langgraph.graph import StateGraph, START, MessagesState

# State definition that extends MessagesState for Chat mode compatibility
class PlanExecute(MessagesState):
    plan: List[str]
    past_steps: Annotated[List[Tuple], operator.add]
    response: str
```

```python
# Pydantic models for structured output
class Plan(BaseModel):
    """Plan to follow in future"""
    steps: List[str] = Field(
        description="different steps to follow, should be in sorted order"
    )

class Response(BaseModel):
    """Response to user."""
    response: str

class Act(BaseModel):
    """Action to perform."""
    action: Union[Response, Plan] = Field(
        description="""Action to perform. If you want to respond to user,
        use Response. If you need to further use tools to get the answer,
        use Plan."""
    )

# Initialize LLM and tools
llm = ChatDatabricks(
    endpoint="databricks-claude-sonnet-4"
)

tools = [TavilySearchResults(max_results=3)]

# Create execution agent
prompt = "You are a helpful assistant."
agent_executor = create_react_agent(llm, tools, prompt=prompt)

# Planning chain
planner_prompt = ChatPromptTemplate.from_messages([
    (
        "system",
        """For the given objective, come up with a simple step by step
        plan. This plan should involve individual tasks, that if executed
        correctly will yield the correct answer. Do not add any superfluous
        steps. The result of the final step should be the final answer.
```

```
        Make sure that each step has all the information needed - do not
        skip steps.""",
    ),
    ("placeholder", "{messages}"),
])
planner = planner_prompt | llm.with_structured_output(Plan)

# Re-planning chain
replanner_prompt = ChatPromptTemplate.from_template(
    """For the given objective, come up with a simple step by step plan.
This plan should involve individual tasks, that if executed correctly will
yield the correct answer. Do not add any superfluous steps. The result of
the final step should be the final answer. Make sure that each step has all
the information needed - do not skip steps.

Your objective was this:
{input}

Your original plan was this:
{plan}

You have currently done the following steps:
{past_steps}

Update your plan accordingly. If no more steps are needed and you can
return to the user, then respond with that. Otherwise, fill out the plan.
Only add steps to the plan that still NEED to be done. Do not return
previously done steps as part of the plan."""
)
replanner = replanner_prompt | llm.with_structured_output(Act)

# Node functions
async def plan_step(state: PlanExecute):
    """Create initial plan based on the last user message"""
    # Get the last user message as the objective
    user_message = next((msg.content for msg in reversed(state["messages"])
    if msg.type == "human"), "")
    plan = await planner.ainvoke({"messages": [("user", user_message)]})
    return {"plan": plan.steps}
```

```python
async def execute_step(state: PlanExecute):
    """Execute the current step in the plan"""
    plan = state["plan"]
    plan_str = "\n".join(f"{i+1}. {step}" for i, step in enumerate(plan))
    task = plan[0]
    task_formatted = f"""For the following plan:
{plan_str}

You are tasked with executing step 1: {task}."""

    agent_response = await agent_executor.ainvoke(
        {"messages": [("user", task_formatted)]}
    )
    return {
        "past_steps": [(task, agent_response["messages"][-1].content)],
    }

async def replan_step(state: PlanExecute):
    """Re-evaluate plan based on completed steps"""
    # Get the original user input for re-planning
    user_message = next((msg.content for msg in reversed(state["messages"])
    if msg.type == "human"), "")

    replan_input = {
        "input": user_message,
        "plan": state.get("plan", []),
        "past_steps": state.get("past_steps", [])
    }

    output = await replanner.ainvoke(replan_input)
    if isinstance(output.action, Response):
        # Add the final response as an AI message
        from langchain_core.messages import AIMessage
        return {"messages": [AIMessage(content=output.action.response)]}
    else:
        return {"plan": output.action.steps}
```

```python
def route_after_replan(state: PlanExecute) -> Literal["agent", "__end__"]:
    """Route to agent for next step execution or end the graph if a final
    response is ready."""
    # Check if the last message is an AI response (indicating completion)
    if state["messages"] and state["messages"][-1].type == "ai":
        # Check if it's a final response (not from agent_executor)
        last_msg = state["messages"][-1]
        if hasattr(last_msg, 'content') and not any(word in last_msg.
        content.lower() for word in ['step', 'plan', 'task']):
            return "__end__"
    return "agent"

# Build the graph
workflow = StateGraph(PlanExecute)

# Add nodes
workflow.add_node("planner", plan_step)
workflow.add_node("agent", execute_step)
workflow.add_node("replan", replan_step)

# Add edges
workflow.add_edge(START, "planner")
workflow.add_edge("planner", "agent")
workflow.add_edge("agent", "replan")
workflow.add_conditional_edges(
    "replan",
    should_end,
    ["agent", "__end__"],
)

graph = workflow.compile()
app = graph
```

Nowadays, all the Deep Research tools come with this capability. Figure 5-9 is the Deep Research planning from Google's Gemini.

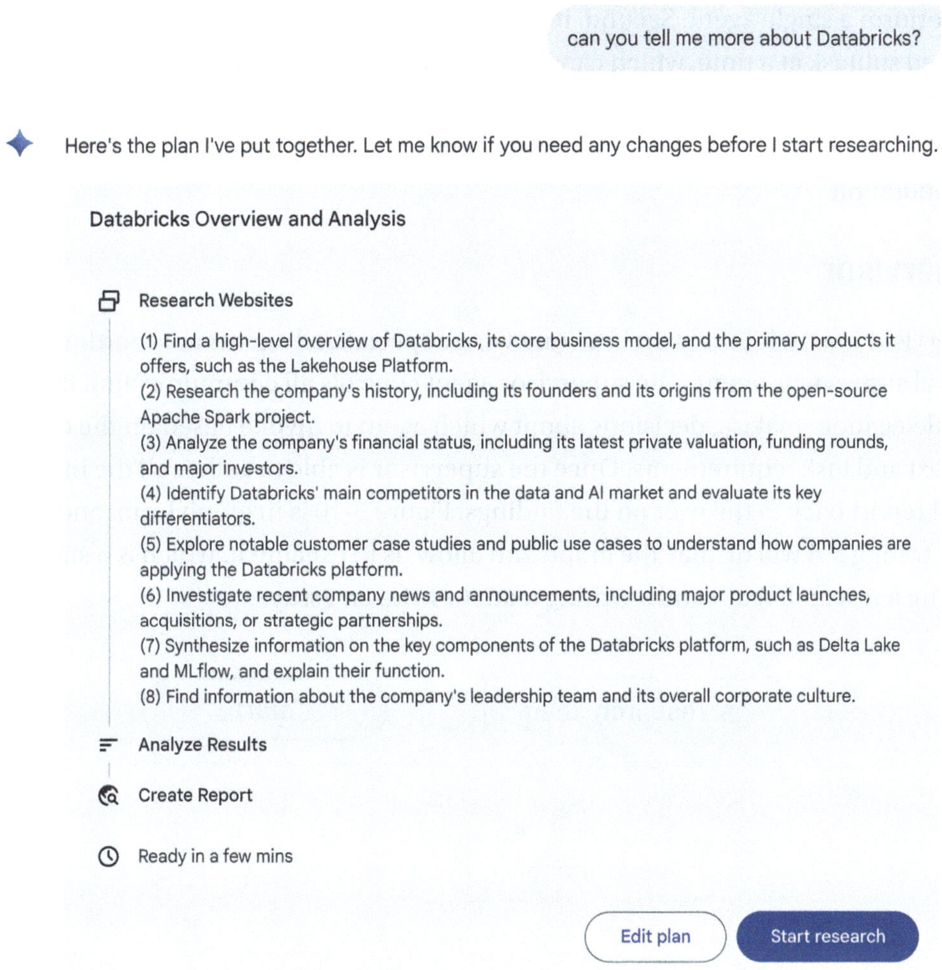

Figure 5-9. *Google Deep Research planning*

Multi-Agent Collaboration

Just as enterprises rely on teams of human specialists, the multi-agent pattern mirrors this structure by connecting a network of specialized AI agents, each focused on a different part of a workflow. A complex task, such as developing software, is broken down into subtasks to be executed by agents with different roles, including a product manager, a designer, a developer, and a QA engineer.

This approach has several advantages. First, as with human teams, multiple agents outperform a single agent. Second, it allows the LLM to focus on one specific, well-defined subtask at a time, which can improve quality. Finally, it provides developers with a powerful mental framework for decomposing complex problems.

There are mainly two different designs for multi-agents: Supervisor and Collaboration

Supervisor

Supervisor is a multi-agent architecture where specialized agents are coordinated by a central **supervisor agent**. The supervisor agent controls all communication flow and task delegation, making decisions about which agent to invoke based on the current context and task requirements. Once the supervisor is able to gather all the information, it will report back to the user on the findings. Figure 5-10 is in graph form, and the code from Listing 5-4 will display the graph and allow us to execute it, which is a supervisor leading a research team and a writing team to write an essay.

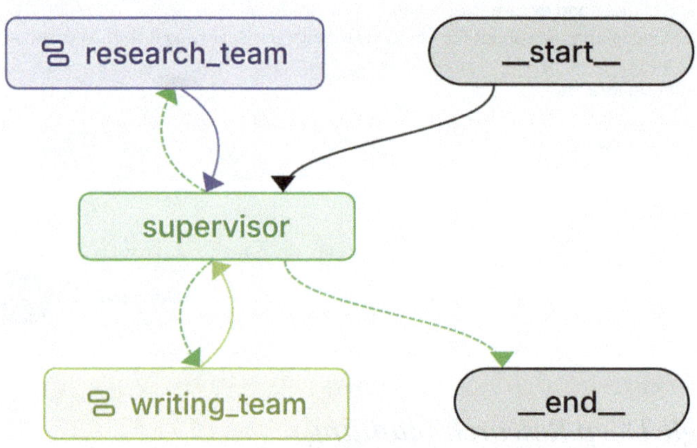

Figure 5-10. *The LangGraph workflow of the Supervisor pattern*

Listing 5-4. Supervisor pattern in LangGraph with ChatDatabricks

```
from pathlib import Path
from tempfile import TemporaryDirectory
from typing import Annotated, Dict, List, Optional, Literal
from typing_extensions import TypedDict
```

```python
from langchain_community.document_loaders import WebBaseLoader
from langchain_core.tools import tool
from langchain_core.messages import HumanMessage
from langchain_core.language_models.chat_models import BaseChatModel
from langchain_tavily import TavilySearch

from databricks_langchain import ChatDatabricks
from langgraph.prebuilt import create_react_agent
from langgraph.graph import StateGraph, MessagesState, START, END
from langgraph.types import Command

# Initialize temporary directory for file operations
_TEMP_DIRECTORY = TemporaryDirectory()
WORKING_DIRECTORY = Path(_TEMP_DIRECTORY.name)

class State(MessagesState):
    next: str

# Initialize LLM
llm = ChatDatabricks(
    endpoint="databricks-claude-sonnet-4-6",
)

# Research team tools
tavily_tool = TavilySearch(max_results=5)

@tool
def scrape_webpages(urls: list[str]) -> str:
    """Use requests and bs4 to scrape the provided web pages for detailed
information."""
    loader = WebBaseLoader(urls)
    docs = loader.load()
    return "\n\n".join([
        f'<Document name="{doc.metadata.get("title", "")}">\n{doc.page_
content}\n</Document>'
        for doc in docs
    ])
```

```python
# Document writing team tools
@tool
def create_outline(
    points: Annotated[List[str], "List of main points or sections."],
    file_name: Annotated[str, "File path to save the outline."],
) -> Annotated[str, "Path of the saved outline file."]:
    """Create and save an outline."""
    with (WORKING_DIRECTORY / file_name).open("w") as file:
        for i, point in enumerate(points):
            file.write(f"{i + 1}. {point}\n")
    return f"Outline saved to {file_name}"

@tool
def read_document(
    file_name: Annotated[str, "File path to read the document from."],
    start: Annotated[Optional[int], "The start line. Default is 0"] = None,
    end: Annotated[Optional[int], "The end line. Default is None"] = None,
) -> Annotated[str, "Contents of the document."]:
    """Read the specified document."""
    with (WORKING_DIRECTORY / file_name).open("r") as file:
        lines = file.readlines()
    if start is None:
        start = 0
    return "\n".join(lines[start:end])

@tool
def write_document(
    content: Annotated[str, "Text content to be written into the
    document."],
    file_name: Annotated[str, "File path to save the document."],
) -> Annotated[str, "Path of the saved document file."]:
    """Create and save a text document."""
    with (WORKING_DIRECTORY / file_name).open("w") as file:
        file.write(content)
    return f"Document saved to {file_name}"
```

```python
@tool
def edit_document(
    file_name: Annotated[str, "Path of the document to be edited."],
    inserts: Annotated[
        Dict[int, str],
        "Dictionary where key is the line number (1-indexed) and value is
        the text to be inserted at that line.",
    ],
) -> Annotated[str, "Path of the edited document file."]:
    """Edit a document by inserting text at specific line numbers."""
    with (WORKING_DIRECTORY / file_name).open("r") as file:
        lines = file.readlines()

    sorted_inserts = sorted(inserts.items(), reverse=True)
    for line_number, text in sorted_inserts:
        if 1 <= line_number <= len(lines) + 1:
            lines.insert(line_number - 1, text + "\n")
        else:
            return f"Error: Line number {line_number} is out of range."

    with (WORKING_DIRECTORY / file_name).open("w") as file:
        file.writelines(lines)
    return f"Document edited and saved to {file_name}"

# Supervisor node factory
def make_supervisor_node(llm: BaseChatModel, members: list[str]):
    options = ["FINISH"] + members
    system_prompt = f"""You are a supervisor tasked with managing a
    conversation between the following workers: {', '.join(members)}. Given
    the following user request, respond with the worker to act next. Each
    worker will perform a task and respond with their results and status.
    When finished, respond with FINISH."""

    class Router(TypedDict):
        """Worker to route to next. If no workers needed, route to
        FINISH."""
        next: Literal[*options]
```

```python
    def supervisor_node(state: State) -> Command[Literal[*members, "__
end__"]]:
        """An LLM-based router."""
        messages = [
            {"role": "system", "content": system_prompt},
        ] + state["messages"]
        response = llm.with_structured_output(Router).invoke(messages)
        goto = response["next"]
        if goto == "FINISH":
            goto = END
        return Command(goto=goto, update={"next": goto})

    return supervisor_node

# Research Team
search_agent = create_react_agent(llm, tools=[tavily_tool])

def search_node(state: State) -> Command[Literal["supervisor"]]:
    result = search_agent.invoke(state)
    return Command(
        update={
            "messages": [
                HumanMessage(content=result["messages"][-1].content,
                name="search")
            ]
        },
        goto="supervisor",
    )

web_scraper_agent = create_react_agent(llm, tools=[scrape_webpages])

def web_scraper_node(state: State) -> Command[Literal["supervisor"]]:
    result = web_scraper_agent.invoke(state)
    return Command(
        update={
            "messages": [
                HumanMessage(content=result["messages"][-1].content,
                name="web_scraper")
```

```python
            ]
        },
        goto="supervisor",
    )

research_supervisor_node = make_supervisor_node(llm, ["search", "web_
scraper"])

# Build research team graph
research_builder = StateGraph(State)
research_builder.add_node("supervisor", research_supervisor_node)
research_builder.add_node("search", search_node)
research_builder.add_node("web_scraper", web_scraper_node)
research_builder.add_edge(START, "supervisor")
research_graph = research_builder.compile()

# Document Writing Team
doc_writer_agent = create_react_agent(
    llm,
    tools=[write_document, edit_document, read_document],
    prompt="""You can read, write and edit documents based on note-taker's
    outlines. Don't ask follow-up questions.""",
)

def doc_writing_node(state: State) -> Command[Literal["supervisor"]]:
    result = doc_writer_agent.invoke(state)
    return Command(
        update={
            "messages": [
                HumanMessage(content=result["messages"][-1].content,
                name="doc_writer")
            ]
        },
        goto="supervisor",
    )
```

```python
note_taking_agent = create_react_agent(
    llm,
    tools=[create_outline, read_document],
    prompt="""You can read documents and create outlines for the document
    writer. Don't ask follow-up questions.""",
)

def note_taking_node(state: State) -> Command[Literal["supervisor"]]:
    result = note_taking_agent.invoke(state)
    return Command(
        update={
            "messages": [
                HumanMessage(content=result["messages"][-1].content,
                name="note_taker")
            ]
        },
        goto="supervisor",
    )

doc_writing_supervisor_node = make_supervisor_node(
    llm, ["doc_writer", "note_taker"]
)

# Build document writing team graph
paper_writing_builder = StateGraph(State)
paper_writing_builder.add_node("supervisor", doc_writing_supervisor_node)
paper_writing_builder.add_node("doc_writer", doc_writing_node)
paper_writing_builder.add_node("note_taker", note_taking_node)
paper_writing_builder.add_edge(START, "supervisor")
paper_writing_graph = paper_writing_builder.compile()

# Top-level supervisor and team coordination
# Orchestrates the research and writing teams via a hierarchical graph
teams_supervisor_node = make_supervisor_node(llm, ["research_team",
"writing_team"])
```

```python
def call_research_team(state: State) -> Command[Literal["supervisor"]]:
    # Pass the full conversation context to the research team
    response = research_graph.invoke({"messages": state["messages"]})
    return Command(
        update={
            "messages": [
                HumanMessage(content=response["messages"][-1].content,
                name="research_team")
            ]
        },
        goto="supervisor",
    )

def call_paper_writing_team(state: State) -> Command[Literal["supe
rvisor"]]:
    # Pass the full conversation context to the writing team
    response = paper_writing_graph.invoke({"messages": state["messages"]})
    return Command(
        update={
            "messages": [
                HumanMessage(content=response["messages"][-1].content,
                name="writing_team")
            ]
        },
        goto="supervisor",
    )

# Build top-level hierarchical graph
super_builder = StateGraph(State)
super_builder.add_node("supervisor", teams_supervisor_node)
super_builder.add_node("research_team", call_research_team)
super_builder.add_node("writing_team", call_paper_writing_team)
super_builder.add_edge(START, "supervisor")

# Compile the graph
graph = super_builder.compile()
app = graph
```

Agent Collaboration

In this scenario, the different agents collaborate on a shared workspace of messages. This means that all their work is visible to others. This has the benefit that other agents can see all the individual steps and their progress. We call this collaboration because the agents have a peer status and communicate with each other. Figure 5-11 is in graph form, and the code in Listing 5-5 will display the graph and allow us to execute it.

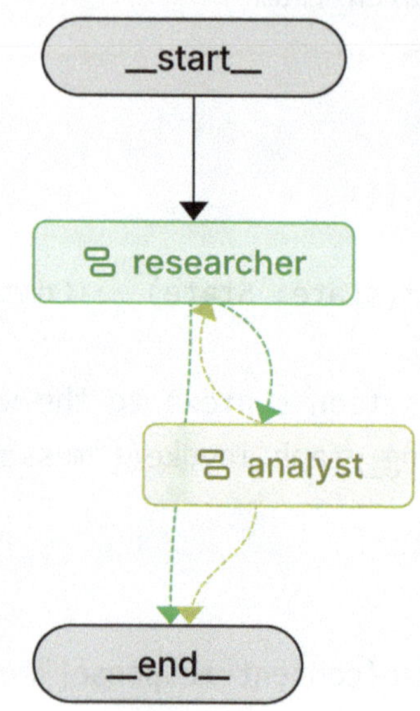

Figure 5-11. *The LangGraph workflow of the Collaboration pattern*

Listing 5-5. Collaboration pattern in LangGraph with ChatDatabricks

```
from typing import Annotated, Literal

from langchain_core.messages import BaseMessage, HumanMessage
from langchain_core.tools import tool
from langchain_tavily import TavilySearch

from databricks_langchain import ChatDatabricks
from langgraph.prebuilt import create_react_agent
from langgraph.graph import MessagesState, END, StateGraph, START
from langgraph.types import Command
```

```python
llm = ChatDatabricks(
    endpoint="databricks-claude-sonnet-4"
)

tavily_tool = TavilySearch(max_results=5)

def make_system_prompt(suffix: str) -> str:
    return f"""You are a helpful AI assistant, collaborating with other
    assistants. Use the provided tools to progress towards answering
    the question. If you are unable to fully answer, that's OK, another
    assistant with different tools will help where you left off. Execute
    what you can to make progress. If you or any of the other assistants
    have the final answer or deliverable, prefix your response with FINAL
    ANSWER so the team knows to stop.

{suffix}"""

# Helper function to determine next node
def get_next_node(last_message: BaseMessage, goto: str):
    if last_message.content and "FINAL ANSWER" in last_message.content:
        # Any agent decided the work is done
        return END
    return goto

# Research agent and node
research_agent = create_react_agent(
    llm,
    tools=[tavily_tool],
    prompt=make_system_prompt(
        """You can only do research. You are working with an analyst
        colleague."""
    ),
)

def research_node(
    state: MessagesState,
) -> Command[Literal["analyst", "__end__"]]:
    result = research_agent.invoke(state)
```

```python
    goto = get_next_node(result["messages"][-1], "analyst")
    # Wrap in a HumanMessage so other agents in the graph treat it as
    user input
    result["messages"][-1] = HumanMessage(
        content=result["messages"][-1].content, name="researcher"
    )
    return Command(
        update={
            # Share internal message history of research agent with
            other agents
            "messages": result["messages"],
        },
        goto=goto,
    )

# Analyst agent and node
analyst_agent = create_react_agent(
    llm,
    tools=[],
    prompt=make_system_prompt(
        """You analyze and summarize research data. You are working with
        a researcher colleague. Provide insights, trends, and conclusions
        based on the research provided."""
    ),
)

def analyst_node(state: MessagesState) -> Command[Literal["researcher",
"__end__"]]:
    result = analyst_agent.invoke(state)
    goto = get_next_node(result["messages"][-1], "researcher")
    # Wrap in a human message for compatibility
    result["messages"][-1] = HumanMessage(
        content=result["messages"][-1].content, name="analyst"
    )
    return Command(
        update={
```

```
            "messages": result["messages"],
        },
        goto=goto,
    )

# Build the graph
workflow = StateGraph(MessagesState)
workflow.add_node("researcher", research_node)
workflow.add_node("analyst", analyst_node)

workflow.add_edge(START, "researcher")

graph = workflow.compile()
app = graph
```

ReAct (Reason + Act)

When a static plan is insufficient, the ReAct pattern enables an agent to solve problems adaptively in real time. Instead of following a fixed script, ReAct agents alternate between two steps: *reasoning* about the problem and *acting* to make progress. After each action, the agent observes the result and uses that new information to decide what to do next.

This cycle of reasoning, acting, and observing allows the agent to handle ambiguity, unexpected errors, and situations where the best path forward is not clear from the start. Figure 5-12 is the graph form, and the code from Listing 5-6 will show the graph and allow us to execute.

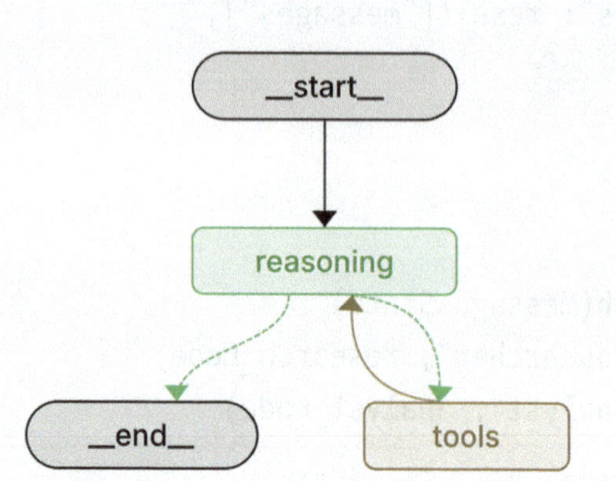

Figure 5-12. *The LangGraph workflow of the ReAct Pattern*

Listing 5-6. ReAct pattern in LangGraph with ChatDatabricks

```python
from datetime import UTC, datetime
from typing import Dict, List, Literal, Optional, Any, cast

from langchain_core.messages import AIMessage, BaseMessage
from langchain_core.tools import tool
from langchain_tavily import TavilySearch

from databricks_langchain import ChatDatabricks
from langgraph.graph import StateGraph, MessagesState, START
from langgraph.prebuilt import ToolNode

llm = ChatDatabricks(
    endpoint="databricks-claude-sonnet-4-6",
)

# ReAct System prompt
SYSTEM_PROMPT = """You are a helpful AI assistant that uses the ReAct
(Reasoning and Acting) approach to solve problems.

You should follow this pattern:
1. **Thought**: Think about what you need to do to answer the question
2. **Action**: Use a tool if needed to gather information
3. **Observation**: Analyze the results from the tool
```

4. Repeat steps 1-3 as needed

5. **Final Answer**: Provide your conclusion when you have enough information

Always start your response with "Thought:" and clearly show your reasoning process.
When you use a tool, explain why you're using it.
When you get results, analyze what they mean.

System time: {system_time}"""

```python
@tool
async def search(query: str) -> Optional[dict[str, Any]]:
    tavily = TavilySearch(max_results=10)
    return cast(dict[str, Any], await tavily.ainvoke({"query": query}))

TOOLS = [search]

class State(MessagesState):
    is_last_step: bool = False
    reasoning_step_count: int = 0

async def reasoning_step(state: State) -> dict[str, Any]:
    """ReAct reasoning step — the model thinks about what to do next.

    This is where the 'Reasoning' part of ReAct happens.
    """
    # Initialize the model with tool binding
    model = llm.bind_tools(TOOLS)

    system_message = SYSTEM_PROMPT.format(
        system_time=datetime.now(tz=UTC).isoformat()
    )

    # Add ReAct-specific guidance based on step count
    step_count = state.get("reasoning_step", 0)
    if step_count == 0:
        guidance = "\n\nThis is your first step. Start with 'Thought:' and
        think about how to approach this problem."
```

```python
    else:
        guidance = f"\n\nThis is step {step_count + 1}. Continue
        with 'Thought:' and decide what to do next based on previous
        observations."

    messages = [
        {"role": "system", "content": system_message + guidance}
    ] + state["messages"]

    # Get the model's response
    response = cast(AIMessage, await model.ainvoke(messages))

    if state.get("is_last_step", False) and response.tool_calls:
        return {
            "messages": [
                AIMessage(
                    id=response.id,
                    content="Thought: I've reached the maximum number of
                    steps. Let me provide a final answer based on what
                    I've learned so far.\n\nFinal Answer: Sorry, I could
                    not complete the full investigation in the specified
                    number of steps, but I can provide the information I've
                    gathered.",
                )
            ],
            "reasoning_step": step_count + 1
        }

    # Return the model's response and increment step count
    return {
        "messages": [response],
        "reasoning_step": step_count + 1
    }

def route_reasoning_output(state: State) -> Literal["__end__", "tools"]:
    """Determine the next node based on the reasoning output.
```

```python
    In ReAct, after reasoning:
    - If the model wants to use tools (Action), go to tools
    - If the model provides a final answer, end
    """

    last_message = state["messages"][-1]
    if not isinstance(last_message, AIMessage):
        raise ValueError(
            f"Expected AIMessage in output edges, but got {type(last_
            message).__name__}"
        )

    # Check if this is a final answer
    if "final answer" in last_message.content.lower():
        return "__end__"

    # If there is a tool call (Action), go to tools
    if last_message.tool_calls:
        return "tools"

    return "__end__"

# Build the ReAct graph
builder = StateGraph(State)

builder.add_node("reasoning", reasoning_step)  # The "Reasoning" part
of ReAct
builder.add_node("tools", ToolNode(TOOLS))     # The "Acting" part of ReAct

builder.add_edge(START, "reasoning")

builder.add_conditional_edges(
    "reasoning",
    route_reasoning_output,
)

builder.add_edge("tools", "reasoning")

graph = builder.compile()
app = graph
```

"Deep Research" is a product of combining these patterns into a sophisticated, multi-layered agentic system. It works like this:

- **Initial Query:** A user asks a complex question.

- **Plan:** The agent first creates a comprehensive research **plan**.

- **ReAct Loop:** The agent enters a **ReAct** loop and executes the first step of its plan. It alternates between **Thought**, **Action**, and **Observation**. This cycle is repeated as it gathers information for each sub-task.

- **Reflect:** After gathering information for a major part of the plan, the agent pauses to **reflect** on the findings. It looks for gaps, contradictions, or potential inaccuracies.

- **Refine and Replan:** Based on the reflection, the agent may modify its original **plan** and enter another **ReAct** loop to address the newly identified needs.

- **Final Synthesis:** Once it has sufficient, verified information, it synthesizes everything into a comprehensive, cited report.

In essence, **ReAct** is the engine that drives the task execution, while **Plan** and **Reflection** provide the high-level strategy and self-correction that elevate a simple AI agent to a "Deep Research" capability.

Figure 5-13 shows the architecture of LangChain's Open Deep Research, which helps us understand these advanced architectures.

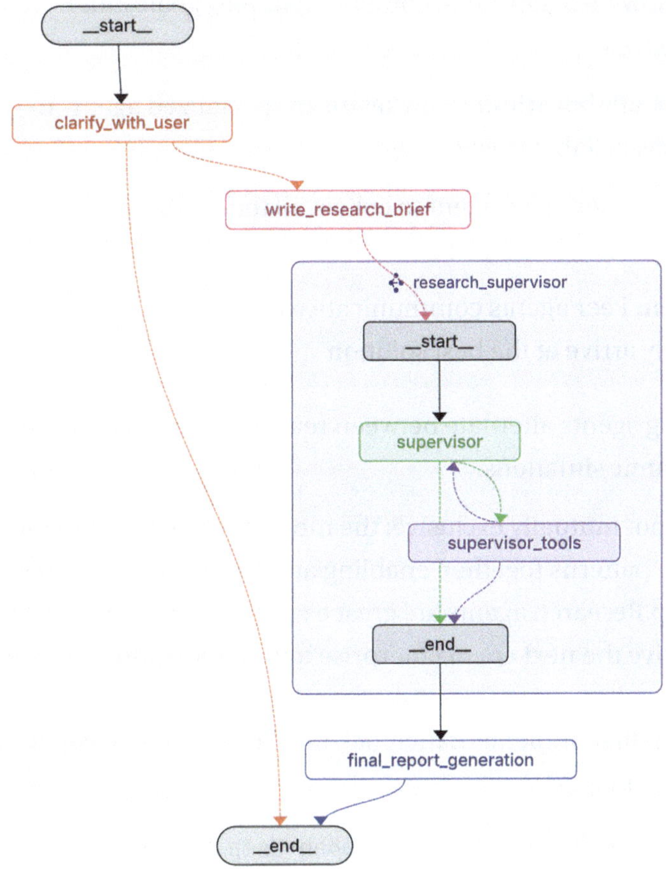

Figure 5-13. *Open Deep Research contains the ReAct pattern*

Conclusion

In this chapter, we defined agentic AI workflows as a powerful alternative to single-pass, zero-shot prompting. These workflows, which allow an LLM to iterate on a task in a series of steps, unlock significant performance gains that can exceed the improvements between foundation-model generations.

We examined five foundational design patterns that serve as the building blocks for intelligent agents:

- **Tool Use:** Equipping agents to interact with functions.

- **Reflection:** Enabling agents to critique and improve their own work.

- **Planning:** Allowing agents to decompose complex goals into manageable steps.

- **Multi-Agent Collaboration:** Using teams of specialized agents to solve problems collaboratively.

- **Supervisor:** A supervising agent to collect all the research insights and create a report

- **Collaboration:** Peer agents communicate with each other to collaboratively arrive at the best solution

- **ReAct:** Letting agents alternate between reasoning and acting to adapt to dynamic situations.

These patterns are not mutually exclusive; the most effective agentic solutions often combine multiple patterns together, enabling automation that is faster, smarter, and more reliable. Deep Research agents are great examples. As these patterns mature, they will continue to drive the next era of enterprise automation and AI-powered applications.

Finally, a Databricks-first implementation of LangChain's Open Deep Research can be found in the following blog post:

```
https://www.tredence.com/blog/databricks-open-deep-research
```

## CHAPTER 6

Databricks As an Agentic Platform: Orchestration, Context, and Control

In the previous chapters, we learned about the new Agent Bricks offerings. Since 2023, with the rise of GenAI and LLMs, Databricks has integrated them into its platform. The Databricks data intelligence platform (see Figure 6-1) combines the lakehouse platform and AI/LLMs to add the "data intelligence" engine that understands the uniqueness of your data and uses that understanding across everything in the platform. However, tools are increasingly designed for AI Agents to work alongside humans. We can no longer maintain a closed-box platform; instead, it must become a hub for hosting AI Agents. In this chapter, we will discuss how to integrate AI components within the Databricks Data Intelligence Platform into our AI apps.

© Jason Yip, Nikhil Gupta and Marcin Wojtyczka 2026
J. Yip et al., *Databricks Data Intelligence Platform*, https://doi.org/10.1007/979-8-8688-2524-8_6

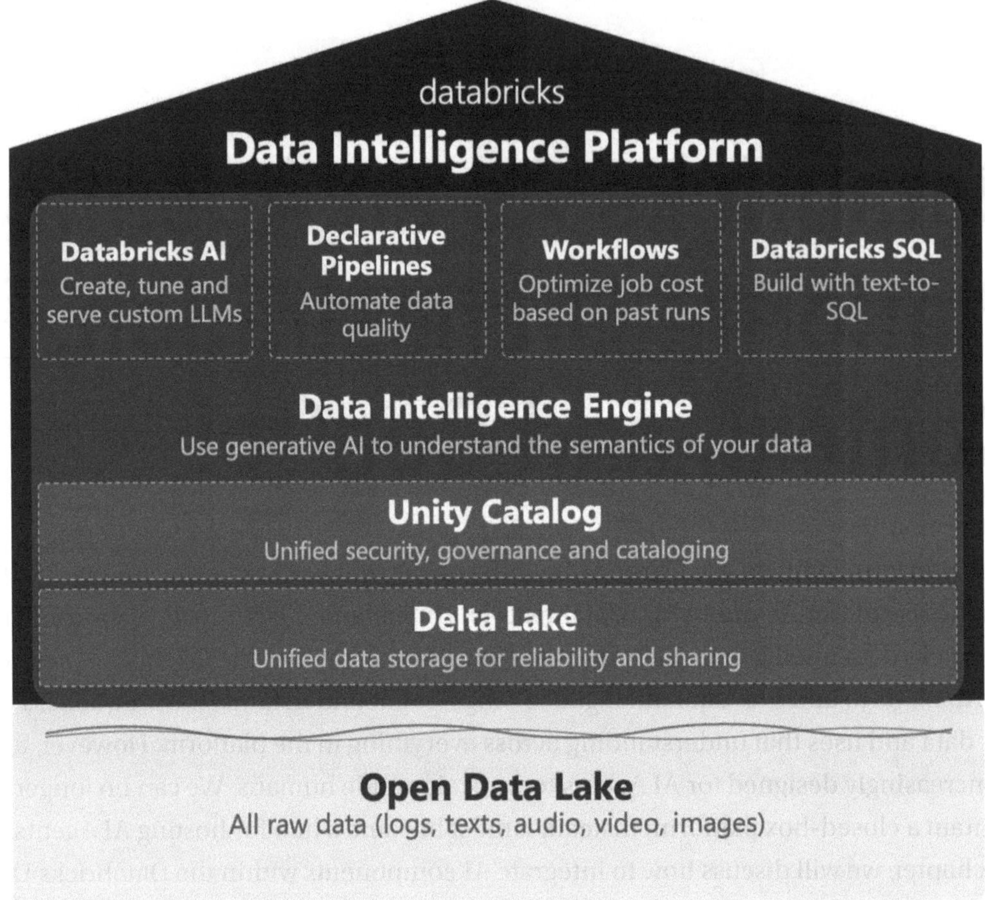

Figure 6-1. *Databricks data intelligence platform*

The Data Intelligence Engine provides the underlying semantic, metadata-aware, and workload-aware capabilities that power features such as the Genie code (Databricks Assistant), agentic automation, AI-driven governance, intelligent search, and AI/BI Genie—enabling the platform to interpret organizational data, reason over it, and support agent workflows across the lakehouse.

Throughout this chapter, we will examine key features of the Databricks data intelligence platform. We will begin by defining the data intelligence platform and how it evolved. Then we will examine key features, such as AI/BI Genie and Vector Search.

Data Intelligence

Data Intelligence is at the heart of the Databricks platform. Many people are using chatbots or copilots to assist with their work. However, most of these are trained on open data sources and have little context around your data.

To put this in perspective, the ideal co-pilot for organizations to be productive while working in Databricks or any other developer tools will need to meet the following requirements:

- Be within a secure environment so internal information is not being used to train the frontier models that will leak confidential information to external users

- Automatically learn about internal information and stay within the organization

- Understand human language and be able to translate into a programming language

- Lightning-fast performance, so problems can be solved in seconds and not minutes

With that in mind, Databricks developed Data Intelligence, which is powered by Mosaic AI Model Serving. Let's look at what areas Data Intelligence can help us with.

1. **Genie Code (Databricks Assistant):** This helps you write queries, create pipelines and ML models, troubleshoot issues, and identify performance bottlenecks in the system, powered by natural language understanding.

2. **AI-Powered Governance:** This helps in a variety of tasks, including generating comments for the metadata and providing lineage; automatic PII detection and masking; and AI security filtering, with the eventual aim to learn how to give advice based on the Databricks AI Security Framework.

3. **Coding Assistant:** The Databricks platform is equipped with AI agents that help you develop data pipelines (via Lakeflow Declarative Pipelines) and machine learning models (via the Data Science Agent). You can also provide Genie Code with skills to help you develop code directly within the platform.

4. **AI/BI Genie:** With the ever-growing data in every organization, it will be impossible for an LLM to keep up with the knowledge. AI/BI Genie enables business users to interact with their data through natural language. It leverages GenAI to understand your data and underlying metadata and gives relevant and accurate answers based on that knowledge.

5. **Automated Job Tuning:** Not all AI is related to a large language model. There are techniques called *deep learning* that can be used to tune the jobs automatically, resulting in less time for human fine-tuning. This is called *predictive I/O*.

Deep Dive into Data Intelligence

In the following sections, we will look at each of these features in detail.

Genie Code (Databricks Assistant)

Genie Code has become a powerful tool right within the Databricks interface. Its function is like the AI plugin in Visual Studio Code or other IDEs, allowing you to write code in natural language. It's also integrated into various parts of the platforms, such as notebooks and jobs, where we can use it for development like Claude code but with deep understanding in Databricks.

Generate Code in Any Language

Not everyone speaks code as their native language. Databricks supports several different languages, including SQL, Python, Scala, and R. There are times you will forget the syntax or simply need to extend a function. The old way involved scanning numerous blog posts or Stack Overflow posts to find your answer, and there were lots of clicks and searches to get the final answer. What if someone is there just to tell you the answer?

Genie Code can generate, explain, and fix SQL and Python code using natural language and is now available across all code editors in the Databricks platform, including notebooks and SQL editors (see Figures 6-2 and 6-3).

```python
show me last 6 months of data in pySpark

from pyspark.sql.functions import col, current_date, add_months

# Assuming your Spark DataFrame is named 'df'
df_last_6_months = df.filter(
    col("date_column") >= add_months(current_date(), -6)
)

display(df_last_6_months)
```

Figure 6-2. *Code generation in cell*

```sql
how many positive vs negative

select rating, ai_analyze_sentiment(Comments), Comments from unitygo.default.amazon_reviews
1  select
2    ai_analyze_sentiment(Comments) as sentiment,
3    count(*) as count
4  from unitygo.default.amazon_reviews
```

Figure 6-3. *Code generation in SQL editor*

Autocomplete Code or Queries

Whether you want IntelliSense or inline code completion, Genie Code can help by reminding you of syntax or wrapping the code for you. This is ideal for developers who know what they want to accomplish but need help with the exact syntax. There are two styles: one via comments (Figure 6-4) and the other via code hints as you type (Figure 6-5).

```python
#last 6 momths of data
df = df[df['reviewTime'] > '2022-06-01']
# DBTITLE 1
```

Figure 6-4. *Generating code based on comments*

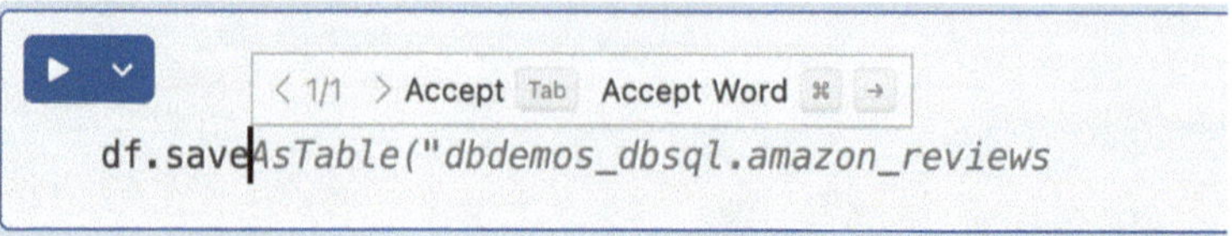

Figure 6-5. *Code completion*

Code Conversion

One of the most common use cases is to convert Python code into PySpark to take advantage of distributed computing, because Python code only executes on a single node, regardless of the cluster size. If you were to use other tools, you first need to copy the code and paste it into another tool, like a chatbot or a search bar. Genie Code has direct access to the notebook and can understand the code and do the migration automatically. The answer can also be replaced with the existing cell with a single click. See Figure 6-6.

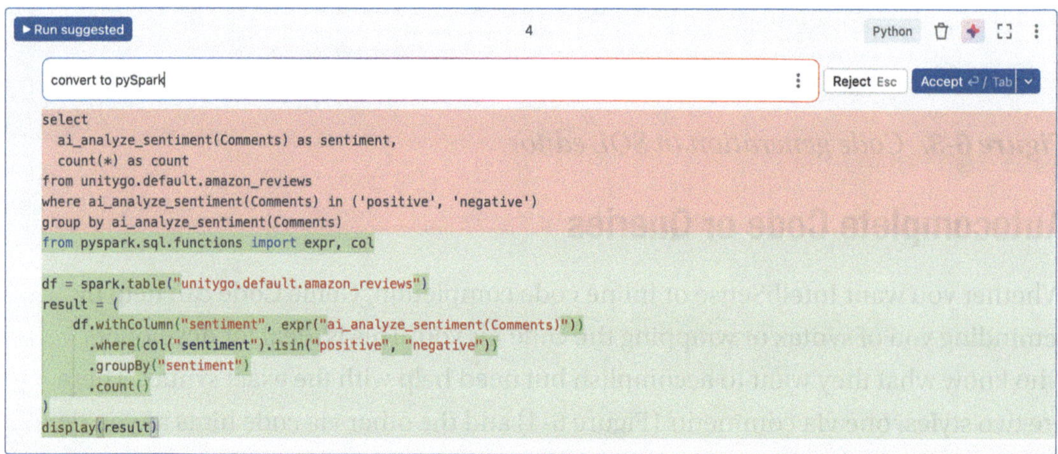

Figure 6-6. *Code conversion*

Code Explanation

Whether you don't understand the code or want to explain your code to a business stakeholder who is interested in the business logic, you can ask Genie Code to do it for you (see Figure 6-7). Having an English description of the code will help you understand it. And if needed, you can always use inline code generation to tweak the business logic.

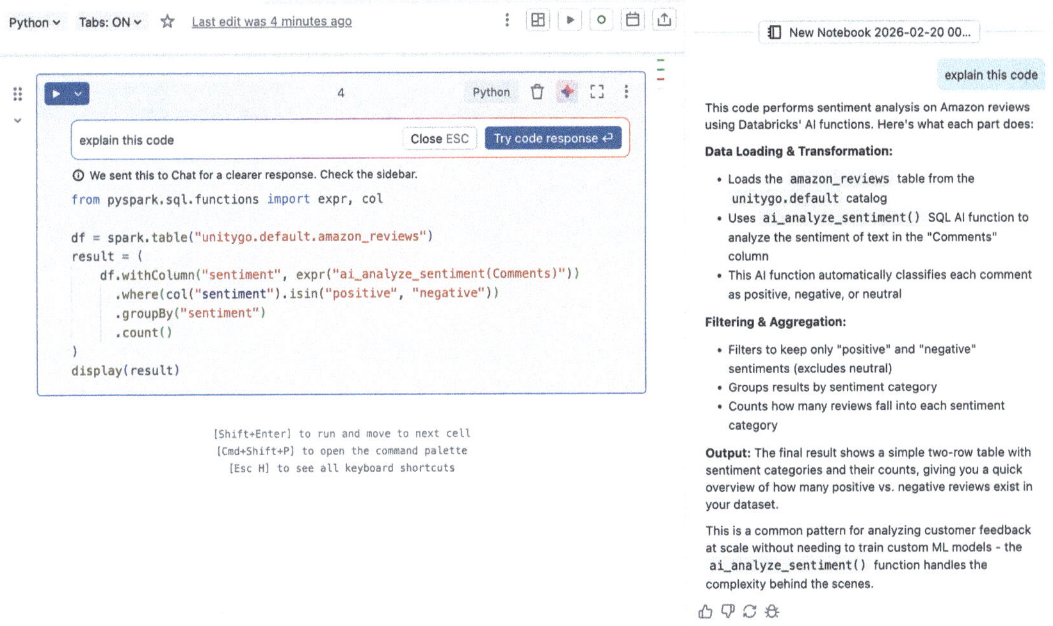

Figure 6-7. *Code explanation*

Code Fixing

While having a debugger is helpful, fixing the code can take a long time if you don't have a good handle on it. Genie Code can explain where the error originates and suggest a fix (see Figure 6-8). Best of all, you can collaborate with the LLM to find the best solution right inside the notebook without leaving the environment.

One thing to note is that Genie Code will only appear when there is an actual error. Some application developers use a `try...except` block to catch exceptions, which is standard practice, but in these scenarios, it will not trigger Genie Code.

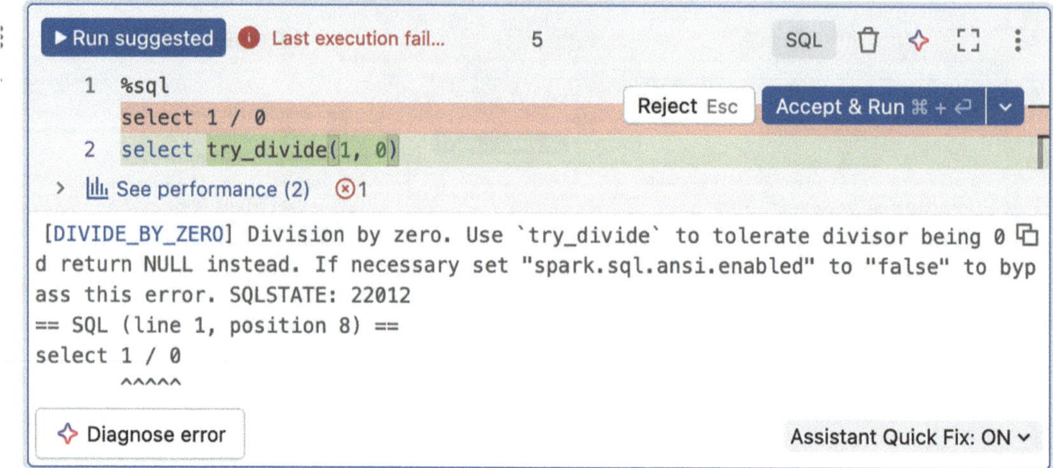

Figure 6-8. *Genie Code is suggesting a code fix*

Agent Mode

Agent mode transforms the Genie Code into an autonomous co-pilot that can independently explore data, generate and run code, and fix errors from a single prompt. It features a planner that allows it to draft and refine multi-step strategies for complex tasks such as training ML models or performing cohort analysis. Throughout the process, the agent remains grounded in Unity Catalog to ensure all automated actions are governed, transparent, and accurate. Figure 6-9 shows the Agent button in Genie Code.

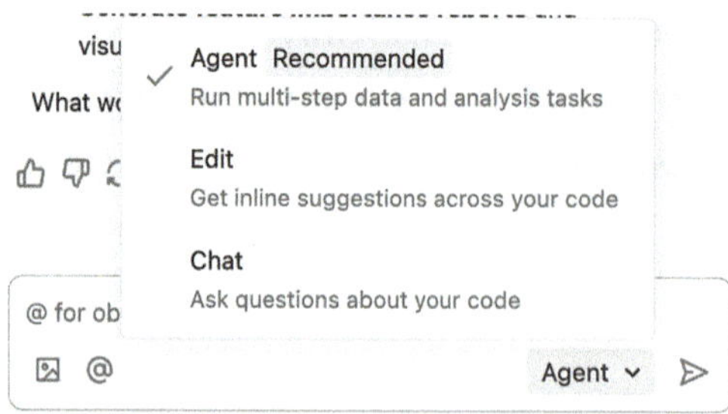

Figure 6-9. *Agent mode in Genie Code*

Data Intelligence would not be possible without data. That's why we can also tag tables or notebooks and provide the agent with more context like what data to use or code samples, as shown in Figure 6-10.

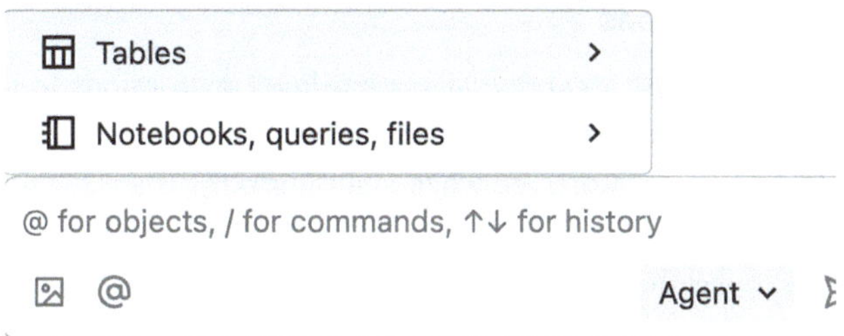

Figure 6-10. *Agent mode can tag tables or notebooks*

Instructions (md) files and skills have been popularized by tools like Claude Code. When Genie Code (🐝) is opened, we can click the settings (⚙) button in the top-right corner and provide our instructions. There are three types of documentation that we can add.

- System-level instructions are specific to the user in the user's home folder

- Workspace instructions are for administrators to create instructions that will impact the whole workspace

- Skills are mini tutorials along with code samples to teach an agent how to perform a specific task

Figure 6-11 shows the instruction panel where we can create our md files.

To understand how to improve the agent response, please refer to the documentation from Databricks: `https://docs.databricks.com/aws/en/notebooks/assistant-tips#tips-to-improve-assistant-responses`

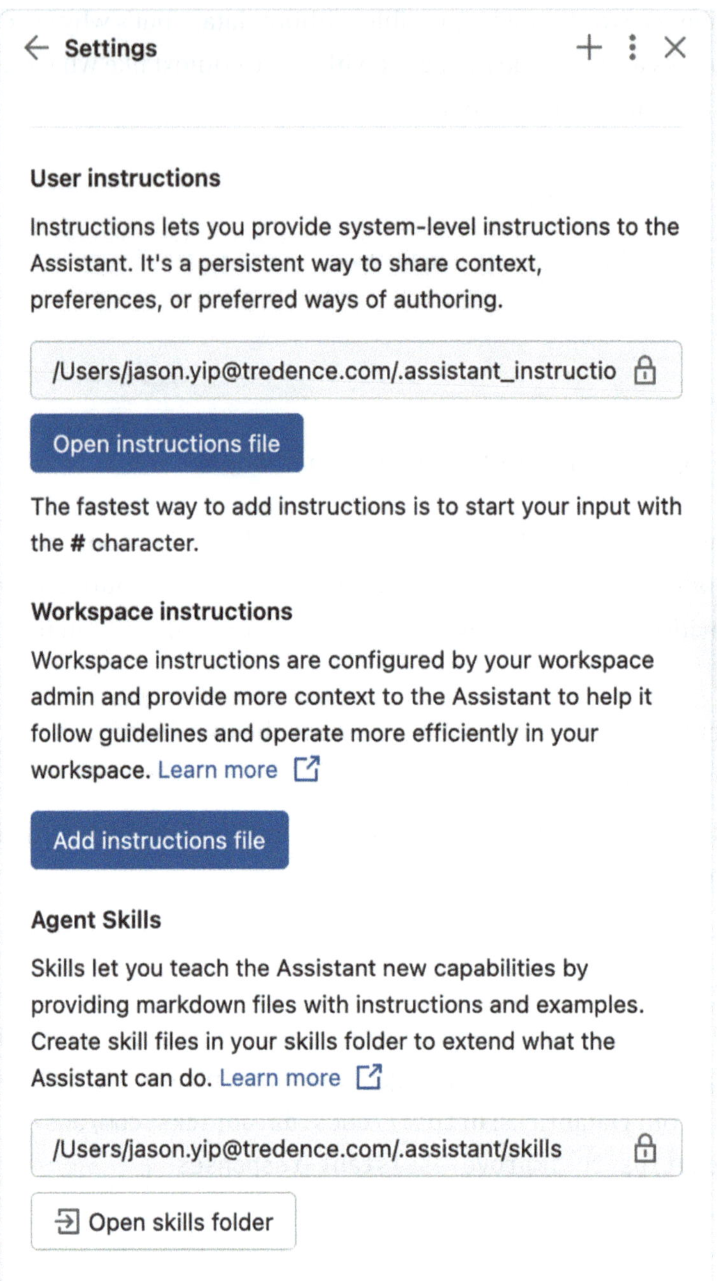

Figure 6-11. *Instruction panel to create md files for agents*

AI-Powered Governance

If you think Unity Catalog is the go-to tool for data governance, then you are on the right track (see Figure 6-12). Lakeflow Declarative Pipeline's data quality monitoring capability, Unity Catalog's lineage information, Lakehouse Federation, and auditing and access control are all perfect elements for data governance. Coupled with its AI power, Unity Catalog will enable organizations to govern more intelligently.

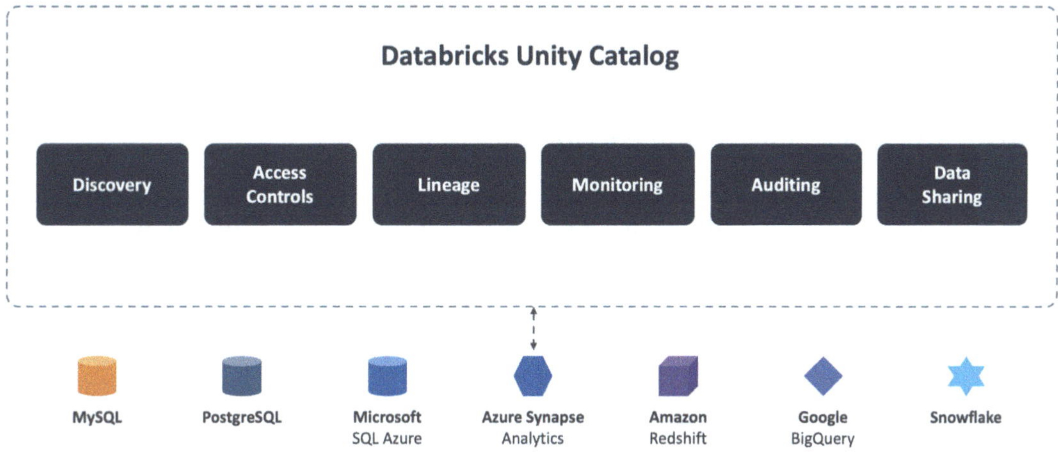

Figure 6-12. *Unity Catalog federated governance*

Below are the AI powers that will help with the governance process.

- **AI-Generated Comments Enhancements**

  Documentation has a love-and-hate relationship with developers. In some cases, there will be initial documentation effort, but as the number of data assets and tables grows, it will become harder to keep the documentation up to date. Although AI-generated comments are not bulletproof, they can analyze the data the way a non-subject-matter expert (non-SME) would—by sampling the data and inferring meaning from the table and column names (see an example in Figure 6-13). Most importantly, the data dictionary can live with the data, rather than maintaining a separate spreadsheet or relying on an external system that requires constant effort to keep up to date.

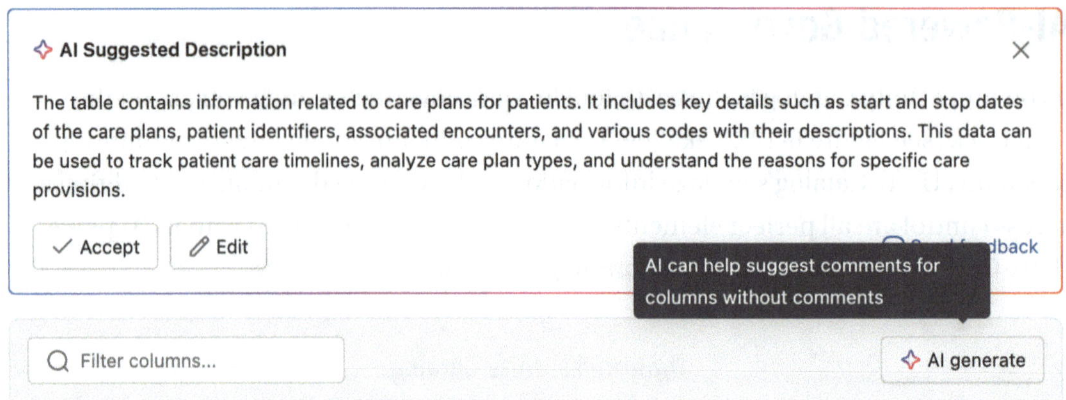

Figure 6-13. *AI-suggested comments for table description*

Transparency is at the heart of Databricks. The following article talks about the making of this AI feature and illustrates how it is not simply looking up in a dictionary:

```
https://www.databricks.com/blog/creating-bespoke-llm-ai-
generated-documentation
```

1. **Lineage**

   Databricks provides lineage in two different ways: Delta Live Tables and Unity Catalog (Figure 6-14). While capturing the lineage is not a result of machine learning or a large language model, it plays a pivotal role as an input to the machine learning model so it can generate meaningful queries in AI/BI Genie and beyond.

Figure 6-14. *Databricks lineage*

- **Data Classification**

  Goodbye regular expression, hello artificial intelligence. To comply with the General Data Protection Regulation (GDPR) and the California Consumer Privacy Act (CCPA), organizations often need to identify columns containing PII and mask them accordingly.

  Previously, without the help of an LLM, regular expressions were often required to extract patterns in email addresses and street addresses, making the process error-prone. Machine learning models were developed to address this problem, but they required an additional layer of processing and inference, either via a batch pipeline or an API.

  Unity Catalog now includes Data Classification features that automatically classify and tag sensitive data in your catalog. We can combine them with policies for automated masking and data filtering.

  Databricks SQL includes two powerful functions for these scenarios: ai_classify and ai_mask.

  ai_classify: What if the LLM is already a very good classifier? Is it possible to ask the LLM to classify whether a column contains PII or not? When we think in this direction, we will have our answer. Consider the query in Listing 6-1.

Listing 6-1. AI Query in serverless SQL

```
SELECT ai_classify('my name is Jason, email address is jason@email.com',
ARRAY('contains PII', 'no PII')) as classification
UNION ALL
SELECT ai_classify('Today''s weather is awesome', ARRAY('contains PII',
'no PII'))
```

The result, shown in Figure 6-15, is as you might expect.

Figure 6-15. *DB SQL AI function: ai_classify*

ai_mask: Similarly, you can mask sensitive values in columns by specifying what you want to mask. While it is not limited to PII, you can mask whatever you want, but from the PII perspective, it is a no-brainer. Listing 6-2 is an example with a name and an email address. The idea is similar to regular expression searches; it will automatically match patterns for you, but in an AI way. The result from Listing 6-2 is shown in Figure 6-16.

Listing 6-2. ai_mask Function for Ease of PII Scanning

```
SELECT ai_mask('my name is Jason, email address is jason@email.com',
ARRAY('name', 'email')) as text
```

Figure 6-16. *DB SQL AI function: ai_mask*

1. **AI Judges**

    Content moderation is one of the hottest topics on the Internet. While content moderation has long been a concern for social media companies, it has gained renewed importance due to hallucinations from LLMs and the risk of generating inappropriate or profane content. Databricks has included AI judges (shown in Figure 6-17) with a toggle in the Playground, or we can apply them using MLflow. Please refer to Chapter 11.

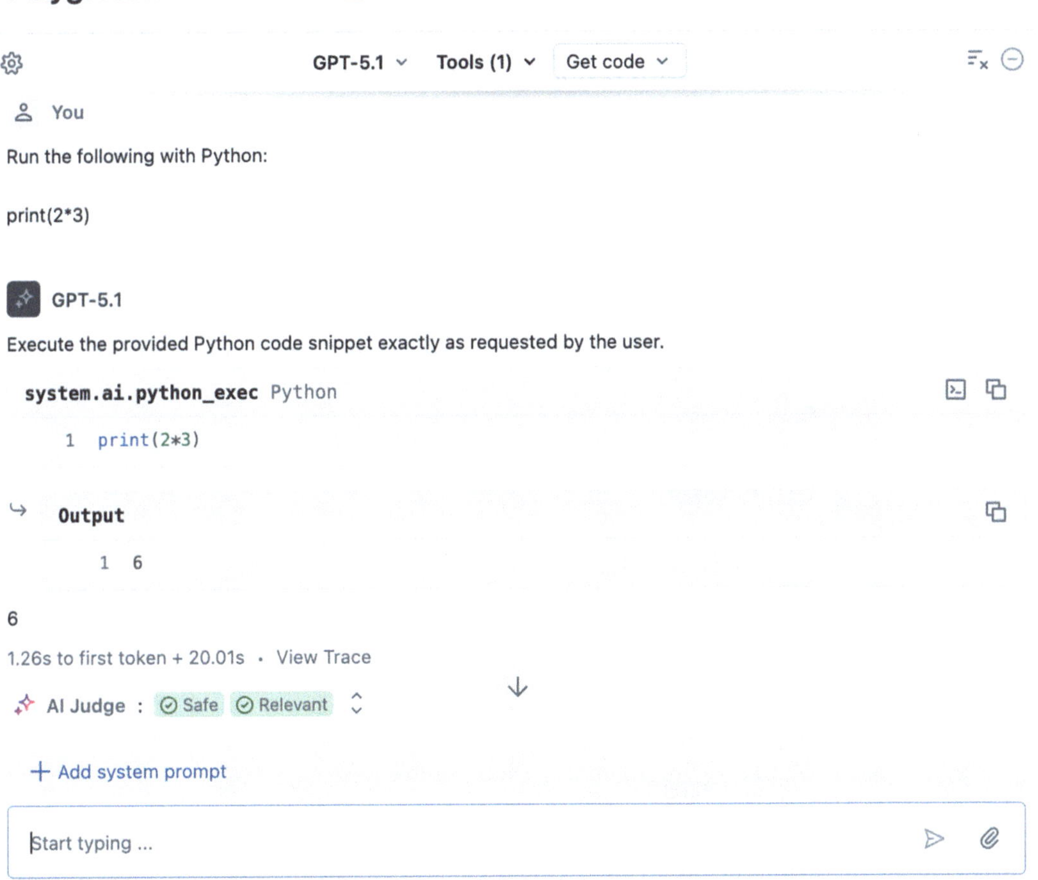

Figure 6-17. *AI judges*

2. **AI Security Framework**

The Databricks AI Security Framework is a comprehensive guide for CISOs and a roadmap for implementing data and AI security within an organization. The whitepaper can be found below, and it contains a lot of valuable information:

```
https://www.databricks.com/resources/whitepaper/
databricks-ai-security-framework-dasf
```

Search and Discovery

Databricks has been working to enhance the user search experience on the platform. The search here refers to the growing number of data assets within the organization.

Intelligent Search

If you are familiar with GitHub's code search, you might think that Databricks is improving its offerings in terms of being able to search code. However, the search is not limited to code; it also includes other objects, such as notebooks and workflows. Figure 6-18 illustrates some of the most popular object types that Intelligent Search can find.

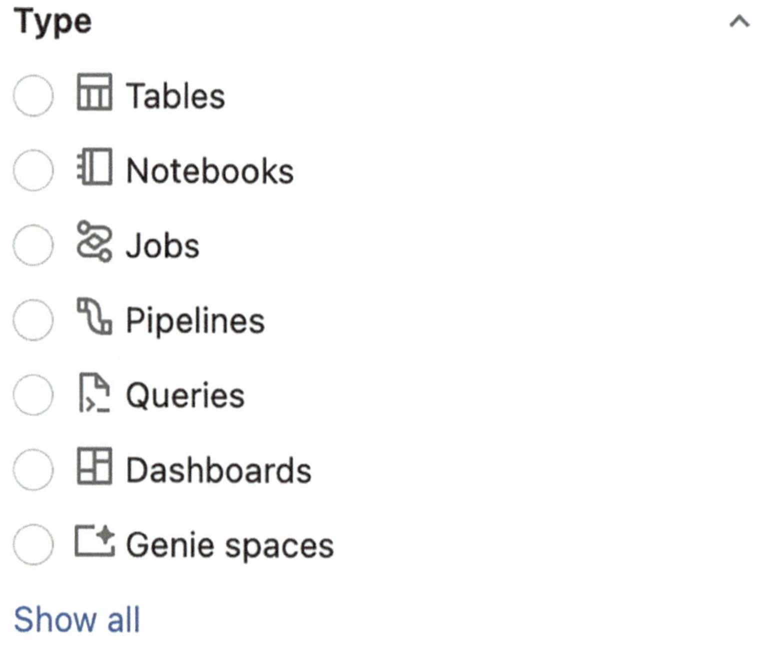

Figure 6-18. Databricks object search

So, what capabilities does the new search experience offer?

1. **Text Search:** This primarily refers to code search. In addition to single-word searches, it supports exact phrase matching by enclosing terms in double quotes. It also supports escape quotes using a backslash.

2. **Semantic Search:** Search with meaning; you can ask questions like "How do I build a financial report?" and it will return relevant financial tables.

3. **Search Engine Style Filter:** You can filter by object types using a search engine style, like `type:table, owner:me`.

4. **Popularity:** Popularity helps ensure others use the returned objects. While it's not always true, the more popular it is, the mmore it implies that the table would contain the right data.

5. **Knowledge Card:** For managed tables only, the search will present a knowledge card for the top search results.

AI/BI Genie

AI/BI Genie is a natural language Q&A experience that allows non-technical business users to ask questions in plain English and receive answers in a table or a visualization, with permissions governed by Unity Catalog. However, a key difference from traditional BI tools is that AI/BI Genie uses agentic reasoning to continuously learn and improve, understanding the nuances of your data and business semantics to deliver useful, contextual answers.

To use Genie, the data should be in Unity Catalog, which provides fine-grained access control to prevent unintended leakage of sensitive data in the Genie environment. Further, Genie is accessible to users with Databricks SQL access. With these requirements in mind, let's move into some of the key aspects of using Genie. Figure 6-19 illustrates the architecture behind AI/BI Genie.

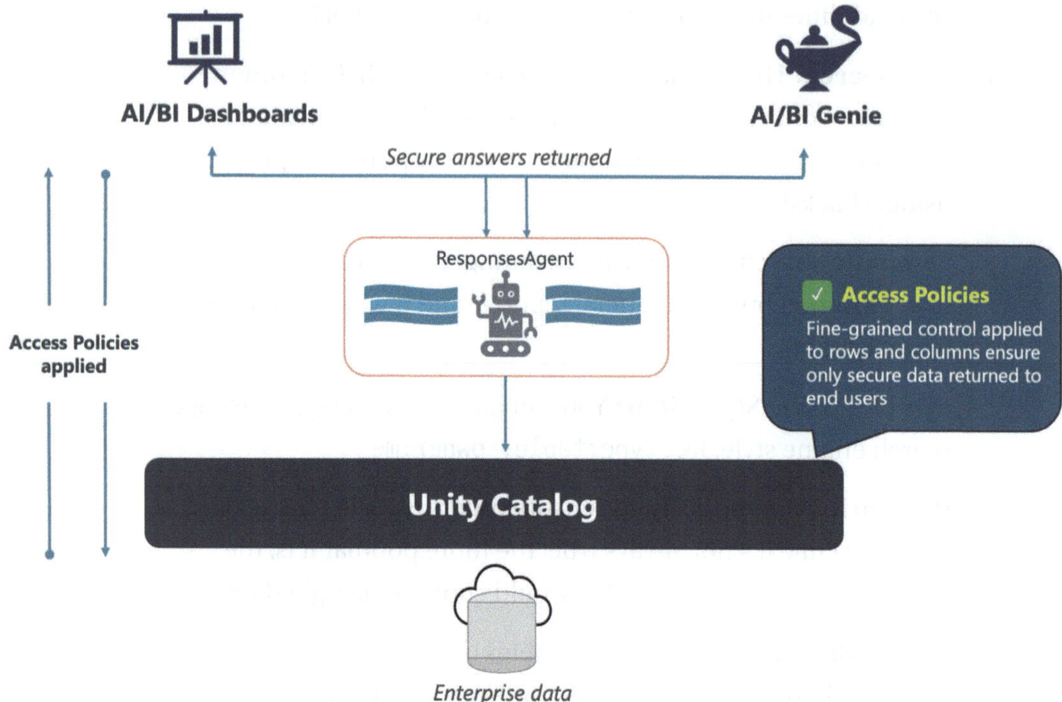

Figure 6-19. *Databricks AI/BI Genie architecture*

How to Set Up Genie

Let's look at an example of a large retail organization that wants its business users across departments to use Genie. As a first step, the data owners and teams within the organizations that know the data best will set up topic- or context-specific Genie spaces (Figure 6-20). For example, POS Genie spaces contain tables that store point-of-sale (POS) data, and a finance space contains all financial data. Note that a Genie space uses table and column names and descriptions to generate an equivalent SQL query from the natural language query, which is then executed against the data in the Unity Catalog.

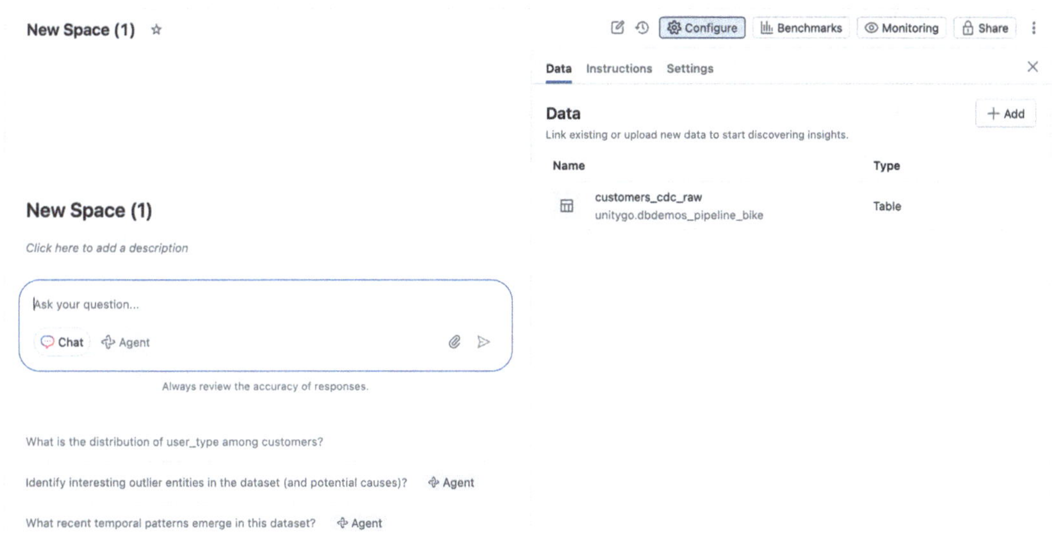

Figure 6-20. *Creating an AI/BI Genie space*

After the Genie space is set up, relevant tables and their associated metadata are imported. It is important to note that your table metadata must be well documented with comments so that Genie can understand columns/tables that may be misnamed and gain more contexts. Further, you can create more focused SQL views and remove unnecessary columns, resulting in cleaner data.

Next, define business-specific terms using the general instructions in your Genie spaces. Here, you can define unique jargon, logic, concepts, and KPIs in the given domain, and this knowledge will be used across all new questions. Further, you can iterate on this over time as you see more questions come in or new KPIs are developed, thus continuously teaching Genie.

Finally, if you already have SQL statements for querying tables in a specific Genie space, you can add them to "Save as Instruction" as well to teach the model how to answer specific questions. You can also keep examining the SQL statements generated by Genie. If you find them a bit off, you can save them, and Genie will learn from them for future questions.

Your Genie space is now ready for use by your end users. Genie is designed to learn over time as it is used increasingly. One way it does this is by asking follow-up questions for clarification when the question is unclear, enabling it to capture more information from user prompts. Further, this new semantic knowledge can be saved as instructions to help Genie learn over time.

Genie works a little differently than the chat interface we are used to interacting with, as it leverages enterprise data. That's why Databricks has provided some tooling for us to document them to "learn" over time properly. As shown in Figure 6-21, **Configure** and **Benchmarks** are the two important capabilities that allow Genie to learn over time. **Configure** provides the space-level metadata that Genie can reference, whereas **Benchmark** serves as validation and establishes baselines to ensure Genie doesn't hallucinate over time.

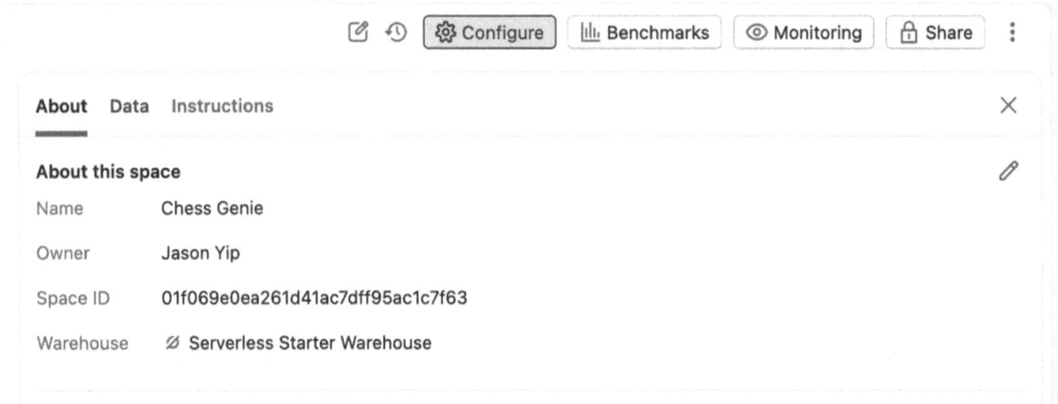

Figure 6-21. *Genie space configurations*

Genie is modeled after how data teams work. When we hire a data engineer, we'd normally do the following steps:

1. Having clear documentation of the data model is critical for Genie or any human to understand the data. Therefore, when we provide the comments of the tables as discussed above, we can enable Genie to learn. We can also provide more context in the **About** tab, as each space is designed to answer different questions.

2. The **Data** tab allows us to override the definitions in Unity Catalog in case we don't have permission to update the metadata or we need to provide more specific instructions to the space as shown in Figure 6-22.

   We can also leverage Prompt matching: **Format assistance** and **Entity matching** help Genie match user language to actual data values, correct misspellings, and generate more accurate SQL

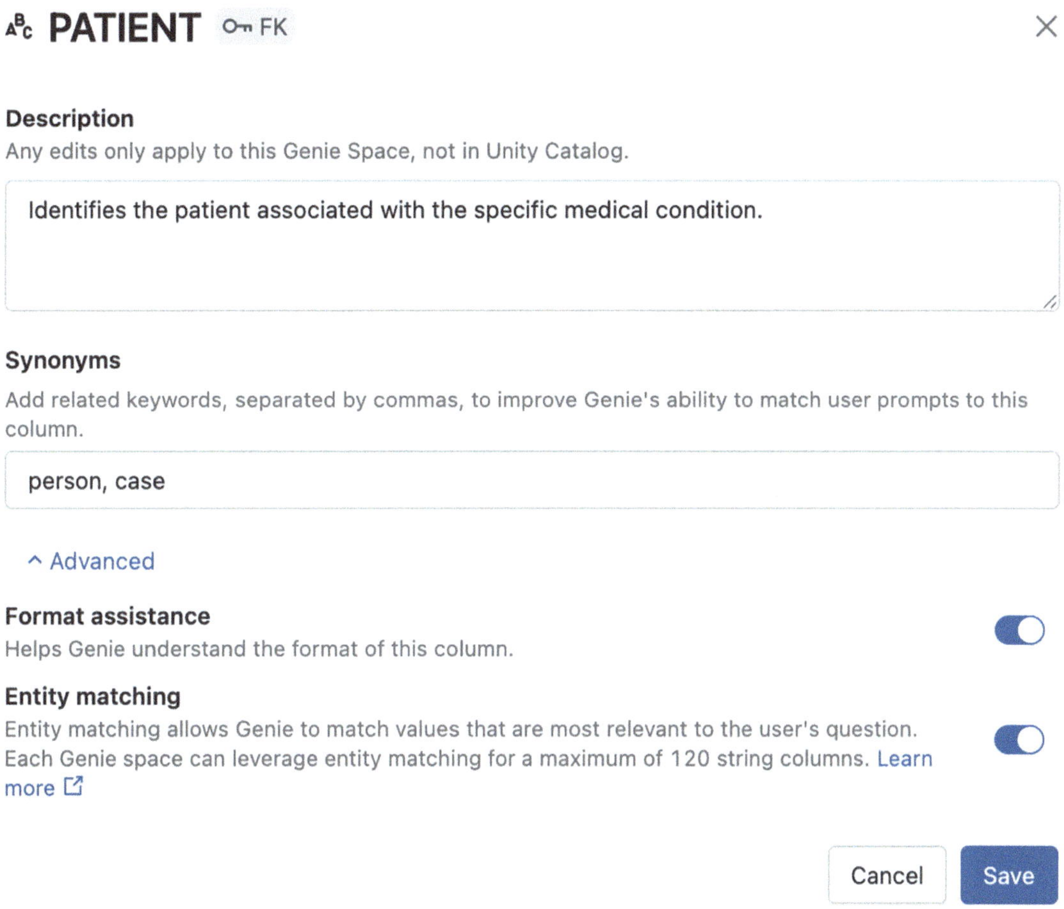
PATIENT
FK
Description
Any edits only apply to this Genie Space, not in Unity Catalog.
Identifies the patient associated with the specific medical condition.
Synonyms
Add related keywords, separated by commas, to improve Genie's ability to match user prompts to this column.
person, case
Advanced
Format assistance
Helps Genie understand the format of this column.
Entity matching
Entity matching allows Genie to match values that are most relevant to the user's question. Each Genie space can leverage entity matching for a maximum of 120 string columns. Learn more
Cancel
Save

Figure 6-22. *Metadata editing for a specific column*

3. The **Instructions** tab is probably the most important of all to allow us to give Genie a lot of help in ensuring all queries come back as expected. As seen in Figure 6-23.

About
Data
Instructions
Text
Joins
SQL Expressions
SQL Queries

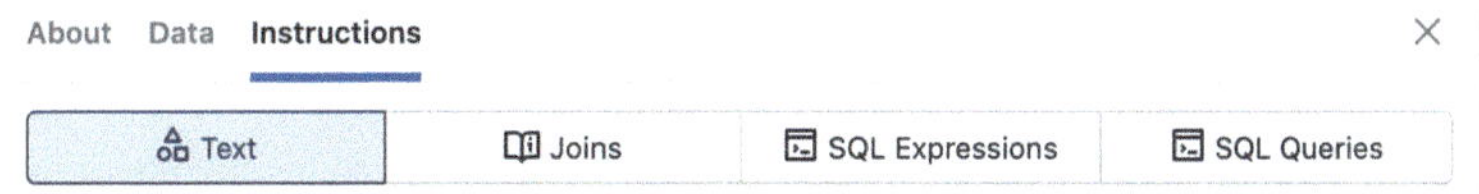

Figure 6-23. *Instructions for Genie space*

a) **Text**, as its name suggested, is free-form text. Similar to md files for agents, we can use it to define custom logic.

b) **Joins** allow us to define relationships between tables. We will discuss table logical relationships, like primary keys and foreign keys, in Chapter 17, but we can also provide them within the Genie space.

c) **SQL Expressions** are Filter, Measure, and Dimension, which allow us to create custom aggregations or a calculated field based on a filter (where condition)

d) **SQL Queries** allow us to specify more complex metrics that Genie can leverage to increase its accuracies, as well as SQL functions, which can also be used in this tab.

For a comprehensive case study on how to use these capabilities to bring Genie to production. Please refer to the below Databricks blog.

```
https://www.databricks.com/blog/how-build-production-ready-genie-spaces-
and-build-trust-along-way
```

Figure 6-24 shows how we can immediately chat with our Genie space and get answers without knowing any code. We can also visualize it from within the space (via Quick actions). The engine will get smarter over time, but the knowledge is there for everyone.

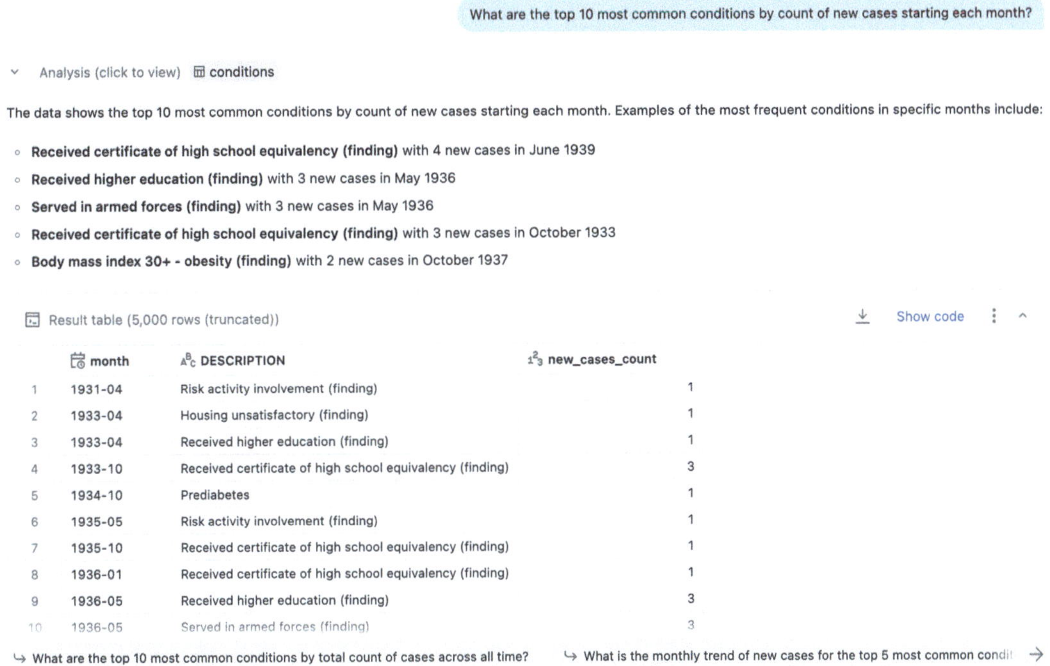

Figure 6-24. *Q/A with Genie in the space*

Genie Deep Research

Enterprise Deep Research has arrived, Genie. This feature is usually available in public models as a premium feature, e.g., Google Gemini or ChatGPT. Databricks' Genie Deep Research can perform the same in-depth reasoning based on the data available within the Genie space. Figure 6-25 shows the toggle can be used for the research agent.

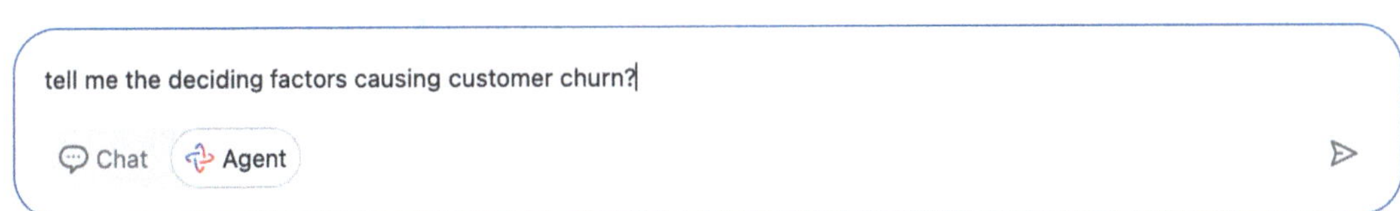

Figure 6-25. *Research agent available in AI/BI Genie*

For example, given a prediction table with customer churn and all the features, we can perform a cohort analysis to determine why customers churn. We can delegate this to the research agent as shown in Figure 6-26.

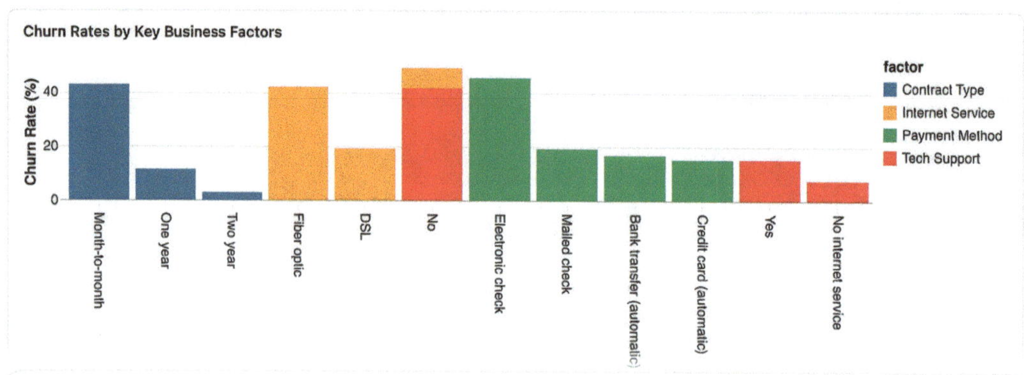

Figure 6-26. *Excerpt from research agent in Genie*

Databricks MCP Servers

According to Anthropic, "The Model Context Protocol is an open standard that enables developers to build secure, two-way connections between their data sources and AI-powered tools. The architecture is straightforward: developers can either expose their data through MCP servers or build AI applications (MCP clients) that connect to these servers."

Source: `https://www.anthropic.com/news/model-context-protocol`

Simply put, MCP is similar to the REST API standard, which is widely used for API tools and is the commonly recognized way to communicate on the internet. MCP implements JSON-RPC 2.0 to allow AI models to execute tools.

MCP Architecture

Like REST APIs, MCP servers are hosted on a server. However, unlike REST APIs, which are usually used by applications to perform specific functions, MCP servers are leveraged by AI models for tool calling in agentic workflows. The architecture is shown in Figure 6-27.

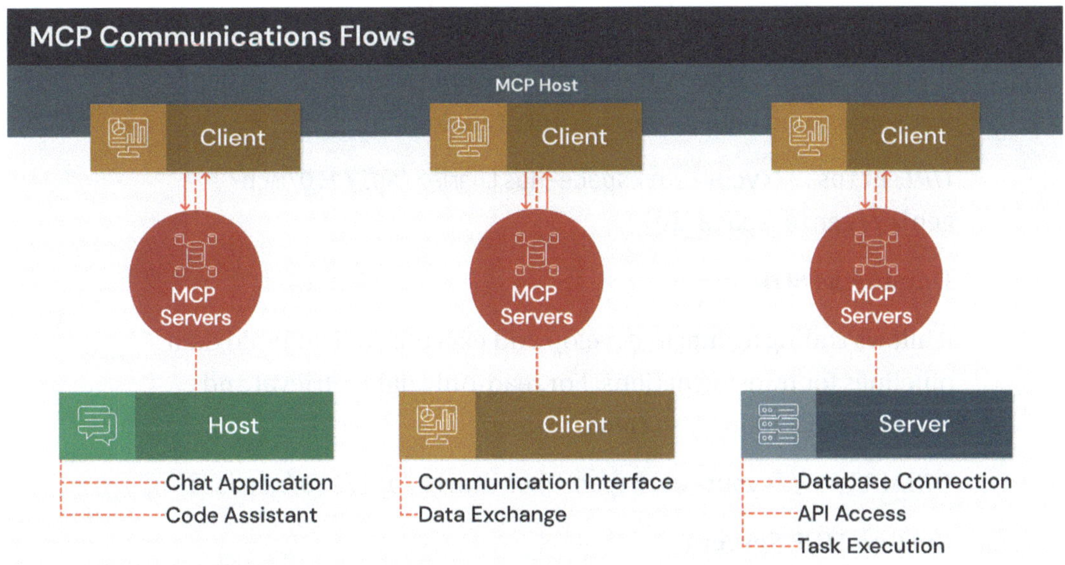

Figure 6-27. *MCP communication flows (source: Databricks) (This image is not included in the book's Creative Commons license.)*

Databricks provides four out-of-the-box MCP servers integrated with Unity Catalog. These are called managed MCP servers. The purpose of these servers is to enable agents to use the platform's components individually rather than through a single software suite.

1. **Vector Search**

   It allows agents to query Databricks Vector Search indexes in the specified Unity Catalog schema.

   URL: `https://<your-workspace-hostname>/api/2.0/mcp/vector-search/{catalog_name}/{schema_name}`

2. **Unity Catalog Functions**

   It allows agents to run Unity Catalog functions in the specified schema.

   URL: `https://<your-workspace-hostname>/api/2.0/mcp/functions/{catalog_name}/{schema_name}`

3. **Genie Space**

   It allows agents to query the specified Genie space to get insights from structured data (tables in Unity Catalog).

   URL: `https://<your-workspace-hostname>/api/2.0/mcp/genie/{genie_space_id}`

4. **Databricks SQL**

   It allows coding agents to develop and execute code to build data pipelines for transformations. For read-only data retrieval and chatbot integrations, use Genie instead.

   URL: `https://<your-workspace-hostname>/api/2.0/mcp/sql`

5. **Custom MCP Servers**

   We can also build custom logic using open-source Databricks libraries, such as the Databricks Function Client and the Databricks UC Toolkit. We can easily host them in Databricks Apps so the application can be governed by Unity Catalog. For more information, please refer to Databricks documentation.

   URL: `https://docs.databricks.com/aws/en/generative-ai/agent-framework/create-custom-tool`

Conclusion

This chapter explores the evolution of Databricks from a traditional data platform into an agentic hub designed for AI agents to work alongside humans. The platform has fully integrated the Data Intelligence Engine, which uses generative AI to understand the unique semantics and metadata of an organization's data.

Core Components of the Data Intelligence Platform

The architecture combines the traditional lakehouse platform with AI/LLM capabilities to provide a unified environment for data and AI.

- **Genie Code (Databricks Assistant) and Agent Mode**: Genie Code provides natural language interfaces to generate, explain, and fix code across multiple languages, like SQL and Python. Agent Mode elevates this to an autonomous co-pilot capable of executing multi-step strategies for complex tasks like training ML models or performing cohort analysis.

- **AI-Powered Governance**: Leveraging Unity Catalog, the platform automates documentation through AI-generated comments and provides intelligent lineage. It also simplifies compliance through AI-driven PII classification and masking using functions like `ai_classify` and `ai_mask`.

- **AI/BI Genie**: This natural language Q&A experience allows non-technical business users to query data in plain English. Genie uses agentic reasoning to learn business nuances over time and includes a Deep Research mode for in-depth reasoning and analysis.

Integration and Connectivity

A significant advancement is the support for the Model Context Protocol (MCP), an open standard that allows LLMs to integrate with external tools and data sources.

Databricks provides several managed MCP servers out-of-the-box, enabling agents to securely interact with:

- **Vector Search:** To query vector indexes within Unity Catalog.

- **Unity Catalog Functions:** To execute specific functions within a schema.

- **Genie Spaces:** To retrieve insights from structured tables.

- **Databricks SQL:** To develop and execute SQL queries for data analysis and transformations.

The Databricks Data Intelligence Platform aims to reduce the barrier between data and insights. By embedding AI agents into every layer—from governance to analysis—the platform ensures that organizational data is not just stored but is actively understood and accessible to every user.

## CHAPTER 7

Quality Tuning and Evaluation with Agent Bricks

So far, we have discussed how to create agents with Agent Bricks; it's a new paradigm for UI-driven AI. It is understandable in a user interface-driven framework that all we are seeing are spinning circles or progress bars. Scientists and engineers alike often want more details, and companies that adopt the latest AI products also need to evaluate the pros and cons of bringing in new technology, since they can no longer analyze the underlying model internals to understand how things work. However, just as not everyone needs to know how a mechanic works to drive a car, we will soon adopt the power of Agent Bricks with full confidence. In this chapter, we will dive into the basic evaluation techniques for agents within Agent Bricks, then discuss how to build custom evaluations, and finally review some of the research techniques powering Agent Bricks.

Evaluating Information Extraction and Custom LLM

Despite their similarity, Information Extraction and Custom LLM presented two different forms of agents and therefore the way to evaluate them will be a bit different. But first, we will dive into the common grounds.

Evaluation-Driven Development with MLflow

Data scientists are used to tracking their experiments with MLflow. The fundamental problem of GenAI is that the evaluation tools are fragmented, or worse still, people don't evaluate their output at all. That's why it's important to understand the lifecycle of GenAI evaluation using MLflow. Like machine learning, GenAI should be iterative and continually improve. That's the role MLflow has been playing over the years, and Databricks has enhanced the GenAI capabilities in the latest MLflow 3.x releases.

Figure 7-1 illustrates a two-loop lifecycle for developing and improving applications and agents, an MLOps lifecycle extended to AI agents. The Inner Loop (Experiment), shown in blue, focuses on rapid iteration and refinement. We can easily track these within MLflow. It starts with App/Agents, which undergo Manual Evaluation (feedback), where human feedback drives initial iterations. This process is then scaled through Systematic Evaluation (expectation), which automates the evaluation to efficiently test and validate changes before deployment. The Outer Loop (Production) is a continuous feedback mechanism that takes place after deployment. It begins with monitoring the live application, collecting real-world usage data, or "traces," to curate real-world datasets. This dataset is then used to fix issues and add new evaluation cases, which feed back into refining and improving the next generation of the App/Agents. Essentially, the inner loop handles initial experimentation and refinement based on evaluation, while the outer loop ensures continuous improvement and robustness using real-world production data.

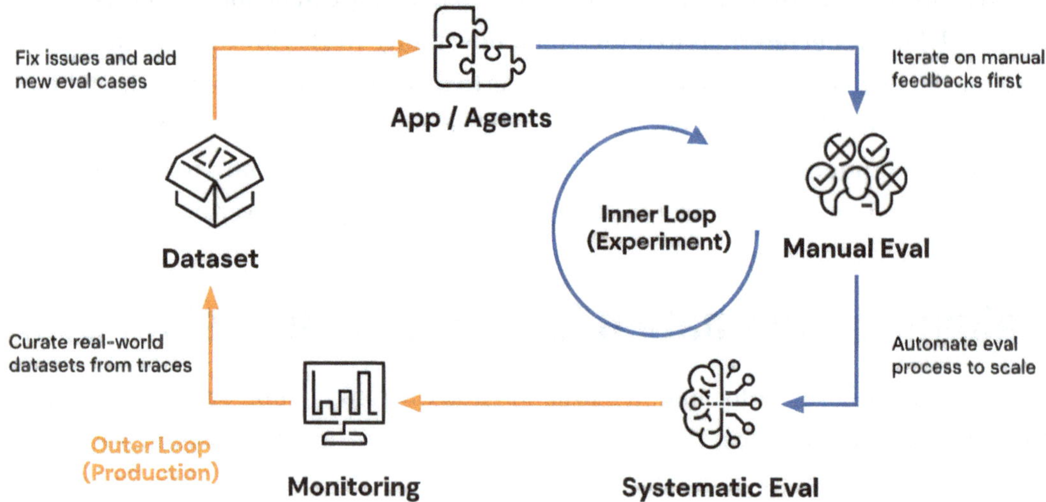

Figure 7-1. *Evaluation-driven development*

Manual Evaluation

In the era of GenAI, everything moves extremely fast, unlike in traditional machine learning, where we evaluate model performance using metrics like F1 scores. People usually don't have a rigid structure for evaluating agents. It's understandable that if we were trying to build domain-specific agents, for example, even as simple as summarizing a transcript, it can take a long time to find a human expert to review and give feedback on it. That's why in Information Extraction and Custom LLM, Agent Bricks acts as your human expert to guide you throughout the manual evaluation process. They are presented as recommendations, as shown in Figure 7-2.

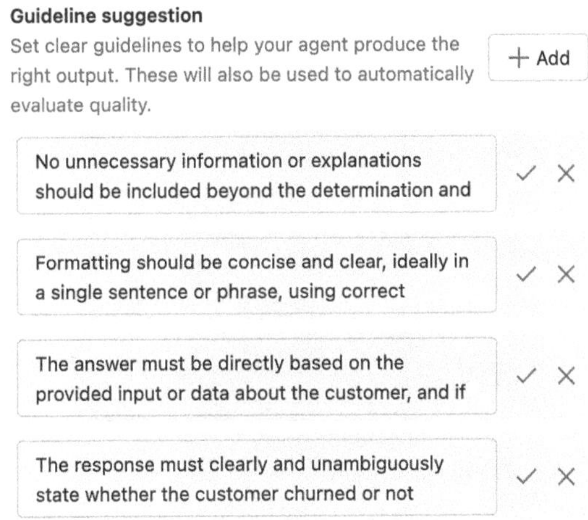

Figure 7-2. *Custom LLM agent guideline*

Databricks has put a research team behind the Agent Bricks user interface. These recommendations would sample the data (up to 50 records) from your dataset and use the latest research techniques like ***chain-of-thought*** and ***tree-of-thought*** to give relevant recommendations. Every time we update the agent, new recommendations will come up. That's what it means by "iterate on manual feedback."

Systematic Evaluation

Since MLflow 3 and above, GenAI capabilities have been taking shape, and we can now track both traditional machine learning and GenAI in a single interface. Information Extraction and Custom LLM provide evaluations using the MLflow framework; this feature is called "Quality Report." Figure 7-3 shows the quality report in Custom LLM.

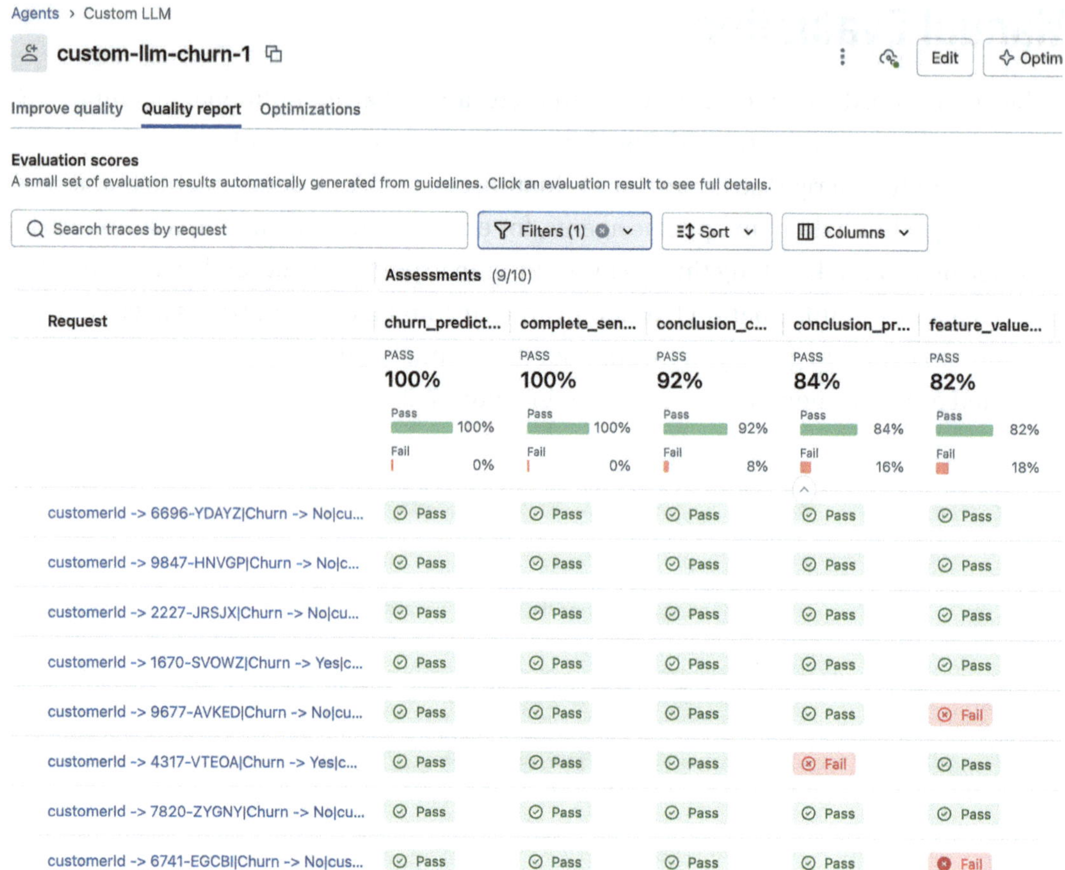

Figure 7-3. *Quality report in Custom LLM*

At first glance, the quality report shows the agents' performance; it provides a quick view of how each agent is doing per request. We can see the overall percentage of each guideline in Custom LLM. As shown in Figure 7-3, Custom LLM provides fine-grained control over what we want to do with the agent line by line, allowing us to tune the output without having to craft a long, complicated prompt. While the agent's tuning can be considered a manual eval, the quality report, on the other hand, is a systematic eval (auto eval) conducted by LLM judges. However, as shown in the diagram above, whether it's manual or auto eval, the process is iterative and continuously improves until we can achieve high quality in our agents.

Monitoring

MLflow has been a part of machine learning experimentation tracking for a long time. In GenAI, in addition to tracking the parameters and metrics, we are also capturing the traces, which include the prompt and the chain of thoughts from the LLM or agent. In Chapter 11, we will discuss extensively the capabilities of MLflow that can capture everything in an agentic workflow. Agent Bricks is one of its consumers only. Figure 7-4 captures a sample input and output from an Agent Bricks endpoint. We can open them from the Agent Bricks quality report or the respective MLflow experiment.

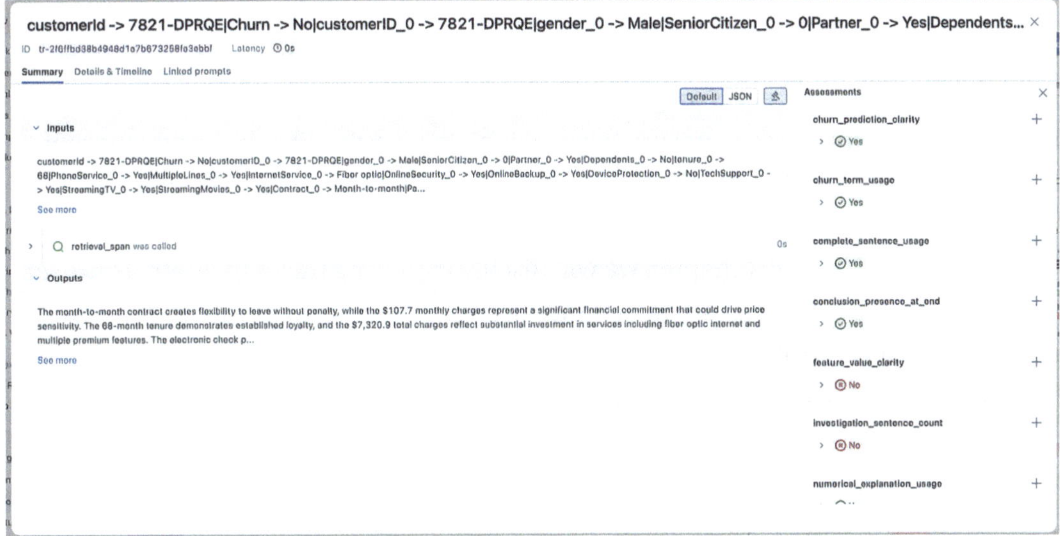

Figure 7-4. *A single trace from an Agent Bricks endpoint*

When we create an Agent Bricks agent, a few components get created:

1. An inference endpoint

2. An MLflow experiment

3. A vector search index (Knowledge Assistant only)

In Figure 7-4, the traces are the initial traces used to develop the agent. However, when developing an application, we can continue to add traces to the same experiment to keep track of performance over time. For example, we can easily use the three components above to create a Databricks app. The most common use is a chat application, using the endpoints of the ChatCompletion API (industry standard) or the

ResponsesAgent (Databricks standard). While a chat application is a typical application, with a little imagination, we can also create applications that don't require a text box, for example, a chess application. For example, Google's DeepMind Game Arena is trying to test LLM's raw intelligence by pairing them up to play games like chess. We can leverage traces in Databricks to capture the chain of thoughts, input, and output, all using the ChatCompletion API. The below blog is a good example:

```
https://www.tredence.com/blog/game-arena-deepmind-on-databricks
```

Databricks Traces

In Figure 7-1, we present the inner loop, which consists of the experiment iterations. However, shipping an agent is only the beginning; we will need to continue to monitor the quality of the inference to ensure that we don't have drifts, like data drift, concept drift, behavior drift, and so on. We are entering the production stage, which is the outer loop.

In the Databricks Apps chapter, we will discuss capturing traces in depth. However, in this chapter, we will focus on understanding the traces created by Agent Bricks and use them to inspect quality in depth.

In Figure 7-4, the left side shows the agent's basic input and output, and the right side shows the ratings from LLM judges based on the criteria we created in Custom LLM, or for each attribute in Information Extraction. These are called assessments.

There are two types of assessments in MLflow:

- **Feedback:** As shown in Figure 7-5, this field allows us to capture human feedback or populate it with LLM judges based on the agent output. In the case of quality reports, feedback will be populated by LLM judges automatically.

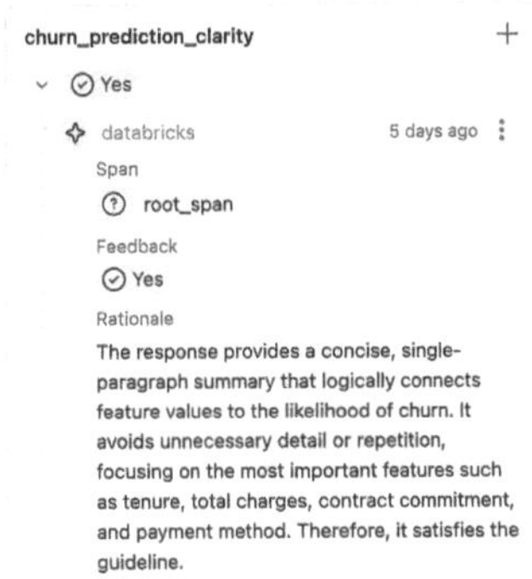

Figure 7-5. *Feedback field*

- **Expectation:** As shown in Figure 7-6, this field captures Ground
  Truth labels or desired outputs for the GenAI application provided
  by human experts. During a review session of an MLflow experiment,
  users can give a thumbs-up or thumbs-down on the response and, in
  the case of thumbs-down, provide a correct answer.

Figure 7-6. *Expectation field*

Table 7-1 is a comparison between Expectations and Feedback.

Table 7-1. *Comparison between Expectations and Feedback*

Aspect	Expectations	Feedback
Purpose	Define what the AI should produce	Evaluate how well the AI performed
Timing	Set before or during development	Applied after AI generates output
Content	Ground truth values	Quality scores, pass/fail judgments
Source	Always from human experts	Can be from humans, LLM judges, or code
Usage	Reference point for evaluation	Actual evaluation results

Dataset

As the name suggests, a dataset is a predefined set of data we can use to evaluate and fix issues for the agent. Normally, these are handled through the MLflow user interface or programmatically using the MLflow Python SDK, but Agent Bricks provides these experimentations in the interface. When developing a GenAI application, it is highly recommended to use evaluation and labeling sessions, as shown in Figure 7-7.

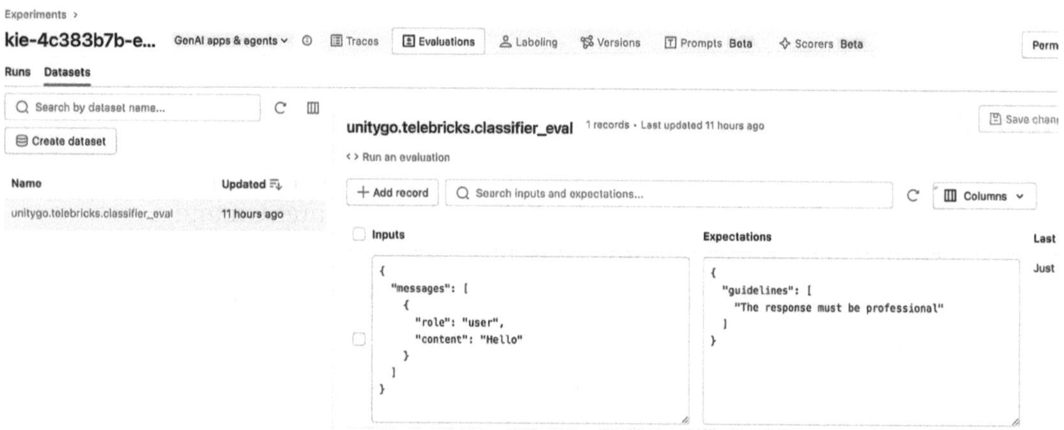

Figure 7-7. *MLflow interface for evaluation and labeling*

In Agent Bricks, additional evaluation and review applications are integrated into the interface without requiring separate development. They follow the same evaluation-driven development approach in Figure 7-1.

New Evaluation

While "Quality report" is built into Agent Bricks, to continue evaluating our agents' quality, we can provide a new dataset. In the case of Agent Bricks, it is simply a table from Unity Catalog. We can simply use the "New evaluation" button, as shown in Figure 7-8, and we will be prompted to choose a table for evaluation. The evaluation will then run again, and a new quality report will be generated.

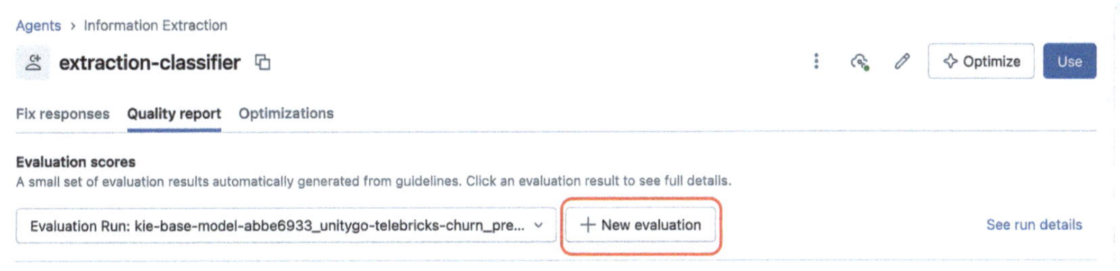

Figure 7-8. *New evaluation in quality report*

Fix Responses

Similar to the Databricks reviews app, Agent Bricks has built-in reviews in the interface. However, one major difference between the reviews app and fixed responses is that Agent Bricks actively incorporates human feedback to improve the agent's quality. As shown in Figure 7-9, we can give the agent a thumbs-up or thumbs-down on the response. In the case of a thumb-down, we can also provide feedback to the agent, and the feedback and expectations will be captured in the traces, as discussed in the last section.

Figure 7-9. *Review session in Agent Bricks*

Knowledge Assistant and Multi-Agent Supervisor

We have spent considerable time discussing the evaluation-driven development approach, which includes Information Extraction and a Custom LLM. Fortunately, the Knowledge Assistant and the multi-agent supervisor share a much simpler approach to quality improvement: leveraging natural-language feedback from human experts, also known as Agent Learning from Human Feedback (ALHF). This technique, a core component of the "improve quality" feature, enables continuous refinement of agents, ensuring they meet specific enterprise needs and deliver high-quality, accurate results.

Agent Learning from Human Feedback (ALHF) in Knowledge Assistant

The quality improvement is powered by a sophisticated technique called Agent Learning from Human Feedback (ALHF). ALHF goes beyond simple binary feedback (like thumbs-up or thumbs-down) by enabling experts to provide detailed, natural-language feedback. This rich feedback is then used to train the agent, like refining the retrieval algorithm, enhancing prompts, filtering the vector database, or even modifying the agentic pattern, from the retrieval mechanisms to the prompt structure and logical reasoning.

While the process is straightforward, it is a crucial step to enhance the agent's quality:

1. **Labeling Sessions**: Subject matter experts are presented with a set of questions in a "Review App." This app is located directly in the "Improve Quality" tab, as shown in Figure 7-10.

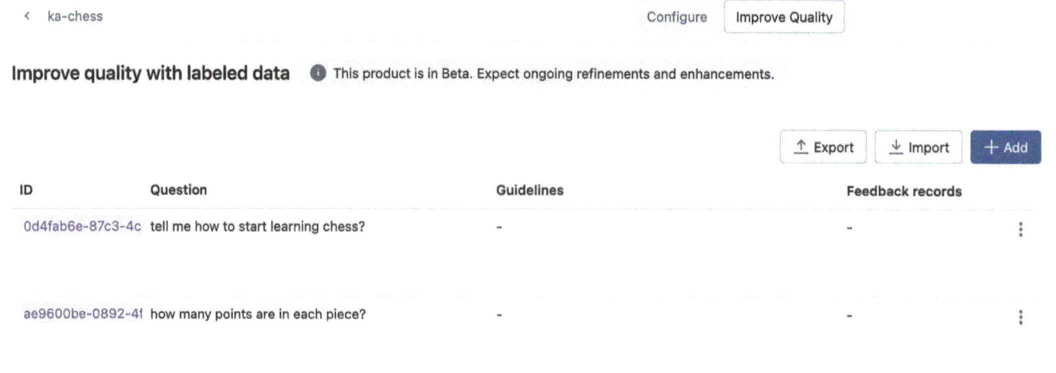

Figure 7-10. *Review questions in Knowledge Assistant*

2. **In-Depth Review:** For each question, reviewers can examine the agent's response, its reasoning, and the sources it cites.

3. **Guideline and Feedback Submission:** Experts can then provide specific feedback and add or refine guidelines to steer the agent's behavior

4. **Continuous Improvement:** This feedback is used to retrain and optimize the agent, resulting in demonstrable improvements in both answer completeness and adherence to expert expectations.

Research by Databricks has shown that ALHF is highly sample-efficient, with significant quality gains achievable with a small number of feedback records. In Figure 7-11, we can see that both answer completeness and feedback adherence show that the agent's performance improved in about four question-and-answer pairs. The full report is available on the Databricks research blog below.

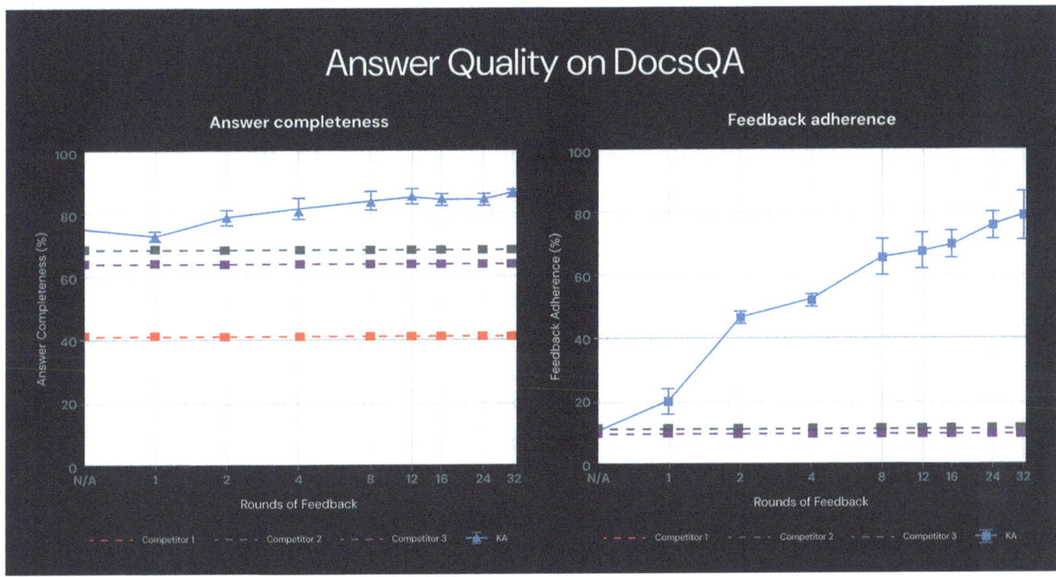

Figure 7-11. *Quality comparison of ALHF in terms of QA iteration. Data from Databricks Labs, DocsQA repository.* `https://github.com/databrickslabs/doc-qa`. *Used under the Databricks License (© 2023 Databricks, Inc.). (This image is not included in the book's Creative Commons license.)*

```
https://www.databricks.com/blog/agent-learning-human-feedback-alhf-
databricks-knowledge-assistant-case-study
```

The **doc-qa** dataset is open-sourced by Databricks Labs:

```
https://github.com/databrickslabs/doc-qa
```

Extending the Paradigm to the Multi-Agent Supervisor

The same fundamental principle of leveraging expert feedback applies to the multi-agent supervisor, but the focus shifts from the quality of a single agent's response to the quality of coordination among multiple agents. The "Improve Quality" tab for the supervisor allows for the collection of human feedback to enhance its orchestration capabilities.

The workflow mirrors that of the Knowledge Assistant. However, the focus now will be different:

1. **Task Scenarios:** Instead of individual questions, experts are presented with complex task scenarios to evaluate how the supervisor delegates and synthesizes results from various sub-agents.

2. **Coordination Review:** Reviewers assess the supervisor's response and overall handling of the multi-agent workflow.

3. **Feedback and Refinement:** Experts provide feedback to improve the supervisor's coordination logic and the behavior of the orchestrated agents.

4. **Optimized Orchestration:** The collected feedback is used to retrain and optimize the supervisor, resulting in more effective and efficient multi-agent systems.

Whether it's a single Knowledge Assistant providing answers from a knowledge base or a multi-agent supervisor orchestrating a complex workflow, Databricks provides a unified, no-code interface for continuous improvement through expert feedback.

Conclusion

In this chapter, we have discussed the evaluation-driven method driving Information Extraction and Custom LLM. The steps are:

1. App/Agents

2. Manual Eval

3. Systematic Eval

4. Monitoring

5. Dataset

These steps ensure an iterative process allows us to systematically evaluate the GenAI agents.

Furthermore, we explored the quality improvement mechanisms for the Knowledge Assistant and multi-agent supervisor, which leverage a simpler, yet powerful approach: Agent Learning from Human Feedback (ALHF). This technique allows human experts to provide detailed, natural-language feedback to refine agent performance across retrieval mechanisms and logical reasoning. This paradigm extends to the multi-agent supervisor, where feedback enhances its ability to orchestrate complex, multi-agent workflows.

These techniques are merely examples of how Agent Bricks can improve on its own, and additional optimization techniques will continue to be applied. For example, by applying automated prompt optimization, Databricks can drastically lower the total cost of ownership compared to using the frontier models as it is.

Ultimately, whether through the evaluation-driven development or the direct, expert-guided refinement of ALHF, Agent Bricks provides a unified, no-code interface for continuous improvement. This integrated framework ensures that agents can be refined to meet specific enterprise needs and deliver high-quality, accurate results.

## CHAPTER 8

Lakebase: The OLTP Engine for Intelligent Applications

Databricks is the most optimized platform for data and analytics, including Online Analytical Processing (OLAP). However, several use cases require sub-second data latency, in which case Lakehouse data must be moved to a low-latency online transactional processing (OLTP) database to serve online applications. In the past, organizations have adopted a strategy of copying data from Delta Lake into a separate OLTP database like Azure SQL Database, Amazon Aurora, or GCP Cloud SQL. The purpose was to enable building applications on top of the OLTP database to serve low-latency queries and write to it.

Although the approach looks simple and straightforward, several practical challenges arise during the synchronization process between Delta Lake and the OLTP database. In this chapter, we will learn more about Lakebase, its capabilities, and how it can solve these challenges.

What Is Databricks Lakebase?

Lakebase is a fully managed PostgreSQL online transaction processing (OLTP) database engine integrated into the Databricks Data Intelligence Platform.

An online transaction processing (OLTP) database is a specialized type of database system designed to handle high volumes of real-time transactional data efficiently. Lakebase allows you to create an OLTP database on Databricks and integrate OLTP workloads with your Lakehouse. This OLTP database enables you to create and manage databases stored in Databricks-managed storage.

© Jason Yip, Nikhil Gupta and Marcin Wojtyczka 2026
J. Yip et al., *Databricks Data Intelligence Platform*, https://doi.org/10.1007/979-8-8688-2524-8_8

Lakebase integrates fully with Databricks Feature Store, SQL Warehouses, Databricks Apps, and Unity Catalog.

While it inherits several core platform capabilities from the original PostgreSQL, the key differentiators of Lakebase including:

- **A Unified Platform for OLAP and OLTP**: Lakebase eliminates the need for separate databases and complex ETL pipelines, allowing for instantaneous, low-latency access to data for applications and AI agents.

- **Simplified Management**: Leverages existing Databricks infrastructure to deploy instances with decoupled compute and storage, managed change data capture with Delta Lake, and support for multi-cloud deployments.

- **Integrated AI and ML Capabilities**: Supports feature and model serving, retrieval-augmented generation (RAG), and other AI and ML integrations.

- **Integrated Authentication and Governance**: Use Unity Catalog to enforce secure access to data.

Getting Started with Lakebase

To prioritize the new experience, Databricks has created a brand-new interface for Lakebase. To find the new Lakebase interface, look for the 6-dot icon next to the profile in the top-right corner, as shown in Figure 8-1.

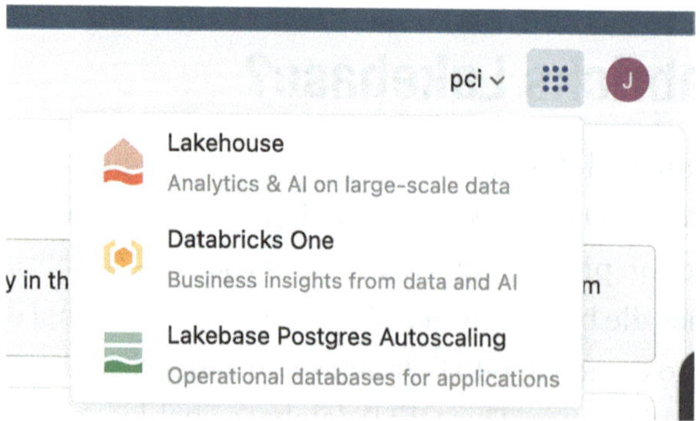

Figure 8-1. *Lakebase Postgres autoscaling switch in Databricks*

Welcome to the Lakebase Interface

The Lakebase interface is brand new and completely redesigned as operational and project-focused. Each project would allow users to create and manage PostgreSQL instances in a single interface, as shown in Figure 8-2. It's a way to allow developers to focus on what matters most. Beyond the operational metrics, there is one notable new feature that is uncommon in traditional database user interfaces: branching.

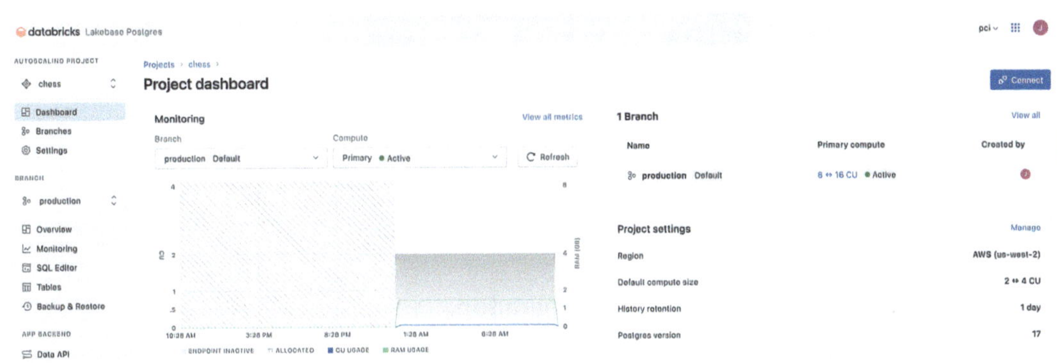

Figure 8-2. *The brand new Lakebase project dashboard*

Storage and Compute Decoupling

To understand what branching all is about, we must first understand one of Lakebase's most groundbreaking technologies: the separation of data and compute without sacrificing performance. In traditional databases, storage and compute are tightly coupled, meaning you must scale both together, even if your workload only demands more CPU or more disk. This leads to over-provisioning, higher costs, and inflexibility.

As shown in Figure 8-3, Lakebase separates storage and compute into two layers:

- **Storage Layer:** All data resides in open formats in low-cost, highly durable object storage (e.g., Delta Lake on cloud storage).

- **Compute Layer**: Multiple compute clusters can access the same data independently, scaling up or down based on workload requirements.

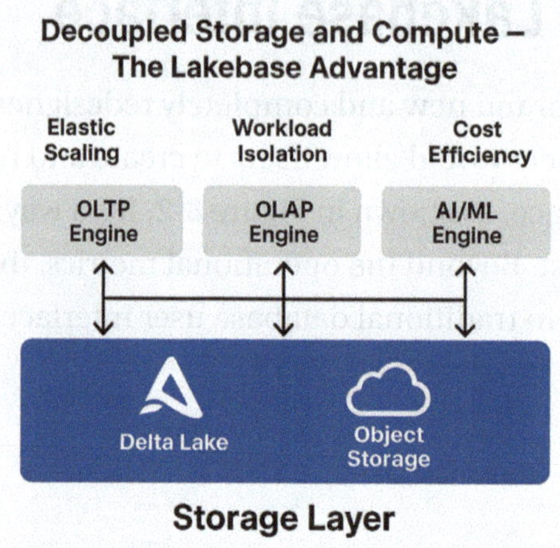

Figure 8-3. *Lakebase's compute and storage separation*

There are several key benefits of this separation:

1. **Cost Efficiency:** Pay only for the compute resources when they are in use, and storage remains inexpensive and persistent.

2. **Workload Isolation:** Analytical queries, real-time ingestion, and machine learning jobs can run on separate compute clusters without interfering with each other.

3. **Elastic Scalability:** Serverless architecture scales compute resources instantly without affecting the storage layer and vice versa.

4. **High Availability:** If a compute cluster fails, the data remains safe and still accessible to other compute clusters.

In the last edition, we discussed CI/CD strategies in depth. And the differentiator here is that Databricks Lakebase has been able to branch the data. We are not talking about tracking the version of the database scripts but about branching the data. However, in theory, the operation does not perform a full data copy in milliseconds, but it uses a technique called "Copy-on-write."

Copy-on-Write

While the full technical details of copy-on-write are beyond the scope of this book. At a high level, copy-on-write creates a pointer to the existing data at a snapshot in time. This pointer ensures we capture a snapshot of the data without interfering with the original "branch" and that when a new write event is requested in the new branch, only that block of data is copied. Figure 8-4 illustrates this process.

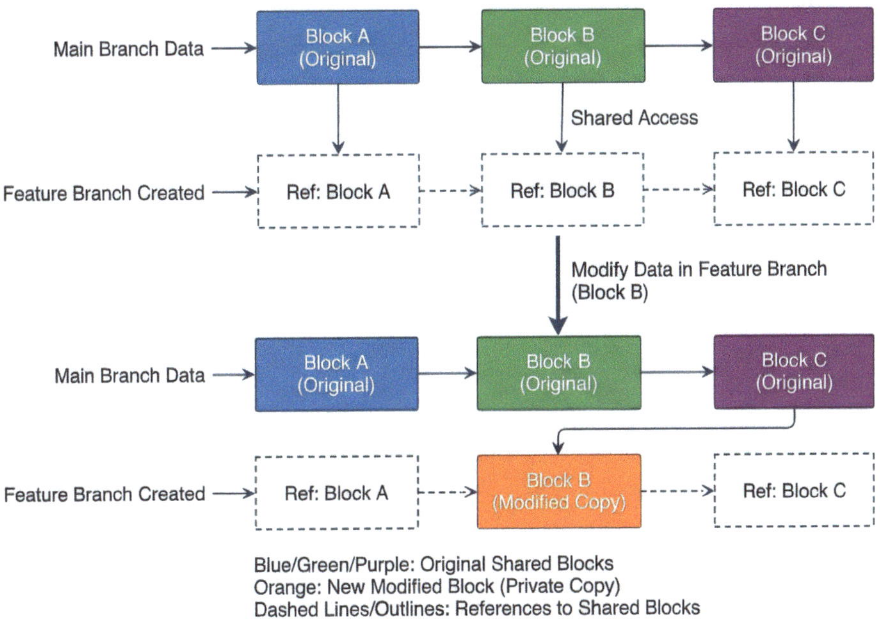

Figure 8-4. *The process of copy-on-write*

What Is Data Branching?

We learned a new concept called data branching, which is conceptually very similar to traditional CI/CD. The purpose of data branching is also very similar. In modern databases, including PostgreSQL, we need to create separate environments for different teams to support development and testing. We must restore a backup to a different location. Not only will it be labor-intensive, but it will also require duplicating the entire database. Imagine bringing the git branch experience from your codebase to your database. This is what data branching is about.

Data Sharing Between Lakebase and Lakehouse

Databricks lets you create and query tables in the Postgres database using the SQL editor or external tools.

To query in Databricks SQL, we can open the SQL editor and attach the compute cluster to Lakebase Postgres. Or, if you are in the new interface, simply click the "Connect" button in the top-right corner to display the connection window, as shown in Figures 8-5 and 8-6.

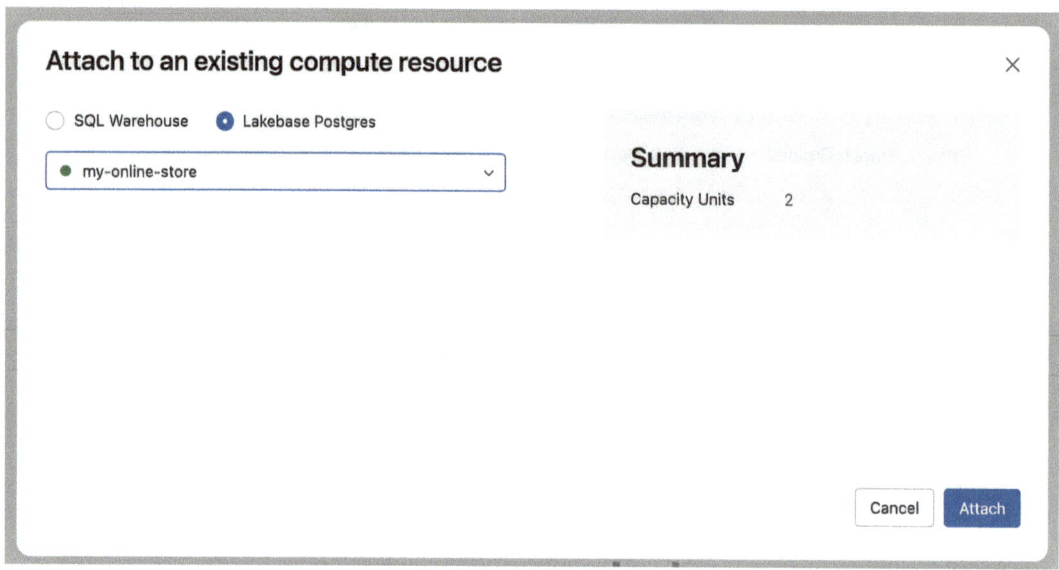

Figure 8-5. *Attaching Lakebase to a query*

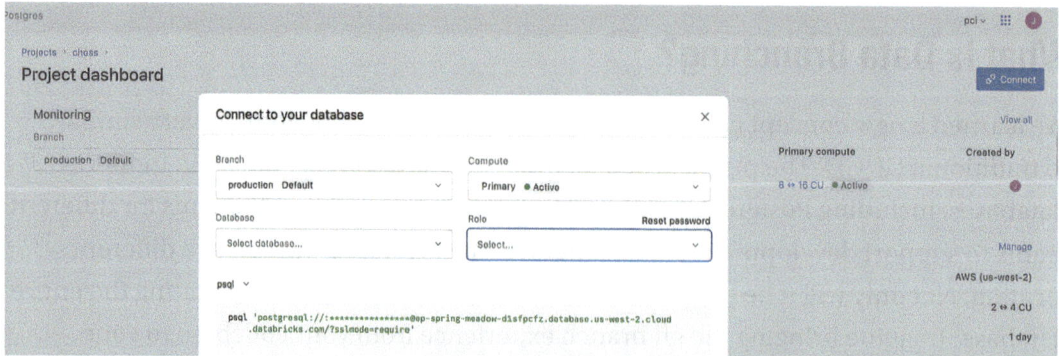

Figure 8-6. *Connecting to Lakebase from the new interface*

As discussed, Lakebase is a fully managed PostgreSQL database hosted by Databricks to abstract management from users. Beyond the Databricks interface, we can also connect external tools to a Lakebase database, just as we can with other PostgreSQL instances. We can find the connection string in the Compute tab inside the Lakebase section, as shown in Figure 8-7.

Figure 8-7. *Connection details of a Lakebase instance*

Lakebase Core Features

In this section, we will explore the core features of Lakebase. We will explain how Databricks' version of PostgreSQL is not only competitive but also an industry-leading offering. Figure 8-8 outlines all the features.

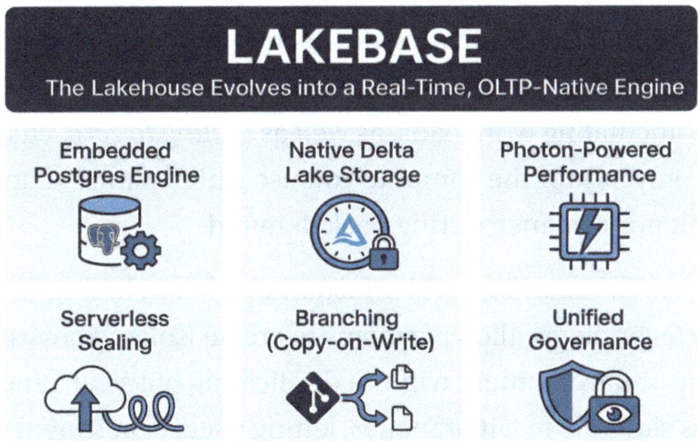

Figure 8-8. *Lakebase core features*

The core features include:

- **Embedded Postgres Engine**: Lakebase leverages the industry-standard PostgreSQL engine to provide full support for ACID transactions and familiar SQL syntax. This ensures that developers can use existing stored procedures and indexes, making it easier to migrate traditional relational workloads to a modern, scalable architecture without losing data integrity.

- **Native Delta Lake Storage**: All transactional data is stored in the open Delta format, which bridges the gap between data lakes and data warehouses. This architecture natively supports "time travel," allowing users to query previous versions of data for auditing or recovery, while strict schema enforcement prevents data corruption and ensures high quality across the pipeline.

- **Photon-Powered Performance**: To achieve high-speed query execution, the system utilizes a Photon-powered engine designed for vectorized execution. By leveraging predicate pushdown and intelligent caching, Lakebase minimizes data processing and maximizes hardware efficiency, resulting in significantly faster performance for both analytical and transactional queries.

- **Serverless Scaling**: The platform offers automatic elastic scaling of compute resources, eliminating the need for manual infrastructure management. This serverless approach allows the system to instantly adjust to fluctuating workloads, as well as *scaling to zero*, ensuring that you only pay for the compute you use while maintaining consistent performance during peak demand.

- **Branching**: This feature enables instant database cloning via copy-on-write technology, allowing teams to create isolated environments for testing or development without duplicating physical data. It functions similarly to Git branches, letting users experiment with data changes in a safe environment before merging them back or discarding them.

- **Unified Governance**: Security and compliance are managed through Unity Catalog, a single unified governance layer that oversees the entire data estate. Unity Catalog includes advanced features like data masking for privacy, lineage tracking to understand data flow, and comprehensive audit logging to meet regulatory requirements and maintain transparency.

Databricks has integrated all the key features that make the Databricks data platform valuable into Lakebase, making it the best-in-class database platform.

Managed Data Synchronization

In Chapter 12 "Real-Time Intelligence with Spark Structured Streaming" we will discuss why we need to clone a Delta table to Lakebase for an online feature store. However, at a high level, a synced table is a read-only PostgreSQL table in Lakebase that automatically receives and reflects data synchronized from a Unity Catalog (Delta) table into the Lakebase instance. Databricks continuously updates this PostgreSQL table to stay aligned with the source Delta table while maintaining read-only semantics.

The synchronization is handled by Lakeflow Spark Declarative Pipelines. A managed pipeline continuously updates the PostgreSQL table with changes from the source table. After creation, synced tables can be queried directly using Postgres tools.

The key characteristics of synced tables are as follows:

- Read-only in Postgres to maintain data integrity with the source

- Automatically synchronized using managed Lakeflow Spark Declarative Pipelines

- Queryable through standard PostgreSQL interfaces

- Managed through Unity Catalog for governance and lifecycle management

Synced tables can be created using the UI, the Python SDK, the CLI, or cURL. We will explore the UI approach of building the synced table, as shown in Figure 8-9.

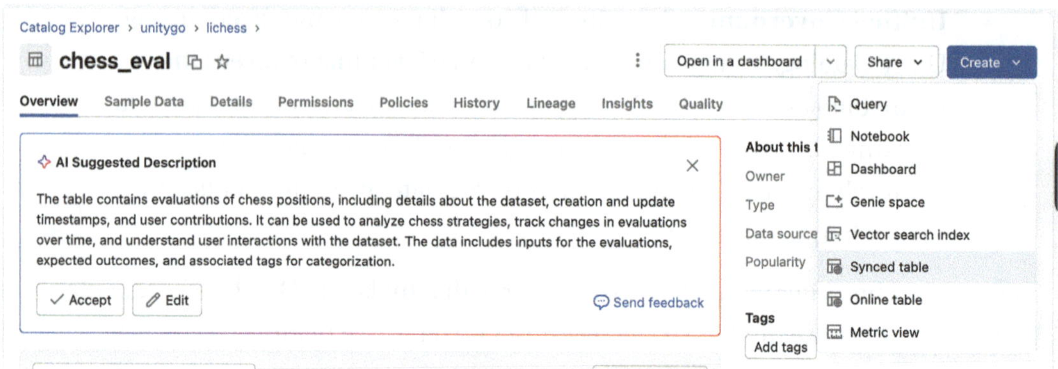

Figure 8-9. *Creating a synced table from a delta table*

As shown in Figure 8-10, there are three sections when creating a synced table: Destination, Synchronization settings, and Pipeline settings.

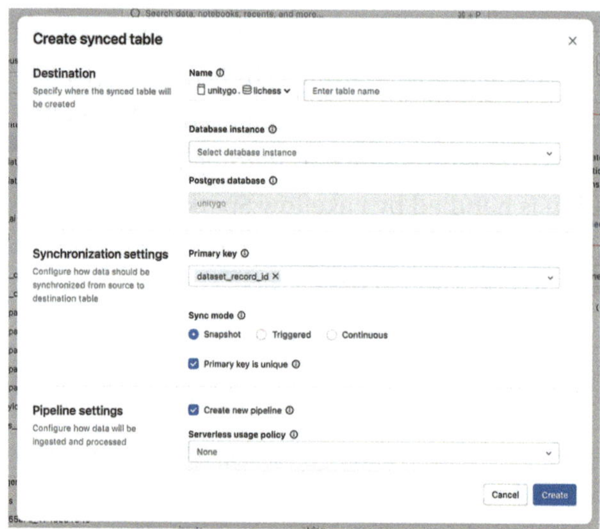

Figure 8-10. *Create a synced table user interface*

While the "Destination" section is self-explanatory, the "Synchronization settings" section requires more attention. While there is no primary key enforcement in Delta tables, database users would understand what a primary key is. The purpose of a primary key is to enable efficient access to rows for reads, updates, and deletes, allowing seamless sync between the two tables.

However, in case there are two rows with the same primary key in the source table due to the lack of primary key enforcement, we can choose a timestamp, e.g., create_time, for deduplication. When a timestamp is specified, the synced table contains only the rows with the latest timestamp value for each primary key, as shown in Figure 8-11.

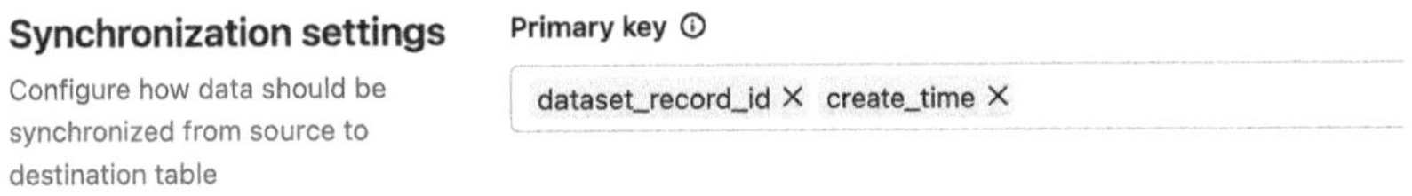

Synchronization settings   Primary key ⓘ

Configure how data should be synchronized from source to destination table

dataset_record_id ✕   create_time ✕

Figure 8-11. *Synchronization settings with multiple keys*

There are a few "sync modes":

- **Snapshot**: This mode is designed for complete data replacement. The pipeline runs once to take a full snapshot of the source table and copies it to the synced table; subsequent runs replace the destination data atomically. According to Databricks documentation, it is considered 10 times more efficient than incremental modes when recreating data from scratch, making it the ideal choice if you are modifying more than 10% of your source table.

- **Triggered**: Triggered mode serves as a balance between cost and data freshness. After an initial full snapshot, it only retrieves and applies changes that have occurred since the last execution. While it helps minimize lag between refreshes, according to Databricks documentation, it can become more expensive than continuous mode if triggered more often than every 5 minutes due to the overhead of repeatedly starting and stopping the pipeline.

- **Continuous**: For applications requiring the most up-to-date data, "Continuous mode" provides real-time incremental updates without requiring manual refreshes. However, if cost is a concern, we should consider using a less frequent "triggered" sync mode.

Lakebase: Before and After
The Operational Complexity

Before Lakebase, there was significant operational complexity in keeping the operational and analytical worlds separate. To meet low-latency application requirements, organizations had to rely on external OLTP databases, which technically worked but posed a significant architectural challenge. Teams often spent time designing and maintaining custom ETL and CDC pipelines required to keep Delta Lake in sync with these external systems. Each of those pipelines introduced a new failure point, additional reconciliation logic, and constant monitoring overhead. Figure 8-12 demonstrates this endless lifecycle.

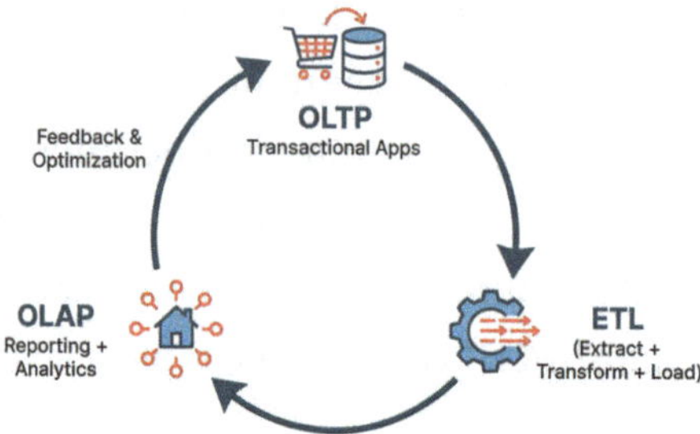

Figure 8-12.   *The lifecycle of ETL between OLTP and OLAP*

Lakebase changes the math by bringing transactional, low-latency access directly into the Databricks platform. It eliminates the boundary between operational and analytical systems that previously required constant synchronization. Because operational and analytical data now coexist, there is no longer a need for complex ETL pipelines. Applications and AI workloads can access fresh data immediately.

The Governance Layer

In addition to data copying, we were fighting governance drift. Because we were operating across multiple platforms, security policies like data masking and audit logging had to be implemented and enforced independently. This not only doubled the work, but it also created inconsistent access controls and increased our compliance risk. On top of that, we also required reliability engineers solely to manage infrastructure, network topology, and backups, which further complicated the team structure.

By moving everything under Unity Catalog, we get a single, unified governance plane. Access control, lineage, and auditing are consistent across Delta Lake and Lakebase, helping close the gap on governance drift. Since Lakebase is native to the platform, the heavy lifting of infrastructure management, like network design and security hardening, is handled by the control plane. This allows teams to focus entirely on business logic. Ultimately, we are not just lowering the total cost of ownership; we are also optimizing the entire system for speed, ensuring data is fresh and secure for real-time applications and inference.

Stateful AI: Lakebase As Short- and Long-Term Storage

By now, most of us have learned basic prompt engineering, the concept of chaining prompts together (like LangChain and LangGraph), and retrieval systems like RAG are also very popular. Additionally, the Agentic AI patterns, such as Planning and ReAct, offer the most advanced and comprehensive ways to enhance the quality of agents' output. However, one critical enhancement we need is to understand past conversations and create stateful AI. In the same way an old friend understands you better than a new acquaintance. We can leverage both Lakebase and Mosaic AI Vector Search to store and retrieve this information, also known as Agent Memory.

Cognitive Memory

The design of modern agent memory systems is inspired by the paper "Cognitive Architectures for Language Agents" by Summers et al. from Princeton University (`https://arxiv.org/abs/2309.02427`). The high-level design is shown in Figure 8-13. Recall that large language models are loosely inspired by neural processes; it makes sense to model memory based on human memory. These memory patterns can also be found in medical publications.

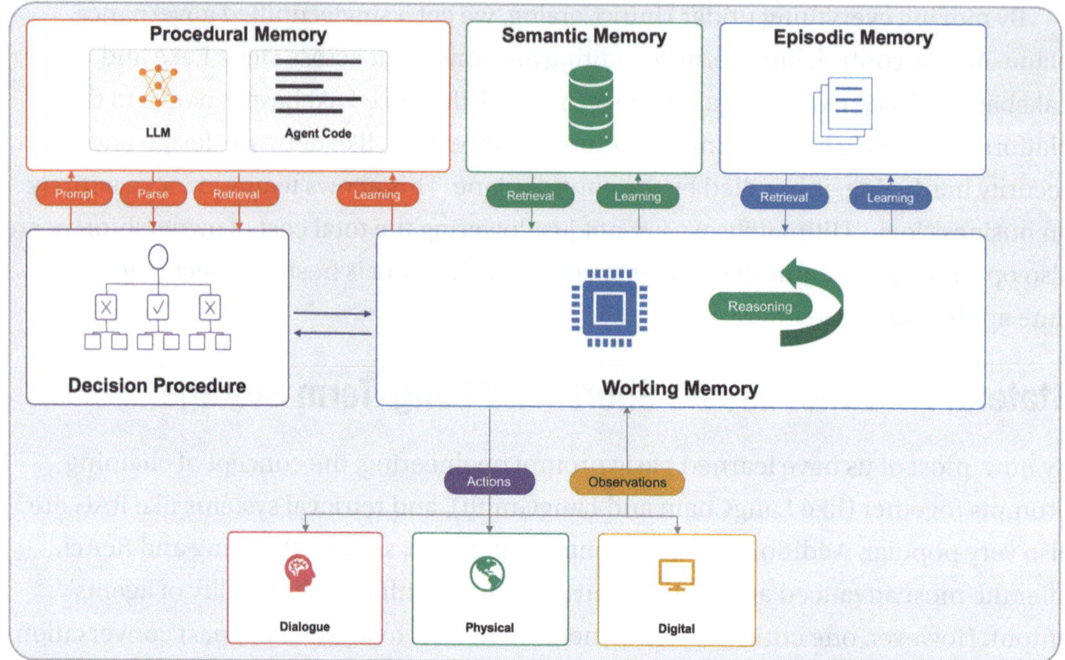

Figure 8-13. *Cognitive architectures for language agents by Summers*

The diagram above illustrates the cognitive architecture of an AI agent, designed to mimic human thought processes. At its center is the "working memory," which acts as the active processing hub where reasoning takes place based on incoming "observations" from the outside world. To process these observations, the working memory interacts with three distinct long-term memory banks: procedural memory, semantic memory, and episodic memory. The system continuously pulls relevant information from these memories to understand its situation and updates them through an ongoing learning process. Table 8-1 summarizes the differences between short-term and long-term memory.

Table 8-1. *Short-term vs. long-term memory*

Memory type	Definition	Example
Short-term memory	Memory that is discarded after an event completes	Conversation history, a chat session
Long-term memory		
Semantic (Declarative)	Facts and concepts	Skill level: intermediate; Chess strategies: Sicilian Defense
Episodic (Declarative)	Experiences	A PGN (game log) of your match where you won or especially impressed with
Procedural (Non-declarative)	Instructions and rules	The rules of chess (Is it a checkmate?)

When users interact with an LLM agent, the working memory collaborates with a "decision procedure" to map out and weigh different choices. This feeds into a dedicated "Planning" phase, which operates in a careful loop: the system first proposes potential plans, evaluates their viability, and selects the best option. Once an optimal path is selected, the system moves to "Execution," translating its internal reasoning into concrete "Actions" directed back out into the environment. This workflow is shown in Figure 8-14.

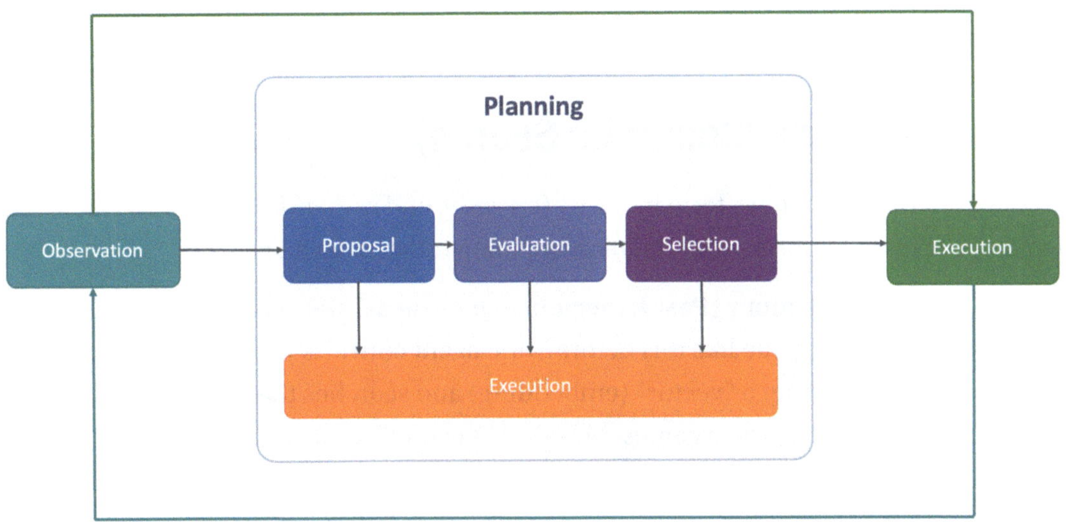

Figure 8-14. *The planning phase of working memory*

Memory Storage

The human brain is such a miracle that it can store a lot of information. However, in the case of agent memory, we need a place to store information temporarily for short-term memory or permanently for long-term memory. Fortunately, Databricks provides two lightning-fast storage options: Lakebase and Mosaic AI Vector Search.

Lakebase (Relational/Key-Value Storage)

Lakebase is primarily used for short-term memory (Checkpoints) and procedural memory. This is because these types of memory require exact, high-speed lookups based on specific IDs.

- **Short-Term Memory:** Every time the agent makes a move in your chess game, the current state (the board, the turn, the chat history) is saved as a "checkpoint." You retrieve this using a thread_id. You don't need to "search" for it; you need the exact latest version.

- **Procedural Memory (Rules):** Instructions like "If a player is in checkmate, end the game" are static. These are stored as structured text or JSON in a standard relational table.

- **Profile (Semantic):** Simple user facts (e.g., user_id: 123, favorite_ opening: "Sicilian") are best stored in a standard JSONB column for quick, direct lookups.

Vector Database (Semantic Search)

A Vector DB is used for long-term memory (episodic and semantic collection) where the agent needs to find information based on meaning rather than an exact ID.

- **Episodic Memory (Past Experiences):** If you ask the agent, "Have I ever lost a game like this before?" the agent converts your current board state into a "vector" (embedding) and searches the database for similar historical games. Mosaic AI Vector Search can efficiently handle millions of records.

- **Semantic Memory:** If you have thousands of chess strategy documents, a Vector DB allows the agent to "retrieve" only the relevant paragraphs about "opening tactics." Similar to learning from practicing chess puzzles and from real-world tactics, they are factual but require semantic understanding. Figure 8-15 illustrates the comparison.

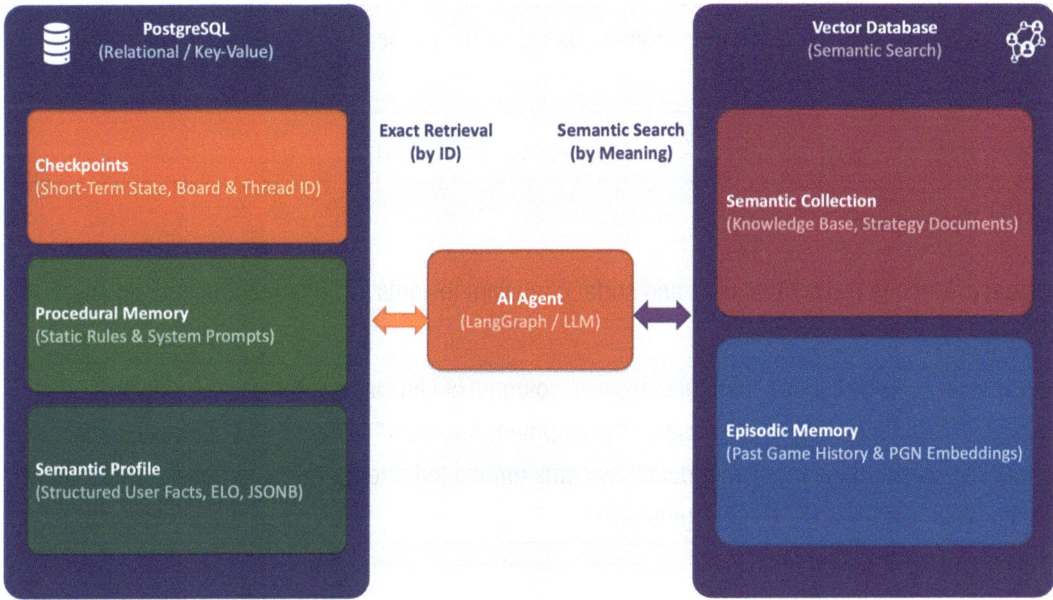

Figure 8-15. *Comparison between memory storage*

Real-Life Use Cases

While cloud databases are not new, the use cases are constantly evolving. The following examples show how organizations in different industries use Lakebase for real-time decision-making and workflow automation.

Databricks Lakebase has many existing customers, including Adobe, T-Mobile, and Rolls-Royce, to name a few. We will discuss how these customers are using Lakebase; all stories are based on Databricks' real case studies.

Below are the use cases by industry:

 E-commerce[1]

Use case: Use pre-calculated customer segments and insights to support workflows such as preferential delivery, offer targeting, and personalized product recommendations.

How it works: Much like Uber Eats, **iFood** uses Databricks to process over 800 billion records to personalize the user journey. They use real-time data and ML models to support preferential delivery routes and target specific offers to customers based on their ordering behavior and geographic location.

Source: https://www.databricks.com/customers/ifood/

✚ Healthcare[2]

Use case: Manage clinical trial data and surface relevant insights through recommendation systems embedded in clinical workflows.

How it works: AstraZeneca manages massive volumes of clinical trial data by building an integrated knowledge graph. They use a "Development Assistant" AI agent that allows scientists to surface insights through recommendation systems embedded directly into their research workflows, helping them identify novel drug targets faster.

🏛 Financial services[3]

Use case: Enable automatic market trading based on streaming data and pretrained models.

How it works: Erste Group uses real-time data processing to automate financial workflows, such as reducing transaction categorization time from minutes to seconds. This high-speed analysis supports market trading and investment decisions by providing a consolidated, real-time view of payment processes and data across 12 countries.

Source: https://www.databricks.com/customers/ceska-sporitelna

---

[1] Databricks, "iFood," Customer Case Study, https://www.databricks.com/customers/ifood.

[2] Databricks, "AstraZeneca," Customer Case Study, https://www.databricks.com/customers/astrazeneca.

[3] Databricks, "Erste Group," Customer Case Study, https://www.databricks.com/customers/ceska-sporitelna.

🏪 Retail[4]

Use case: Use a chatbot that incorporates recent conversation history and real-time data (for example, shopping cart contents) to personalize responses and drive engagement.

How it works: FordDirect developed an AI Copilot chatbot designed to transform how dealers interact with customers. The bot incorporates recent conversation history and real-time inventory data to personalize responses, helping drive engagement and higher conversion rates during the car-buying process.

🏭 Manufacturing[5]

Use case: Track and manage machine telemetry and IoT data to support low-latency decision-making and automated maintenance workflows.

How it works: Rolls-Royce tracks machine telemetry from over 13,000 jet engines in real time. They use this IoT data to create "Digital Twins" that support low-latency decision-making, allowing them to predict maintenance needs precisely on time and even use "swarm robots" for automated maintenance tasks.

Conclusion

Databricks Lakebase evolves the Lakehouse into a real-time, OLTP-native engine by integrating a fully managed PostgreSQL database directly into the platform. By decoupling storage and compute, it provides a scalable, cost-efficient solution for sub-second-latency workloads without the operational complexity of external databases.

In this chapter, we discussed the following:

- **Architectural Efficiency**: Eliminates complex ETL pipelines by allowing operational and analytical data to coexist under a single Unity Catalog governance layer.

---

[4] Databricks, "FordDirect," Press Release, https://www.databricks.com/company/newsroom/press-releases/databricks-unveils-new-mosaic-ai-capabilities-help-customers-build.

[5] Databricks, "Rolls-Royce," Customer Case Study, https://www.databricks.com/customers/rolls-royce.

- **Core Functionality**: Features copy-on-write branching for instant database cloning and Photon-powered performance for high-speed transactional queries.

- **Seamless Synchronization**: Uses Lakeflow Spark Declarative Pipelines to automatically sync Delta tables with Lakebase via Snapshot, Triggered, or Continuous modes.

- **Stateful AI:** With Lakebase as the storage layer, we can build agents that maintain context, such as which user made the request and what they've asked before, creating a more coherent experience.

Proven Impact: Already adopted by industry leaders like AstraZeneca, iFood, and Rolls-Royce, and many more, for real-time AI, IoT telemetry, and personalized e-commerce.

## CHAPTER 9

Building Security-First, Production-Grade Databricks Apps

In the previous chapters, we learned about the Databricks Lakehouse, which essentially means storing all your data in open storage in an open format with Unity Catalog providing a single governance layer. Databricks provides features that support all use cases, including data engineering, data science, streaming, and warehousing. Since 2023, with the rise of GenAI and LLMs, Databricks has integrated them into its platform. The Databricks data intelligence platform (see Figure 9-1) combines the Lakehouse platform and AI/LLMs to add the "data intelligence" engine that understands the uniqueness of your data and uses that understanding across everything in the platform.

Insights developed on a data platform must be shared with developers, often by moving data or model artifacts to an entirely separate application hosting environment. This process introduces several challenges:

1. **Infrastructure Overhead**: Developers must provision, manage, and secure new infrastructure, such as container orchestration platforms or virtual machines, which falls outside the core data environment and requires specialized expertise.

2. **Governance**: Authentication and access control must be reimplemented independently within the application, for example, by using OAuth or SAML, creating a separate security perimeter that must be manually synchronized with the data platform's policies.

3. **Data Egress and Latency**: Moving or duplicating data to serve the application introduces latency, increases costs, and adds to the security team's burden of monitoring yet another database to prevent unauthorized access.

This entire process, from initial infrastructure setup to final security sign-off, slows the time-to-value, turning what should be a rapid iteration cycle into a prolonged, resource-intensive project. Figure 9-1 shows a typical application architecture with multiple layers. We can easily imagine that different teams would be responsible for the process.

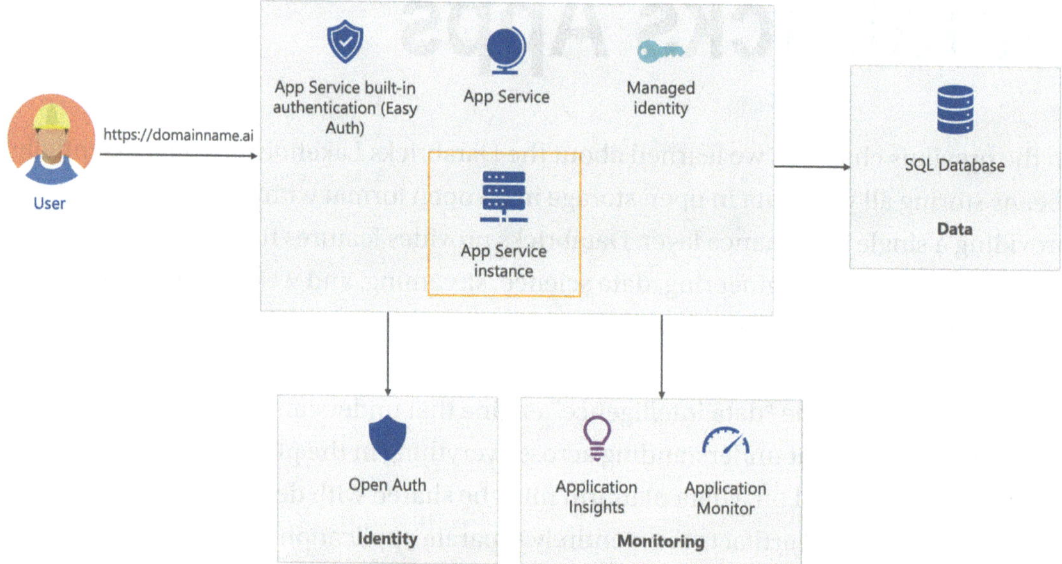

Figure 9-1. *Typical application architecture*

Building Apps Where Your Data Lives

Databricks Apps is trying to resolve the above problem by reimagining how applications are built. It introduces a new philosophy: "Bring the app to where your data is."

Instead of moving data out to a separate application environment, this model brings the application's compute directly to the data, hosting it within the secure, governed perimeter of the Databricks Unity Catalog. A Databricks App is a web application that runs as a containerized service on the Databricks serverless platform. Figure 9-2 shows the new architecture of modern data applications.

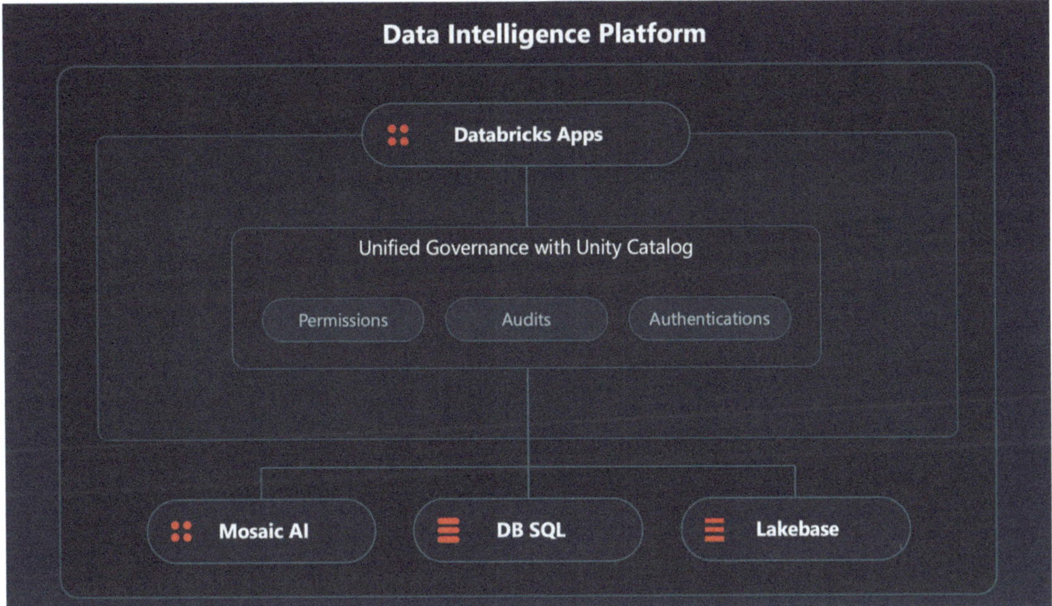

Figure 9-2. *Databricks Apps high-level architecture*

By colocating the application with the data, it dramatically simplifies the development lifecycle reduces operational overhead, and the governance layer of Unity Catalog, which is the first of its kind, allows developers to take advantage of the same security practices they have been using for their data.

Core Architecture: A Serverless, Containerized Foundation

The underlying architecture of Databricks Apps is built on a serverless foundation that is fully managed and containerized. Figure 9-3 shows an internal view of Databricks Apps.

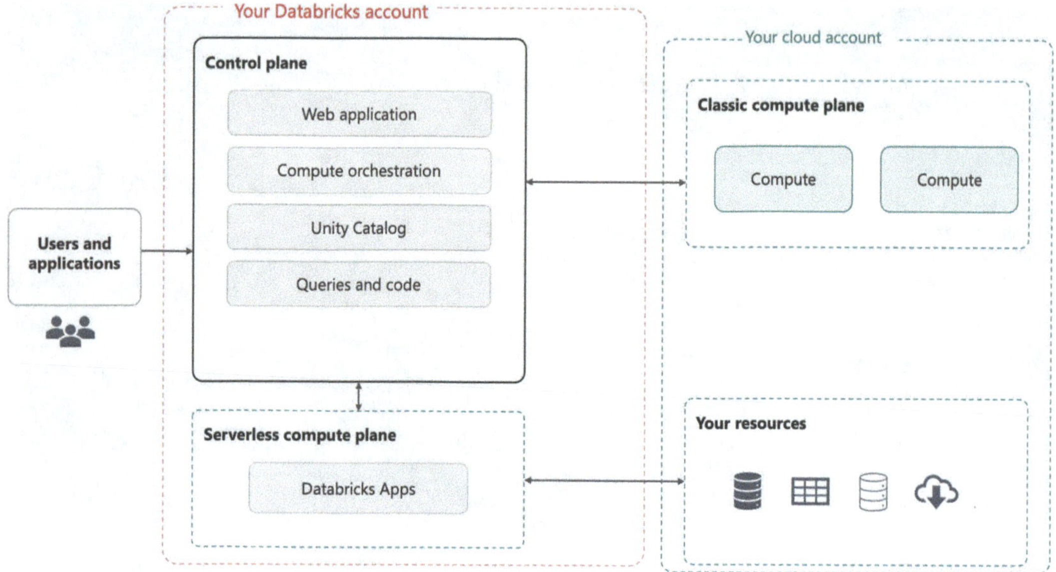

Figure 9-3. *Databricks Apps internal architecture*

The Serverless Compute Plane

Databricks Apps execute on the Databricks serverless compute plane, an infrastructure layer entirely managed by Databricks. Developers might be familiar with Azure Function Apps or AWS Lambda. "Serverless" compute at Databricks is similar: developers no longer need to provision, configure, scale, patch, or manage servers or containers.

This model offers several advantages:

1. **Accelerated Deployment**: Databricks maintains "warm pools" of instances, enabling near-instantaneous compute startup. Compared to serverless functions, Databricks serverless significantly reduces the "cold start" problem. The idea is similar to traditional cluster pools, except that Databricks now manages the pool for users.

2. **Cost Optimization**: Billing is based on compute time when the App is actively running, a pay-for-use model that is more efficient than maintaining perpetually running application servers.

3. **"Versionless" Infrastructure**: Databricks automatically handles all runtime upgrades and security patches, ensuring applications run on a secure, up-to-date platform without developer intervention. This is a critical compliance requirement.

Isolation and Security by Design

While each application runs in its own isolated, containerized environment, Databricks Apps also include a least-privileged service principal, so all permissions must be granted explicitly. We will cover the security model in detail later in this chapter. Figure 9-4 shows the unique credential generated for each App.

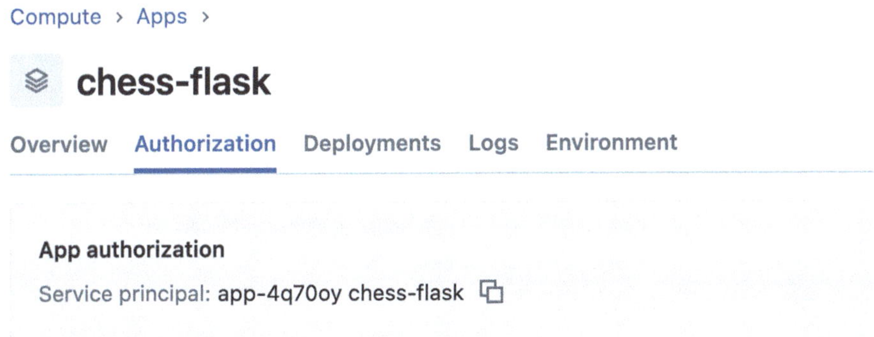

Figure 9-4. *Service principal for the App*

There are several advantages to this isolation design:

1. **Dedicated Resources**: Each App is allocated dedicated compute resources and encrypted storage for the duration of its execution. This infrastructure is never shared between different Apps. Upon termination, all resources are securely wiped.

2. **Network Segmentation**: Every App operates within a logically isolated private network with no public IP addresses, preventing data leakage between workloads.

3. **End-to-End Encryption**: All attached storage is encrypted at rest (AES-256), and all network traffic is encrypted in transit (TLS 1.2 or higher). This is often required to meet the highest financial compliance requirements.

4. **Principle of Least Privilege**: The application's runtime environment operates with minimal privileges. Access to Databricks resources is mediated through short-lived, securely passed tokens, ensuring access is temporary and strictly scoped.

The Runtime Environment

Databricks Apps include prebuilt templates for Dash, Flask, Gradio, Shiny, Streamlit, and Node.js. Figure 9-5 shows the templates provided. While it's convenient to use these templates, many users assume that Databricks Apps are limited to them, which can unnecessarily constrain what they build.

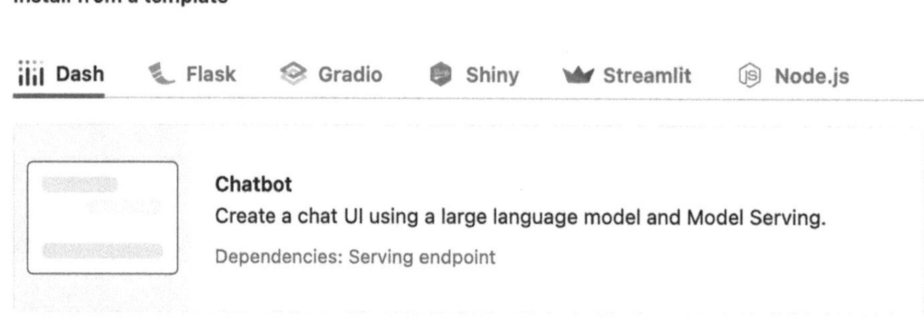

Figure 9-5. *Databricks Apps templates*

Understanding the specific runtime is critical for development and dependency management, as well as for thinking outside the box. As of this writing, the runtime environment consists of:

- **Operating System**: Ubuntu 22.04 LTS

- **Python Environment**: Python 3.11, running in a dedicated virtual environment.

- **Node.js Environment**: Node.js version 22.16 for apps developed using JavaScript frameworks.

- **System Resources**: Each app can utilize up to 2 virtual CPUs (vCPUs) and 6 GB of memory.

With the above in mind, we can very easily migrate our existing applications to Databricks Apps, with the added security provided by Unity Catalog.

Do you want to use the MEAN or MERN stack? You can keep your MongoDB instance, or use Lakebase, Databricks' PostgreSQL database offering, as your application database.

Is your application built on a Python backend with a React frontend? You can host it on the Databricks Apps platform.

By understanding the runtime environment, we understand that Databricks Apps is not only a rapid development platform but also a secure-by-design platform. In the next section, we will do a deep dive on Databricks security.

Unity Catalog As the Governance Layer

Enterprises have been using Unity Catalog to govern their data for some time, and Databricks is extending the offering to Apps. A Databricks App is a security-first application by design, inheriting its governance model directly from the platform's core. If we follow the security settings provided by the Apps platform, we will not introduce unnecessary risk in our application.

Distinguishing Permissions from Authorization

To fully grasp the security model, it is essential to distinguish between two concepts:

1. **Permissions**: These control access *to the application asset itself.* These are workspace-level privileges that define who can interact with the App in a management capacity. The primary levels are IS OWNER (full control), CAN MANAGE (edit settings), and CAN USE (run the App). See Figure 9-6.

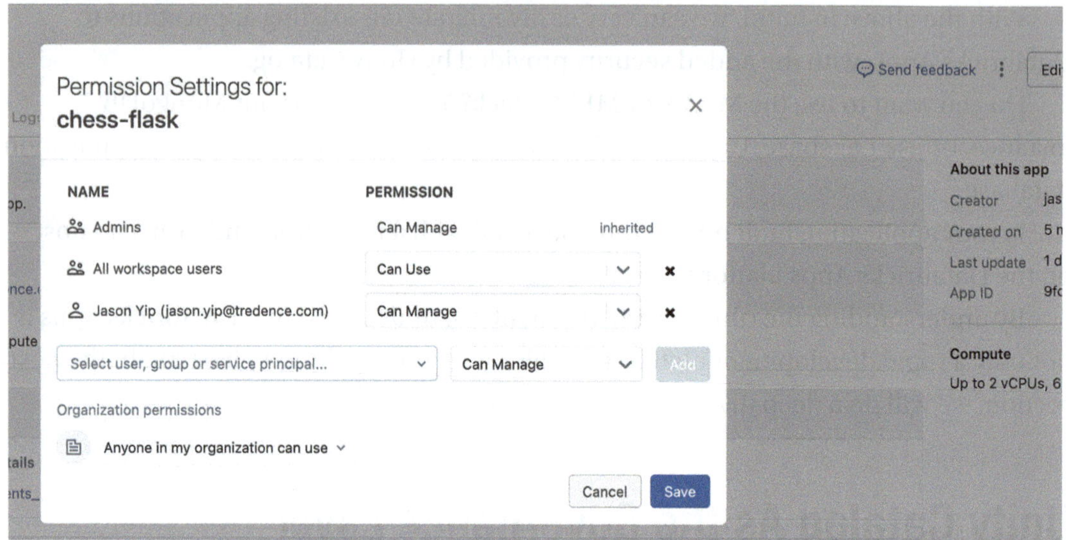

Figure 9-6. *Databricks Apps permissions*

2. **Authorization:** This controls which data and Databricks resources the App can access during execution. There are two roles: one is the App's default service principal, and the other acts on behalf of the user. These are set up during the App creation, as seen in Figure 9-7.

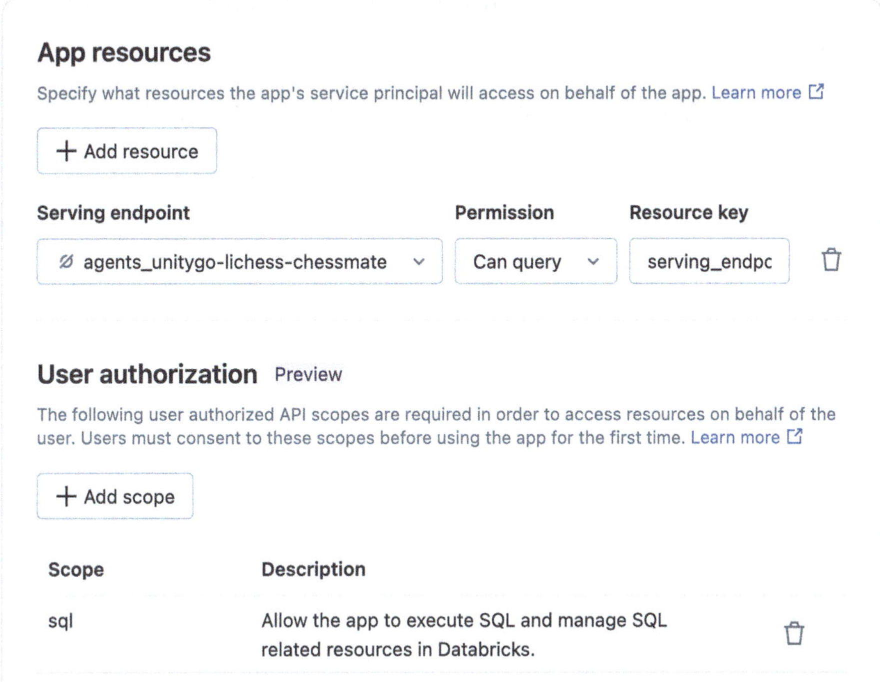

Figure 9-7. *Authorization of App resources*

In the next section, we will discuss the two security models we can use to govern resource access.

App Authorization (Service Principal Model)

When a new Databricks App is created, the platform automatically provisions a dedicated **service principal** for it. This service principal acts as the App's own unique, non-human identity within the Databricks ecosystem.

Just like any other user, we can explicitly grant this service principal the permissions it needs for the App's operational requirements; in other words, anyone with access to the App can use these resources. For example, you might grant the service principal the following permissions:

- CAN USE on a specific SQL warehouse to ensure the App can always execute queries.

- SELECT on a global configuration table in Unity Catalog.

- CAN QUERY on a model serving endpoint.

Figure 9-8 shows that we can find the service principal and grant it permissions as a normal user. This is explicit granting.

Grant on unitygo.rag

Granted privileges will be inherited by applicable objects (e.g. tables, views) in this schema. Learn more

Principals

app-4q70oy chess-flask ✕

Privilege presets

Custom

Prerequisite	Read	Create
☐ USE SCHEMA	☐ EXECUTE	☐ CREATE FUNCTION
	☐ READ VOLUME	☐ CREATE MATERIALIZED VIEW
Metadata	☐ SELECT	☐ CREATE MODEL

Figure 9-8. *Granting permission to a service principal*

The red arrow in Figure 9-9 illustrates that a service principal is directly accessing the sources, and we need to grant the service principal, not the user, access to the resources if we use this model.

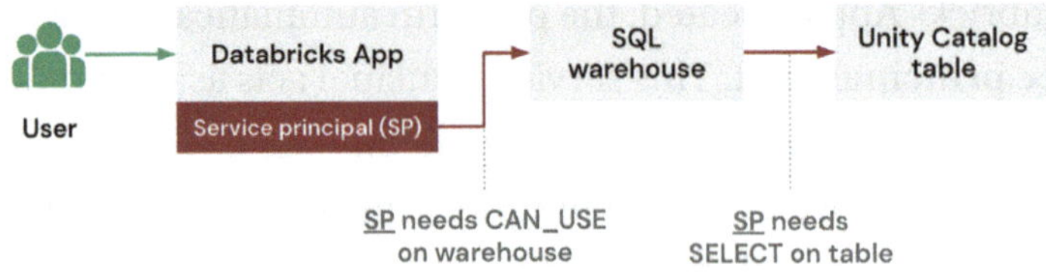

Figure 9-9. *Illustration of the App authorization model*

When the App operates under this model, all API calls and data queries are executed using the identity and permissions of the App's service principal. All users interacting with the App share this single, consistent set of permissions. This model is ideal for operations that are independent of the user's context, such as writing to a centralized logging table, fetching metadata, or performing automated background tasks. This ensures the application is robust and can always function.

User Authorization ("On-Behalf-Of" Model)

The "On-Behalf-Of" (OBO) model is the key to enterprise-grade, fine-grained data security. An App can be configured to request the end user's identity token upon authentication. When a user logs into the App, Databricks forwards their access token to the App's runtime. The App can then use this token to make API calls and data queries *on behalf of that specific user*. Figure 9-10 shows that the service principal is no longer being used to access resources. Instead, if the user has access to the resources, their identity will be used.

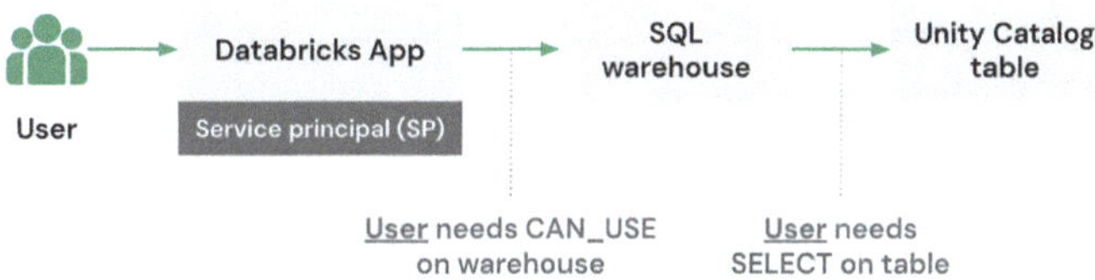

Figure 9-10. *Illustration of the on-behalf-of model*

This mechanism ensures that all of Unity Catalog's most granular security controls are automatically and dynamically enforced for each user session:

- **Row-Level Security**: If a table has a row-level filter (e.g., current_ user() = sales_rep_email), the App will *only* see and return the rows that the specific user is authorized to access.

- **Column-Level Masking**: If a column containing PII is masked (e.g., "pii_email" is masked for all users outside the "HR" group), the App will automatically display the masked data to unauthorized users.

- **Attribute-Based Access Control (ABAC)**: All granular permissions defined in Unity Catalog are respected. If a user does not have SELECT privilege on a particular table, any query the App attempts to run against that table on their behalf will fail, as it should.

To prevent misuse, the OBO model is protected by scope-based access. The App's manifest must explicitly declare the specific access scopes it requires (e.g., permission to access SQL warehouses). The user is prompted to consent to these scopes, ensuring they are aware of the actions the App can perform with their identity. Figure 9-11 shows the scopes available to allow the App to access resources on a user's behalf. Unlike service principal access, we assume the user has already granted access to the resources, so this App is only accessing on the user's behalf and not requesting access for them.

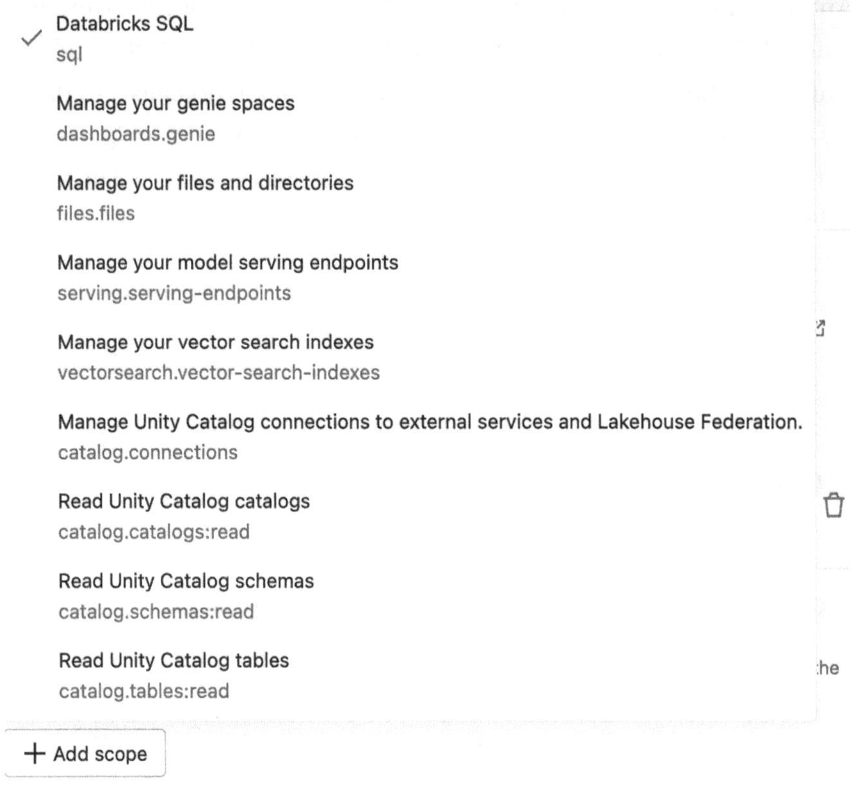

Figure 9-11. *Scopes available to manage on behalf of users*

Table 9-1 summarizes the differences between the two models.

Table 9-1. *Differences Between Service Principal and On–Behalf-Of User Model*

Feature	App Authorization (Service Principal)	User Authorization (On-Behalf-Of)
Identity Used	The App's dedicated, auto-generated service principal.	The identity of the end user interacting with the App.
Permission Source	Permissions explicitly granted to the App's service principal.	The end user's existing permissions as defined in Unity Catalog.
Data View	Consistent and uniform for all users of the App.	Dynamic and personalized for each user.
Granularity	Coarse-grained; all users share the same access level.	Fine-grained; enforces user-specific access controls.
User Policy Enforcement	No. Bypasses user-specific row filters and column masks.	Yes. Automatically enforces row-level security, column masking, and ABAC.
Ideal Use Cases	System-level actions, logging, background jobs, uniform data views.	Multi-tenant Apps, applications with sensitive data, personalized data views.

From Code to Production

While Databricks Apps are flexible, following a standardized project structure is a best practice. The structure differs slightly between Python and Node.js applications.

For Python Applications

This structure centers on a Python application, and dependencies are managed using a requirements.txt file, as shown in Listing 9-1.

Listing 9-1. Python App folder structure

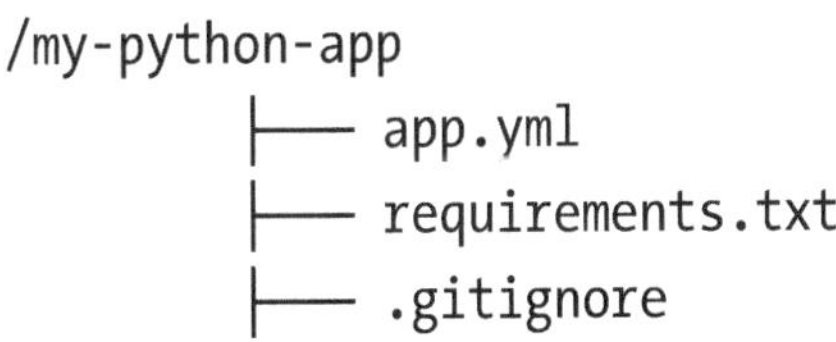

```
/my-python-app
        ├── app.yml
        ├── requirements.txt
        ├── .gitignore
```

```
├── README.md
└── src
     └── app.py
```

1. **app.yml**: This is the App's configuration manifest that defines how the App is started. For example: `command: ['streamlit', 'run', 'app.py']` will translate to "streamlit run app.py"

2. **requirements.txt**: This file lists all the **Python** dependencies your application needs (e.g., streamlit, flask, databricks-sdk, pandas). This file is read by Databricks when deploying your App.

3. **src/app.py**: This directory contains your main **Python** application source code. For example, src/app.py would be the entry point for your Streamlit, Flask, or Gradio application.

For Node.js Applications

This structure uses a `package.json` file for dependencies and scripts. The default run command is `npm run start`. The folder structure can be found in Listing 9-2:

Listing 9-2. Node.js folder structure

```
/my-node-app
          ├── app.yml
          ├── package.json
          ├── .gitignore
          ├── README.md
          └── src
               └── app.js
```

1. **app.yml**: This is the App's configuration manifest; this is the same as the Python example above.

2. **package.json**: This file lists all **Node.js** dependencies and defines scripts (like the start script) that npm will use.

3. **src/app.js**: The entry point for your **Node.js** (e.g., Express) application or the entry point for a compiled React/Angular App.

Environment variables can also be defined in the `app.yml` file, as seen in Listing 9-3. A security best practice is to avoid exposing secrets in clear text in the YAML file; retrieve them from App resources by specifying their resource key using **valueFrom**. The resource key can be found on the App resources page, as seen in Figure 9-12.

Listing 9-3. Environment variables in app.yml

```
env:
  - name: WAREHOUSE_ID
    value: sql_warehouse_id
  - name: SECRET_KEY
    valueFrom: secret
```

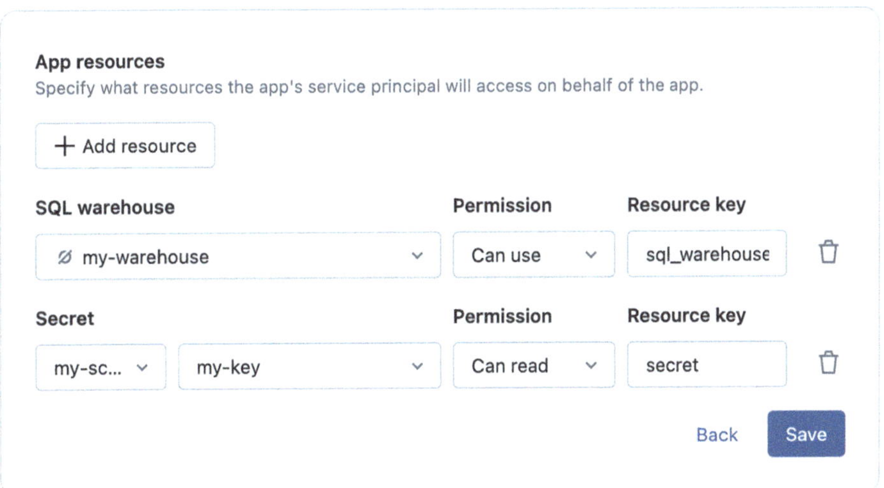

Figure 9-12. *An App resources page where we can specify a secret as a resource*

This structure provides a clean, scalable, and automation-friendly foundation for your Databricks App.

Deployment Using Declarative Automation Bundles

DABs are the infrastructure-as-code (IaC) framework for the Databricks platform. They promote portability by separating an application's code from its environmental configuration.

A developer can write code that is environment-agnostic (e.g., by referring to a resource key such as my_sql_warehouse). The `app.yml` manifest then maps this key to a *specific* resource ID for each target environment (e.g., dev_warehouse_id in dev and prod_warehouse_id in prod). This allows the *exact same codebase* to be promoted across environments without modification.

The App-specific features are a subset of the DABs where we can specify jobs or LakeFlow pipelines in the yml file, allowing everything to be promoted to a higher environment.

Figure 9-13 illustrates the end-to-end process.

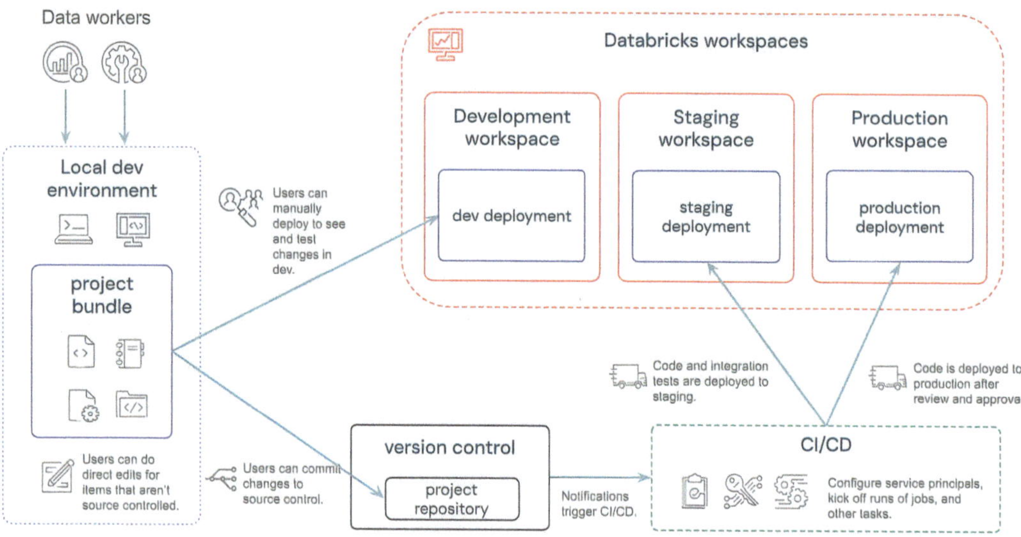

Figure 9-13. *High-level view of a development and CI/CD pipeline with bundles*

A CI/CD Workflow Example

A typical CI/CD pipeline (using tools like GitHub Actions or Azure DevOps) follows these steps:

1. **Local Development**: A developer clones the Git repository and uses the Databricks CLI to run the App locally (`databricks apps run-local`) to simulate the environment.

2. **Commit and Push**: Changes are committed to a feature branch.

3. **CI Trigger**: A pull request triggers the CI pipeline.

4. **Validate and Test**: The pipeline installs dependencies, runs unit tests, and validates the bundle configuration (`databricks bundle validate`).

5. **Deploy**: Upon merging to the main branch, the CD pipeline deploys the bundle to the target environment (`databricks bundle deploy -t prod`).

6. **Run**: The pipeline can then start or update the application using "`databricks bundle run <app_name>`," which triggers a rolling update of the running container.

Auditing and Observability

A secure platform must be auditable. Databricks Apps provides a two-tiered approach to logging and monitoring.

1. **Application Logs**: These are the standard output (stdout) and standard error (stderr) streams generated by the running application container. They are visible in a live-streaming view in the Apps management UI (or via a /logz endpoint) and are invaluable for real-time, interactive debugging. However, these logs are App-specific and are permanently deleted when the App is stopped or restarted. To keep the logs persistently, we can connect the App to application performance monitoring tools like New Relic and write the logs to Unity Catalog volumes.

2. **Platform Audit Logs**: For security, compliance, and forensics, Databricks captures a comprehensive and immutable audit trail of all significant platform activities. These persistent logs are made available automatically as part of the system tables ("system. access.audit").

Integrating with External Monitoring Tools

For organizations that require persistent application-level logs, the logs can be delivered in a couple ways:

1. **Application Performance Monitoring (APM) tools**: Use New Relic, Datadog, or similar application performance monitoring tools to collect and analyze logs, metrics, and traces.

2. **Custom Log Persistence**: Write logs periodically to Unity Catalog volumes or tables for long-term storage and analysis.

While audit logs are stored in system tables, we can configure the application to forward its logs to an external system. For example, Databricks' Splunk integration add-in can be used to fulfill this purpose: `https://splunkbase.splunk.com/app/5416`

Strategic Use Cases: Beyond Dashboards

The true strategic value of Databricks Apps is their ability to operationalize complex data and AI workflows far beyond the capabilities of traditional BI. Below, we examine strategic use cases that demonstrate the platform's capabilities.

- **Interactive ML Model Interfaces**: Building custom front ends for MLflow models where stakeholders can input new data, adjust parameters, and receive real-time predictions and explanations (e.g., SHAP values). This shortens the feedback loop between model development and business validation.

- **Front Ends for AI Agents and RAG Systems**: As organizations build Retrieval-Augmented Generation (RAG) systems, Databricks Apps provide an ideal, secure interface for users to interact with these agents via chat or structured forms, all within a governed, auditable workspace.

- **Data Collection and Annotation Tools**: Creating "human-in-the-loop" applications for subject matter experts to upload, validate, label, or annotate data, writing the results securely back to Delta tables for immediate use in model retraining.

- **Business Process Automation**: Serving as the UI for custom operational tools, such as alert triage systems or data quality approval workflows, that are deeply integrated with Lakehouse data.

- **Chess Game**: A fun chess game between humans and AI. Figure 9-14 shows a chess game between humans and AI agents built on Databricks Apps.

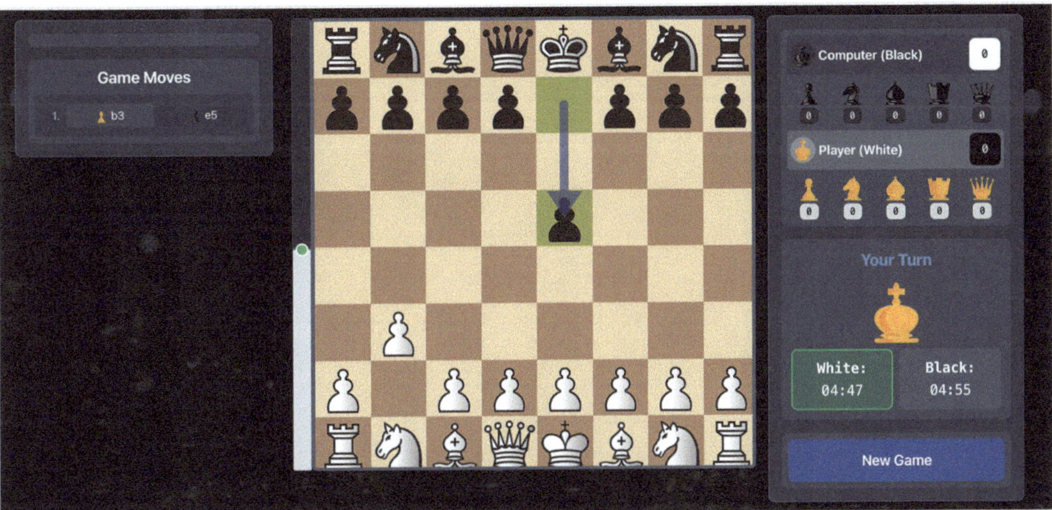

Figure 9-14. *Databricks chess App*

Conclusion: The Future of the Full-Stack Data Intelligence Platform

Databricks Apps are not merely a feature for building dashboards. They represent a fundamental and well-architected shift, transforming the Lakehouse from a passive data platform into an active, intelligent, and interactive application platform.

By providing a unified architecture, serverless operations, enterprise-grade security, and a modern DevOps-ready framework, Databricks Apps solve the "last mile" problem, enabling organizations to finally close the loop between data insight and business value.

In conclusion, Databricks Apps are the essential bridge that connects the platform's powerful back-end capabilities—like Agent Bricks—with end users, providing the key to unlocking the full business value of an organization's data and AI assets in a secure, governed, and scalable manner.

## CHAPTER 10

AI Runtime: Elastic Compute for AI Workloads

When it comes to graphics processing units (GPUs), people undoubtedly think of Copilot or ChatGPT. No doubt that large language models are powered by a massive number of GPUs, and we will also discuss AI models in this chapter. One lesser-known limitation is the limited support for training traditional machine learning models on serverless CPU compute. The reason is that serverless is using the Spark Connect architecture. As shown in Figure 10-1, the Spark Connect architecture decouples the Spark driver and the application layer or the execution environment. While the client-server architecture allows the core Spark engine and the execution environment to continue to evolve independently, the drawback is that the application must now access Spark via API, and it no longer has access to the low-level RDD API.

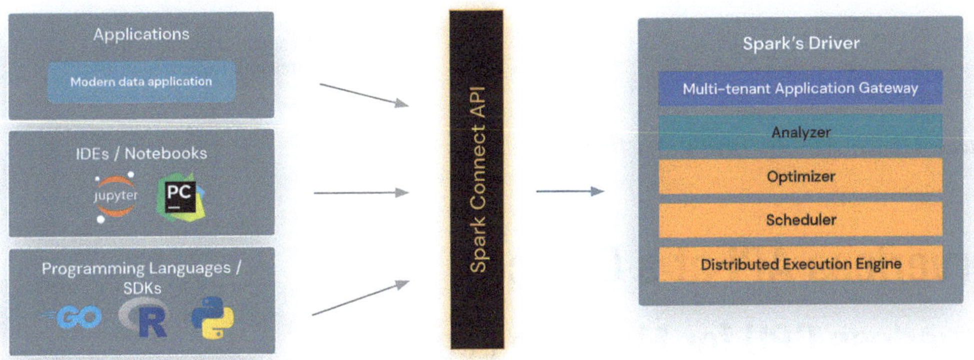

Figure 10-1. *Spark Connect architecture*

AI (Serverless GPU) Runtime

When it comes to serverless GPUs, we would probably compare them to virtual machines with GPUs installed. However, the software stack is equally important because, without suitable software, it will be hard to manage an array of machines ourselves, not to mention GPU communication, which makes the development of large language models so difficult. Fortunately, with the Spark Connect architecture and newly developed runtime and API, data scientists and AI researchers can now build custom models more easily. Figure 10-2 shows the workflow of serverless GPU development. We will examine how to apply serverless GPU computing across the AI spectrum, including machine learning, deep learning, and LLM fine-tuning.

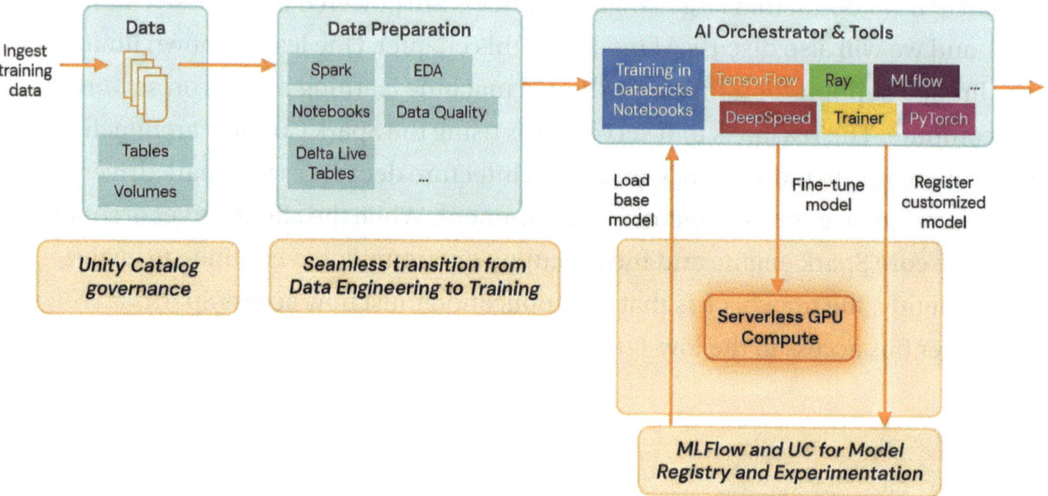

Figure 10-2. AI runtime workflow

Serverless API at a Glance

Serverless CPU for Machine Learning

Distributed machine learning is possible in Spark Connect, but it will require new API support. Spark 4.0 introduces support for Spark ML, the standard Spark MLlib DataFrame API. The good news is that serverless automatically handles the Spark version based on the environment, as shown in Figure 10-3, making upgrades very easy without a cluster restart. From serverless environment version 4 onwards, Spark ML support has been added.

Figure 10-3. *Serverless environment can be found in the right sidebar*

Just as the Databricks runtime version details are published, the environment details are also published. For example, the details of environment 4 can be found on the Databricks website:

```
https://docs.databricks.com/aws/en/release-notes/serverless/environment-
version/four
```

Table 10-1 summarizes which types of workloads are suitable for serverless CPU. This will help you test the CPU in your workflow before deciding whether to use a GPU.

Table 10-1. *Machine learning workloads supported by serverless CPU*

Workload	Will it run on Serverless CPU?
Scikit-Learn/Pandas	Yes, but requires `pip install scikit-learn`
Deep Learning	Yes, but not recommended without GPUs
Spark MLlib (env <= 3)	No, it does not have Spark ML support
Spark MLlib (env >= 4)	Yes, use Spark ML `pyspark.ml`. DO NOT use `pyspark.ml.connect`

Listing 10-1 is a very basic logistic regression model that can run on a serverless CPU using Spark ML. This model takes a dataset from scikit-learn and classifies breast tumors as malignant or benign. It's a demonstration of how we can do this easily without relying on a machine learning runtime or a dedicated cluster.

Listing 10-1. Sample Logistic Regression that runs on a serverless CPU

```python
from pyspark.ml.classification import LogisticRegression
from pyspark.ml.feature import VectorAssembler
from sklearn.datasets import load_breast_cancer

pdf = load_breast_cancer(as_frame=True).frame
pdf["label"] = pdf["target"]
feature_cols = [col for col in pdf.columns if col not in ("label",
"target")]
spark_df = spark.createDataFrame(pdf)

assembler = VectorAssembler(inputCols=feature_cols, outputCol="features")
dataset = assembler.transform(spark_df).select("features", "label")

lor = LogisticRegression(maxIter=20, regParam=0.0, elasticNetParam=0.0)
lor_model = lor.fit(dataset)
transformed_dataset = lor_model.transform(dataset)
```

Ray on Databricks Serverless GPU

As shown in Figure 10-2, serverless GPU provides a runtime for managing GPU
workloads. While serverless GPUs are designed for deep learning and large language
models, many platforms support this kind of workload. We will provide one more
example of machine learning before diving into deep learning capabilities.

What Is Ray?

Apache Spark, the execution engine behind Databricks, is very good at data processing.
However, it lacks comprehensive machine learning support. Ray and Apache Spark
are complementary frameworks. Ray excels at logical parallelism, handling dynamic,
compute-intensive tasks like machine learning and reinforcement learning. Apache
Spark specializes in data parallelism, efficiently processing large datasets for tasks
like ETL and data analytics. Together, they provide a powerful combination for both
data processing and complex computation. Figure 10-4 illustrates a Ray cluster, which
consists of the head node (global control store) and worker nodes.

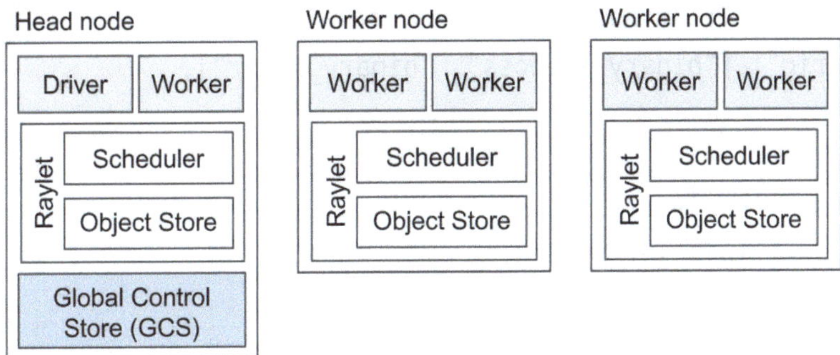

Figure 10-4. *A Ray cluster*

While we can run Ray on Databricks, it requires setting up a Ray server. Serverless GPU abstracted this complexity into its runtime by providing the `serverless_gpu.ray` module. This module provides integration with Ray for distributed computing on serverless GPU infrastructure. It includes:

1. Ray cluster setup and management on multiple GPUs

2. Utilities for Ray head node detection and connection

3. Support for Ray distributed training and inference

There are two modes in serverless: a single-node GPU running on the driver and distributed GPU training via the Python API, which we will dive into throughout this chapter. To demonstrate the ease of use of Ray on serverless GPU, we will adopt the example from the `lightgbm_ray` repo in Listing 10-2.

Listing 10-2. Sample Ray on LightGBM model runs on a serverless GPU

```
from lightgbm_ray import RayDMatrix, RayParams, train
from sklearn.datasets import load_breast_cancer
import mlflow

train_x, train_y = load_breast_cancer(return_X_y=True)
train_set = RayDMatrix(train_x, train_y)

evals_result = {}
bst = train(
    {
```

```
        "objective": "binary",
        "metric": ["binary_logloss", "binary_error"],
    },
    train_set,
    evals_result=evals_result,
    valid_sets=[train_set],
    valid_names=["train"],
    verbose_eval=False,
    ray_params=RayParams(num_actors=2, cpus_per_actor=2))
mlflow.set_registry_uri("databricks-uc")
with mlflow.start_run():
    mlflow.lightgbm.log_model(
        bst.booster_,
        "model",
        input_example=train_x[:5],
        registered_model_name="unitygo.ml.lgbm_breast_cancer"
    )
    mlflow.log_metrics({"final_train_error": evals_result["train"]["binary_
    error"][-1]})
```

The above code shows that we can use Ray on serverless GPUs to train a machine learning model. Trying to run the code on a serverless CPU will result in an error: "Failed to connect to GCS." A more thorough example of distributed training using Ray can be found at the below repo:

```
https://github.com/databricks-industry-solutions/ray-framework-on-
databricks/blob/main/Distributed_Training/xgboost/01c-train-with-
serverless-GPUs-Out-of-Core.ipynb
```

Serverless GPU API

AI runtime is an instant-on, pay-as-you-go service, so developers don't need to worry about cluster maintenance and idle time. It takes a few seconds to connect to the AI runtime, and we can immediately see the availability of the GPU as shown in Figure 10-5.

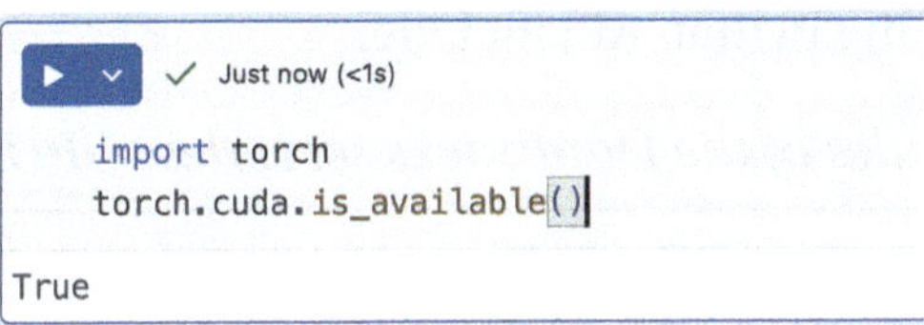

Figure 10-5. *GPU availability in serverless GPU*

Serverless GPU also natively supports critical GPU performance enhancements, like Remote Direct Memory Access (RDMA). RDMA enables low-latency, GPU-to-GPU communication across nodes in a cluster. As a result, we can achieve a linear decrease in model training time using RDMA. For instance, if training a model on one node takes 100 seconds, then training the same model on two nodes should take 50 seconds. Training the model on four nodes should take 25 seconds, and so on. Figure 10-6 compares traditional Ethernet and RDMA.

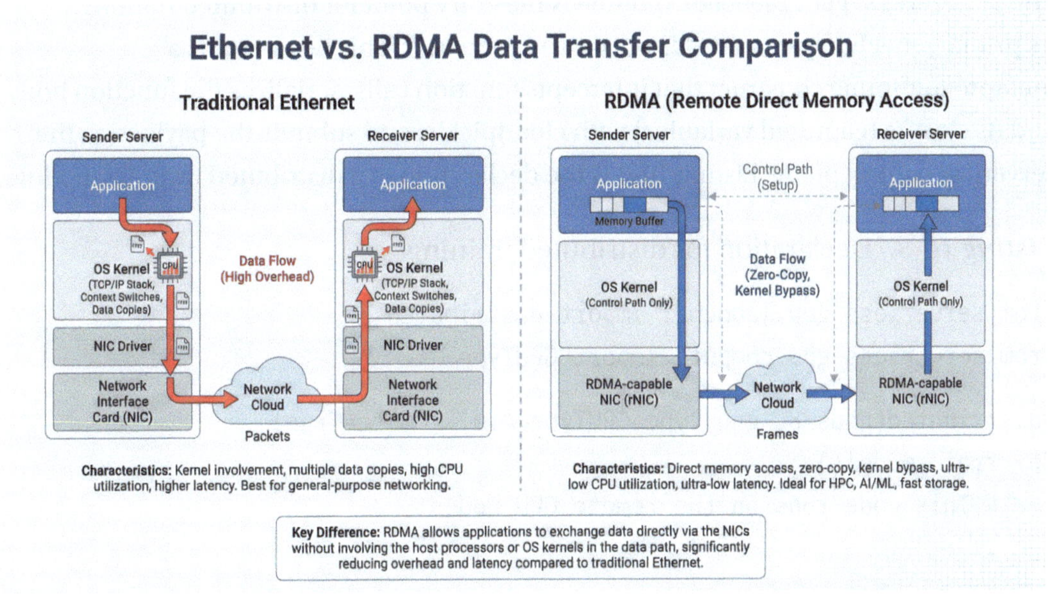

Figure 10-6. *Ethernet vs. RDMA data transfer comparison*

The hardware architecture is further simplified by software-level abstraction, accessible via the Python API. For full documentation, please refer to the serverless GPU API documentation below by Databricks.

```
https://api-docs.databricks.com/python/serverless_gpu/index.html
```

The core modules of the Python API include:

Table 10-2. *Core modules of the Databricks serverless GPU Python API*

Module	Description
serverless_gpu.compute	GPU compute type definitions and utilities. Represented by ENUMs (GPUType.H100/GPUType.A10)
serverless_gpu.launcher	Main launcher module for distributed serverless GPU compute.
serverless_gpu.runtime	Runtime utilities for distributed serverless GPU compute.
serverless_gpu.ray	Ray integration for distributed serverless GPU compute.

The Launcher Module: @distributed and Orchestration

The serverless_gpu.launcher module is the entry point for distributed training. Its primary mechanism is the @distributed decorator. This decorator is a metaprogramming construct that intercepts function calls, serializes the function body and its closure (captured variables) with cloudpickle, and submits the payload to the serverless control plane. Listing 10-3 is the declaration of a distributed training module.

Listing 10-3. Declaration for distributed training

```
from serverless_gpu.launcher import distributed
from serverless_gpu.compute import GPUType

@distributed(gpus=8, gpu_type=GPUType.A10, remote=True)
def train_model():
    # This code runs on the remote GPU nodes
    ...
```

If remote is set to False, the code will run locally on the driver as a single-node cluster. This is meant for debugging syntax issues before distributing the code to multiple GPUs.

To execute the train_model function, we need to call distributed() again. Note that train_model() can contain parameters; in other words, we can define train_model(x: int), and we'd pass the parameter into the distributed function as shown in Listing 10-4.

Listing 10-4. Launching the model training code

```
train_model.distributed()

train_model.distributed(1)
```

The Runtime Module: The Core of Distributed Processing

Adopted from the MosaicML composer library, the `runtime` module represents the core of distributed computing concepts it manages. In a distributed training job, there is no single "program" running the whole job. Instead, multiple independent processes run in parallel, often on different physical nodes. To coordinate their actions—whether to synchronize gradients, partition a dataset, or broadcast model weights—these processes must share a common understanding of the "state" of the "world."

This state is defined by a set of integer identifiers and counts, commonly referred to as the "Distributed Configuration Parameters."

World Size

- **Definition**: The total number of processes participating in the entire distributed training job across all nodes.

- **Usage**: It defines the scope of the communication group. Every process needs to know the world size to understand how many peers it needs to coordinate with. This value is used by the data loader to split the dataset into `world_size` of non-overlapping chunks, ensuring each process receives a unique subset of the data for training.

Global Rank (Rank)

- **Definition**: A unique identifier assigned to each process across the entire cluster, ranging from 0 to `world_size - 1`.

- **Usage**:

- **Unique Identification**: It serves as the primary, unique ID for each process, regardless of the node on which it runs.

- **Collective Communication**: It is crucial for collective communication operations (e.g., all-reduce for gradient synchronization), enabling processes to send data to and receive data from specific peers across the network.

- **Main Process**: The process with global rank 0 is often designated as the "main" or "master" process, responsible for coordinating tasks such as logging, saving model checkpoints, and data loading.

Local Rank

- **Definition**: A unique identifier assigned to each process within its specific physical node (machine), ranging from 0 to the number of GPUs/processes on that node minus 1.

- **Usage**:

  - **Device Assignment**: It is primarily used to assign a specific GPU to a specific process within a single machine. For example, a process with a local rank of 0 will use `cuda:0`, local rank 1 uses `cuda:1`.

  - **Intra-Node Operations**: It helps manage operations that are specific to resources within a single node. It is used to identify a process in the intra-node collective communication operations.

  - **Simplified Launching**: Utilities like `torch.distributed.launch` use `local_rank` as a command-line argument to automatically manage device assignment and spawn processes on each node.

Listing 10-5 provides a simple example of training a multilayer perceptron (MLP) neural network using PyTorch DDP, in which we will see the runtime module in action.

Listing 10-5. Training a simple multilayer perceptron (MLP) neural network using PyTorch DDP

```python
import torch
import torch.distributed as dist
from torch.nn.parallel import DistributedDataParallel as DDP
from torch.utils.data import DataLoader, DistributedSampler, TensorDataset
import torch.nn as nn
import torch.optim as optim
from serverless_gpu import distributed, runtime
from serverless_gpu.compute import GPUType
import mlflow

def setup():
    dist.init_process_group("nccl")
    torch.cuda.set_device(runtime.get_local_rank())

def cleanup():
    dist.destroy_process_group()

class SimpleMLP(nn.Module):
    def __init__(self, input_dim=10, hidden_dim=64, output_dim=1):
        super().__init__()
        self.net = nn.Sequential(
            nn.Linear(input_dim, hidden_dim),
            nn.ReLU(),
            nn.Dropout(0.2),
            nn.Linear(hidden_dim, hidden_dim),
            nn.ReLU(),
            nn.Dropout(0.2),
            nn.Linear(hidden_dim, output_dim)
        )
    def forward(self, x):
        return self.net(x)

@distributed(gpus=2, gpu_type=GPUType.A10, remote=True)
def train(num_epochs=3, batch_size=64) -> None:
    setup()
```

```python
    rank = runtime.get_global_rank()
    local_rank = runtime.get_local_rank()
    device = torch.device(f"cuda:{local_rank}")

with mlflow.start_run():
    model = SimpleMLP(input_dim=784, hidden_dim=256, output_dim=10).
    to(device)
    model = DDP(model, device_ids=[local_rank])

    x = torch.randn(5000, 10)
    y = 3 * x.sum(dim=1, keepdim=True) + torch.randn(5000, 1) * 0.1
    dataset = TensorDataset(x, y)
    sampler = DistributedSampler(dataset)
    dataloader = DataLoader(dataset, sampler=sampler, batch_
    size=batch_size)

    optimizer = optim.Adam(model.parameters(), lr=0.001)
    loss_fn = nn.MSELoss()

    for epoch in range(num_epochs):
        sampler.set_epoch(epoch)
        model.train()
        total_loss = 0.0
        for step, (xb, yb) in enumerate(dataloader):
            xb, yb = xb.to(device), yb.to(device)
            optimizer.zero_grad()
            loss = loss_fn(model(xb), yb)
            if rank == 0:
                mlflow.log_metric("mse_loss", loss.item(), step=epoch
                * len(dataloader) + step)

            loss.backward()
            optimizer.step()
            total_loss += loss.item() * xb.size(0)

        if rank == 0:
            avg_loss = total_loss / len(dataloader.dataset)
            mlflow.log_metric("avg_loss", avg_loss)
```

```
        ckpt = {
            "MODEL_STATE": model.state_dict(),
            "EPOCHS_RUN": epoch,
        }
        torch.save(ckpt, "/tmp/model.pt")

    cleanup()

train.distributed()
```

Ray: The "Spark Engine" for Distributed AI Training

We covered Ray at the beginning of this chapter mainly to introduce serverless GPU support for distributed machine learning model training, as well as single-node GPU training. However, Ray has been evolving into a standard for distributed AI training, which is why serverless GPU provides native support for Ray without any warm-up time or cluster maintenance.

As shown in Figure 10-7, Ray has a few components, including:

- **Ray Data** provides scalable, framework-agnostic data loading and transformation, allowing users to process petabyte-scale datasets and apply transformations efficiently, minimizing CPU/GPU idle time. Ray data supports `read_databricks_tables` by leveraging Databricks SQL.

- **Ray Train** orchestrates multi-node, multi-GPU training seamlessly, integrating with popular deep learning frameworks like PyTorch and TensorFlow.

  It supports complex parallelism strategies (data, pipeline, and tensor parallelism) necessary to distribute large model shards across many machines. It also provides fault tolerance, allowing training to resume from checkpoints if a node fails, which is vital for long-running, expensive LLM training jobs.

- **Ray Tune** efficiently parallelizes these hyperparameter searches across the entire cluster, identifying the best model configurations faster by distributing each experiment independently. While we normally don't pre-train a large language models, this is still useful in machine learning.

- **Ray Serve** is a specialized library designed for high-performance, scalable LLM serving in production. We can think of it as a programmable alternative to Databricks Model Serving. But since Databricks offers in-house model serving, the only use case here is a custom deployment outside of Databricks.

- **Ray RLlib** is an open-source library for reinforcement learning (RL), offering support for production-level, highly scalable, and fault-tolerant RL workloads, while maintaining simple and unified APIs for a large variety of industry applications.

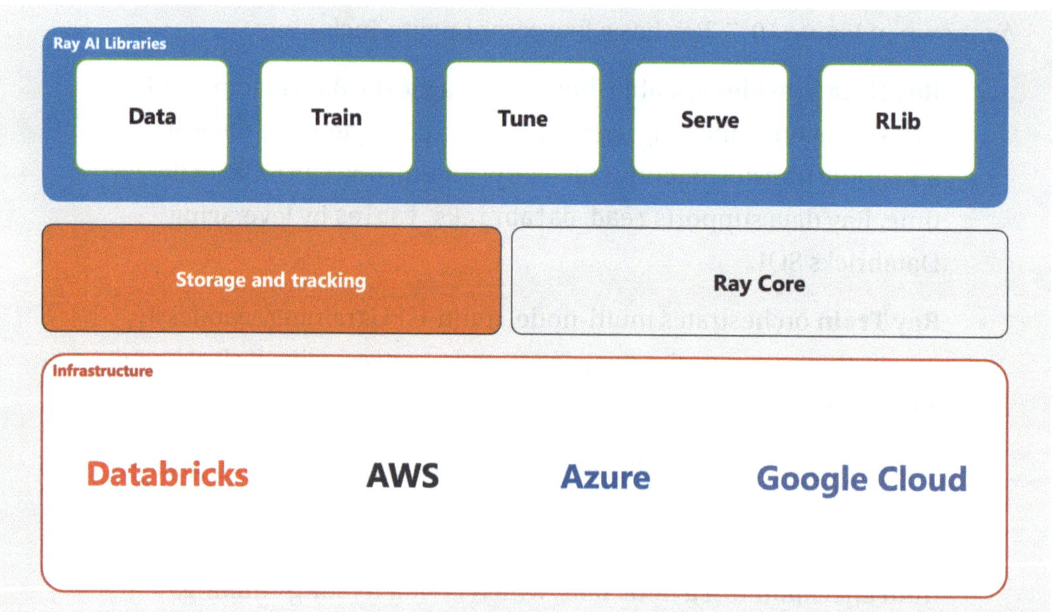

Figure 10-7. Ray architecture overview

With the rich API Ray already provides, there is really no need to add additional APIs to support Ray on serverless GPUs. Nevertheless, there is one important compatibility: it can be used the same way as a launcher. In other words, it supports both using as a decorator and the `distributed()` function. As in Listings 10-3 and 10-4, we can migrate our Ray code as shown in Listing 10-6.

Listing 10-6. Skeleton code to train with Ray in serverless GPU

```python
from serverless_gpu.ray import ray_launch

@ray_launch(gpus=8, gpu_type="H100", remote=True)
def ray_function():
    ...

ray_function.distributed()
```

The Distributed Training Frameworks

When navigating distributed training, Ray serves as a coordinator between nodes. However, it still requires popular training libraries like PyTorch and TensorFlow. But because of the popularity of the PyTorch family, we will do a deeper dive into the decision tree in using the training frameworks. While they are not all full-blown frameworks, we will just call them that for simplicity.

To successfully fine-tune on Databricks serverless GPU, the process begins with an assessment of the instance type, typically NVIDIA A10s (24 GB) or H100s (80 GB), relative to your model's memory requirements. For example, since `gpt-oss-20b` is a larger model, your best choice for standard serverless nodes (such as the 24GB A10) is to prioritize an efficiency strategy with 4-bit quantization. We recommend using tools like Unsloth or QLoRA at this stage; these libraries compress the model effectively, ensuring it fits within the memory constraints of a single node without requiring complex engineering workarounds. The `Unsloth` library is a Python package that optimizes large language model (LLM) fine-tuning by leveraging custom GPU kernels and efficient matrix operations to accelerate and reduce memory usage on consumer-grade hardware. In simple terms, it uses clever optimizations to allow us to run a model on a single GPU. The process is shown in Figure 10-8.

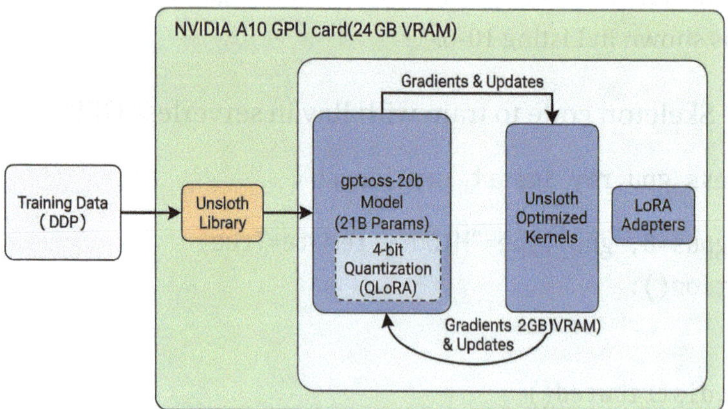

Figure 10-8. *Fine-tuning a model of up to 30B on a single A10 GPU*

Once your model is optimized for the hardware, the next step is to orchestrate training with Ray on Spark. Because Databricks automatically handles the underlying infrastructure, you can simply initialize a Ray cluster directly on top of your existing Spark workers. This setup allows you to seamlessly distribute your training code. If your model requires more memory, Ray can coordinate the communication needed to shard the model across multiple nodes using Fully Sharded Data Parallel (FSDP) or DeepSpeed. This approach lets you leverage the power of a distributed cluster while keeping your workflow straightforward. Figure 10-9 illustrates this process.

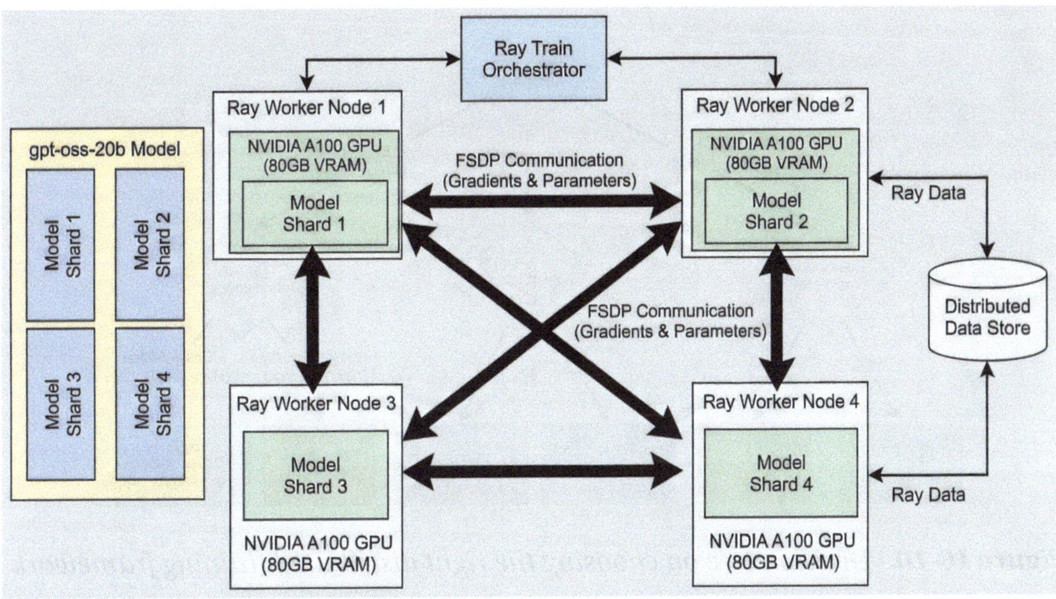

Figure 10-9. *Distributed model fine-tuning with FSDP*

Figure 10-10 summarizes the entire decision process. While we only discussed the PyTorch family in this chapter, the decision process is not limited to PyTorch or Ray. Databricks serverless GPU is a distributed compute environment that allows us to use any library, as long as we distribute the data and libraries across all nodes. For more best practices of using serverless GPUs, please refer to the Databricks documentation:

```
https://docs.databricks.com/aws/en/compute/serverless/sgc-best-practices
```

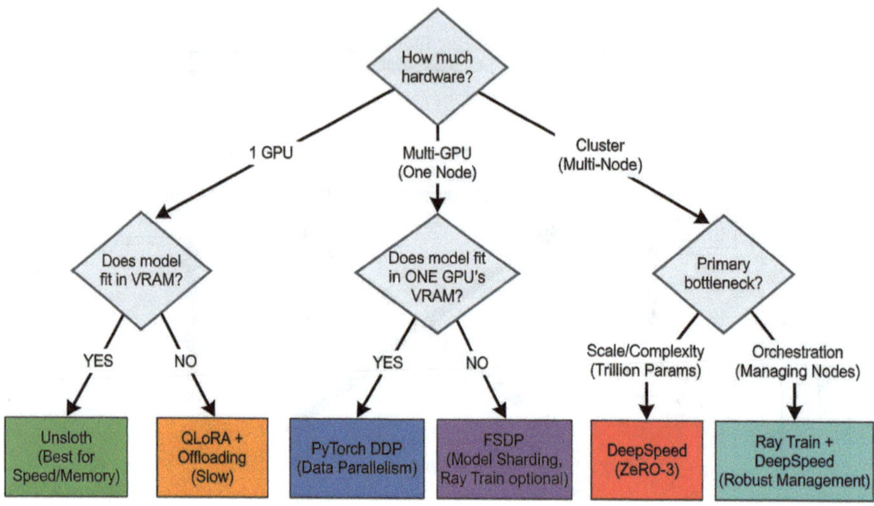

Figure 10-10. *Decision tree on choosing the right distributed training framework*

Multimodal Fine-Tuning: Where the Ray Stack Shines

In the above use case, we have presented fine-tuning `gpt-oss-20b`, which is a text-based model. We presented Ray Train as a robust orchestration tool in large-scale training. However, the real value of Ray lies in combining Ray Data and Ray Train for fine-tuning, since we rarely need to perform distributed hyperparameter tuning. While Databricks provides model serving, only Ray train and Ray data will help with the fine-tuning process by coordinating the nodes and distributing the data across them. In particular, Ray Data provides numerous data loaders that we can leverage. Listing 10-7 is a full implementation of Ray fine-tuning of Qwen's Vision-Language Model (VLM) using FSDP.

Listing 10-7. Full implementation of Ray fine-tuning of Qwen's Vision-Language Model using FSDP

```
%pip install ray
from serverless_gpu.ray import ray_launch

# Config
MODEL_NAME = "Qwen/Qwen2-VL-2B-Instruct"
CHECKPOINT_DIR = "/dbfs/checkpoints/qwen2-vl-demo"
EXPERIMENT_NAME = "/Users/demo/ray-data-fsdp-demo"
```

```python
@ray_launch(gpus=4, gpu_type='A10', remote=True)
def train_func():
    """Ray Train + FSDP distributed Vision-Language Model (VLM)
    training."""

    import os     import io     import torch     import mlflow     import
    ray     import ray.data     import numpy as np
    from PIL import Image
    from functools import partial
    from datasets import load_dataset
    from torch.utils.data import Dataset as TorchDataset, DataLoader
    from transformers import Qwen2VLForConditionalGeneration, AutoProcessor
    from transformers.models.qwen2_vl.modeling_qwen2_vl import
    Qwen2VLDecoderLayer
    from peft import LoraConfig, get_peft_model
    from ray import train as ray_train
    from ray.train.torch import TorchTrainer, TorchConfig
    from ray.train import ScalingConfig, RunConfig
    from torch.distributed.fsdp import FullyShardedDataParallel as FSDP
    from torch.distributed.fsdp import ShardingStrategy, MixedPrecision,
    FullStateDictConfig, StateDictType
    from torch.distributed.fsdp.wrap import transformer_auto_wrap_policy

    # ========= Data Loading ===========================
    def create_ray_dataset():
        hf_dataset = load_dataset("HuggingFaceH4/llava-instruct-mix-vsft",
        split="train[:1000]")
        items = []
        for item in hf_dataset:
            d = dict(item)
            if d.get("images"):
                d["images"] = [np.array(img) if hasattr(img, "convert")
                else img for img in d["images"]]
            items.append(d)
        return ray.data.from_items(items)
```

```python
# ========= Dataset & Collate =============================

class VLMDataset(TorchDataset):
    MAX_TEXT_LEN, MAX_IMG_SIZE = 1500, 448

    def __init__(self, items, processor):
        self.items, self.processor = items, processor

    def __len__(self): return len(self.items)

    def __getitem__(self, idx):
        item = self.items[idx]
        msgs = [{**m, "content": m["content"][:self.MAX_TEXT_
        LEN] + "..."
                    if m["role"] == "assistant" and len(str(m["content"]))
                    > self.MAX_TEXT_LEN
                    else m["content"]} for m in item["messages"]]

        text = self.processor.apply_chat_template(msgs, tokenize=False,
        add_generation_prompt=False)
        img = item.get("image") or Image.new("RGB", (224, 224),
        "white")

        if img and max(img.size) > self.MAX_IMG_SIZE:
            ratio = self.MAX_IMG_SIZE / max(img.size)
            img = img.resize((int(img.size[0]*ratio), int(img.
            size[1]*ratio)), Image.Resampling.LANCZOS)

        return {"text": text, "image": img}

def collate_fn(batch, processor):
    texts, images = [b["text"] for b in batch], [b["image"] for b
    in batch]
    try:
        enc = processor(text=texts, images=images, padding=True,
        truncation=True, max_length=2048, return_tensors="pt")
    except ValueError:
```

```python
        enc = processor(text=texts, images=images, padding=True,
            truncation=False, return_tensors="pt")
    enc["labels"] = enc["input_ids"].clone()
    return enc

# ========= Training Worker ============================

def train_worker(config: dict):
    rank, local_rank = int(os.environ.get("RANK", 0)), int(os.environ.
    get("LOCAL_RANK", 0))
    device = torch.device(f"cuda:{local_rank}")

    # MLflow on rank 0
    if rank == 0:
        os.environ["MLFLOW_EXPERIMENT_NAME"] =
        config["experiment_name"]
        os.environ["MLFLOW_TRACKING_URI"] = "databricks"
        mlflow.start_run()

    # Load model + LoRA
    model = Qwen2VLForConditionalGeneration.from_pretrained(
        MODEL_NAME, torch_dtype=torch.bfloat16, attn_
        implementation="sdpa"
    )
    processor = AutoProcessor.from_pretrained(MODEL_NAME)

    model = get_peft_model(model, LoraConfig(
        r=16, lora_alpha=16, lora_dropout=0.05,
        target_modules=["q_proj", "k_proj", "v_proj", "o_proj"],
        bias="none", task_type="CAUSAL_LM"
    ))
    model = model.to(torch.bfloat16)
    model.train()
    model.gradient_checkpointing_enable(gradient_checkpointing_
    kwargs={"use_reentrant": False})

    # FSDP wrap
    model = FSDP(
```

```python
    model,
    sharding_strategy=ShardingStrategy.FULL_SHARD,
    mixed_precision=MixedPrecision(param_dtype=torch.bfloat16,
    reduce_dtype=torch.bfloat16, buffer_dtype=torch.bfloat16),
    auto_wrap_policy=partial(transformer_auto_wrap_policy,
    transformer_layer_cls={Qwen2VLDecoderLayer}),
    device_id=device,
    use_orig_params=True,
)
print(f"Rank {rank}: FSDP ready")

# Process Ray Data shard
train_items = []
for batch in ray_train.get_dataset_shard("train").iter_
batches(batch_size=10):
    for i in range(len(batch["messages"])):
        raw_imgs = batch.get("images",
        [None]*len(batch["messages"]))[i] or []
        img = None

        if raw_imgs:
            img_data = raw_imgs[0] if isinstance(raw_imgs, list)
            else raw_imgs
            if isinstance(img_data, np.ndarray):
                img = Image.fromarray(img_data).convert("RGB")
            elif isinstance(img_data, dict) and img_data.
            get("bytes"):
                img = Image.open(io.BytesIO(img_data["bytes"])).
                convert("RGB")
            elif hasattr(img_data, "convert"):
                img = img_data.convert("RGB")

        # Format messages
        formatted = []
        for j, msg in enumerate(batch["messages"][i]):
            if msg["role"] == "user":
                content = []
```

```
            if img and j == 0: content.append({"type": "image",
            "image": img})
            txt = msg["content"] if isinstance(msg["content"],
            str) else next((x["text"] for x in msg["content"]
            if x.get("type")=="text"), "")
            content.append({"type": "text", "text": txt})
            formatted.append({"role": "user", "content":
            content})
        else:
            formatted.append({"role": "assistant", "content":
            str(msg["content"])})

    train_items.append({"messages": formatted, "image": img})


dataloader = DataLoader(VLMDataset(train_items, processor), batch_
size=2, shuffle=True,
                        collate_fn=lambda b: collate_fn(b,
                        processor), num_workers=0)

# Training loop
MAX_STEPS, GRAD_ACCUM = 20, 4
optimizer = torch.optim.AdamW([p for p in model.parameters() if
p.requires_grad], lr=2e-4)
global_step, running_loss = 0, 0.0
optimizer.zero_grad()

while global_step < MAX_STEPS:
    for batch_idx, batch in enumerate(dataloader):
        if global_step >= MAX_STEPS: break

        inputs = {k: v.to(device) if torch.is_tensor(v) else v for
        k, v in batch.items()}
        loss = model(**inputs).loss / GRAD_ACCUM
        running_loss += loss.item()
        loss.backward()

        if (batch_idx + 1) % GRAD_ACCUM == 0:
            torch.nn.utils.clip_grad_norm_(model.parameters(), 1.0)
```

```python
                optimizer.step()
                optimizer.zero_grad()
                global_step += 1

                if global_step % 10 == 0 and rank == 0:
                    print(f"Step {global_step}: loss = {running_
                    loss/10:.4f}")
                    mlflow.log_metrics({"train_loss": running_loss/10},
                    step=global_step)
                    running_loss = 0.0

                ray_train.report({"loss": loss.item() * GRAD_ACCUM,
                "step": global_step})

        # Save (all ranks participate, only rank 0 writes)
        adapter_path = f"{config['checkpoint_dir']}/final_adapter"
        os.makedirs(adapter_path, exist_ok=True)

        with FSDP.state_dict_type(model, StateDictType.FULL_STATE_DICT,
                                  FullStateDictConfig(offload_to_cpu=True,
                                  rank0_only=True)):
            state_dict = model.state_dict()  # ALL ranks must call this
            (collective)

            if rank == 0:
                torch.save(state_dict, f"{adapter_path}/adapter_model.bin")
                processor.save_pretrained(adapter_path)
                mlflow.end_run()

    # ========= Run Training ============================

result = TorchTrainer(
    train_loop_per_worker=train_worker,
    train_loop_config={"experiment_name": EXPERIMENT_NAME, "checkpoint_
    dir": CHECKPOINT_DIR},
```

```
    scaling_config=ScalingConfig(num_workers=4, use_gpu=True,
    resources_per_worker={"GPU": 1}),
    run_config=RunConfig(name="ray-data-fsdp", storage_
    path=CHECKPOINT_DIR),
    torch_config=TorchConfig(backend="nccl"),
    datasets={"train": create_ray_dataset()},
  ).fit()

  return result

train_func.distributed()
```

An example image from the dataset is shown in Figure 10-11. A sample Q/A pair is shown in Listing 10-8.

Listing 10-8. Sample image description response from model

```
Question: Describe this image in detail.
Model Response: The image depicts a serene coastal scene with a sandy
path leading towards the beach. The path is bordered by tall, green grass
and some wildflowers, suggesting a natural, possibly remote location. The
beach itself is sandy and appears to be a quiet, unpopulated area. In the
background, the ocean stretches out to the horizon, with several sailboats
anchored in the water. The boats vary in size and design, indicating
a range of activities, from leisurely sailing to possibly fishing or
sightseeing. The sky overhead is overcast, with clouds scattered across
it, adding a sense of calmness and tranquility to the scene. The overall
atmosphere of the image is peaceful and inviting, with the calm ocean and
the quiet beach providing a sense of solitude and relaxation.
```

Figure 10-11. *Sample image from dataset*

Conclusion

This chapter established AI Runtime (serverless GPU) as a powerful and tunable platform for modern data and AI applications. By leveraging the Spark Connect architecture, which decouples the Spark driver from the execution environment, serverless computing offers a foundation for both CPU and GPU workloads.

In this chapter, we have done a comprehensive study of serverless machine learning, deep learning, and LLM fine-tuning, including:

- **Serverless Efficiency:** Serverless GPU is an instant-on, pay-as-you-go service that eliminates cluster maintenance and reduces idle time, allowing developers to immediately check GPU availability with tools like `torch.cuda.is_available()`.

- **Performance:** Serverless GPU natively supports critical performance enhancements, such as Remote Direct Memory Access (RDMA), which enables low-latency GPU-to-GPU communication and a reduction in model training time as more nodes are added.

- **Ecosystem Integration:** Serverless architecture seamlessly integrates with distributed computing frameworks:

  - **Spark MLlib:** Support for Spark ML is available from environment version 4 onwards, making it suitable for traditional machine learning tasks on serverless CPUs.

  - **Ray Integration:** Serverless GPU provides native support for Ray, abstracting away the complexity of setting up a Ray server through the `serverless_gpu.ray` module. Ray excels at dynamic, compute-intensive tasks like machine learning and deep learning, which is complementary to Spark's data parallelism strengths.

- **LLM Fine-Tuning:** For Large Language Models (LLMs), the choice of framework depends on the hardware config:

  - **Single-GPU:** Tools like Unsloth (for best speed/memory) or QLoRA + Offloading (when the model doesn't fit in VRAM) are recommended.

  - **Multi-GPU/Cluster:** Frameworks such as PyTorch DDP (Distributed Data Parallel), FSDP (Fully Sharded Data Parallel), DeepSpeed (for scale and complexity), and Ray Train + DeepSpeed (for orchestration) are used to distribute training across nodes.

In essence, Databricks serverless GPU provides the necessary runtime, API, and underlying architecture to simplify and scale demanding AI workloads, from traditional machine learning to cutting-edge LLM fine-tuning, allowing developers to focus on model development rather than infrastructure management.

## CHAPTER 11

MLflow 3 and the GenAI Agents

Since launching in 2018, MLflow has been one of the major open-source tools in managing machine learning lifecycles. In the previous edition, we discussed the machine learning lifecycle and how to use MLflow to track experiments. Since MLflow 3.0, the community as well as Databricks has successfully evolved MLflow to support various GenAI use cases, from the basics like prompt versioning to agent tracing to custom judges.

In the old days, we'd Google for usage statistics about a tool. But in the GenAI era, we can still use Google's AI mode and get our answers. Listing 11-1 is a response from Google's AI mode based on Gemini 3 to the question "What is the popularity of MLflow?"

Listing 11-1. Popularity of MLflow according to Gemini 3

```
MLflow is currently the industry-standard open-source platform for managing
the machine learning (ML) lifecycle, with widespread adoption across both
traditional AI and modern Generative AI (GenAI) sectors.
```

```
Key Usage Statistics (as of 2025)
```
- **Monthly Downloads:** Surpassed **30 million monthly downloads in 2025**, a significant increase from 13 million in late 2022.
- **Organization Adoption:** Trusted by more than 5,000 organizations worldwide.
- **User Base:** Reached 10 million individual users by late 2022.
- **Community Support:** Features over 850 community contributors and a following of 45,000+ on social platforms.

In this chapter, we will go through all GenAI features in MLflow. While the features will continue to evolve, this chapter will set a strong foundation for you to build and evaluate GenAI apps, optimize prompts, collect feedback, and track fine-tuning experiments. The features include:

- **Tracing:** Capturing the chain of thoughts from Agents

- **Evaluation:** Evaluate agent output with criteria

- **Optimization:** Using DSPy-inspired function for prompt optimization

- **Feedback:** Use the no-code review app, MLFlow Review App, to collect user feedback

- **Experiment Tracking:** Tracking fine-tuning metrics

- **Model Management:** Deploy and serve fine-tuned models

Tracing

If you are familiar with the LangChain stack, you'd have heard about LangSmith. If you have a LangGraph or LangChain pipeline, LangSmith Studio provides a very nice way to develop, visualize, and troubleshoot your agent. However, if you were to develop in Microsoft's AutoGen, you would not get the same experience. MLflow has streamlined this experience for developers by integrating most of the tools on the market and allowing us to view a single trace. The list is quite extensive, but to name a few, MLflow tracing supports OpenAI, Anthropic, LangChain, LangGraph, LlamaIndex, AutoGen, and more. The latest list can be obtained from the MLflow website:

```
https://mlflow.org/docs/latest/genai/tracing/
```

There are two types of tracing: auto tracing and manual tracing. And within manual tracing, there are three types of APIs: function decorators, span tracing, and low-level API. We recommend going with function decorators most of the time and span tracing for advanced use cases. Leave the low-level API for coding agents to figure out and use auto-tracing for testing purposes. We will go through examples below to understand the reasoning behind each approach.

Auto Tracing

MLflow supports auto-tracing for the majority of open-source GenAI libraries in the form of `mlflow.<library>.autolog()`. The most popular library by far is OpenAI. Auto-tracing automatically breaks down a single request into a hierarchy of "spans." For example, in a RAG (Retrieval-Augmented Generation) application, it will capture separate spans for the retrieval step, the prompt augmentation, and the final LLM call. Listing 11-2 shows the syntax for auto-tracing OpenAI calls. Please note that while using the OpenAI SDK, we can also use it to call Databricks Foundation Model APIs (FMAPI) and not the premium OpenAI models, which makes the development experience much more streamlined.

Listing 11-2. Auto-tracing for OpenAI library

```
import mlflow
mlflow.openai.autolog()
```

It is critical to remember that we need to set the Databricks token for the API to work, but we can retrieve a token via code and never hardcode any secrets in clear text, as shown in Listing 11-3.

Listing 11-3. OpenAI-compatible Chat Completions call (via the DatabricksOpenAI client) to Databricks Foundation Model APIs (FMAPI)

```
from databricks_openai import DatabricksOpenAI

# In Databricks notebooks, DatabricksOpenAI auto-configures auth and
base_url
client = DatabricksOpenAI()

response = client.chat.completions.create(
    model="databricks-claude-opus-4-5",
    messages=[
        {"role": "system", "content": "You are a chess master."},
        {"role": "user", "content": "What is the next best move? 1rb5/4r3/3
        p1npb/3kp1P1/1P3P1P/5nR1/2Q1BK2/bN4NR w - - 3 61"},
    ],
```

```
    max_tokens=5000,
)

print(response.choices[0].message.content)
```

For better organization, we can optionally specify the experiment location. Otherwise, the traces would appear in an experiment named after the notebook's name. The syntax is the same as how we do in machine learning experiments, as shown in Listing 11-4.

Listing 11-4. Setting the experiment location for auto-tracing

```
mlflow.set_tracking_uri("databricks")
mlflow.set_experiment(f"/Shared/llm-chess-game")
```

Figures 11-1 and 11-2 show the trace right below the cell that executed the Chat Completion as well as the Experiment tab. The code can be found in Listing 11-3.

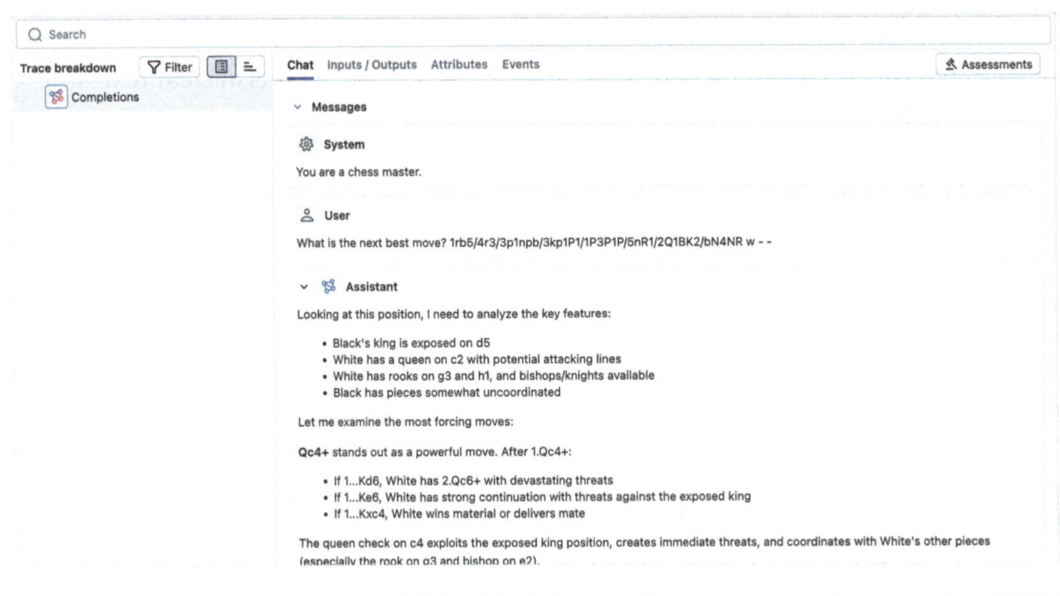

Figure 11-1. *Traces found in notebook with chat completion call*

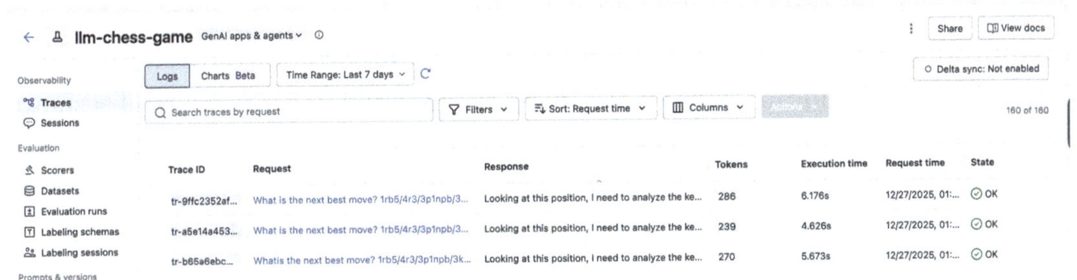

Figure 11-2. *Traces can also be found in the Experiment tab*

A hierarchy of spans can be seen in more complex use cases, like deep research agents with multiple steps of execution, as seen in Figure 11-3.

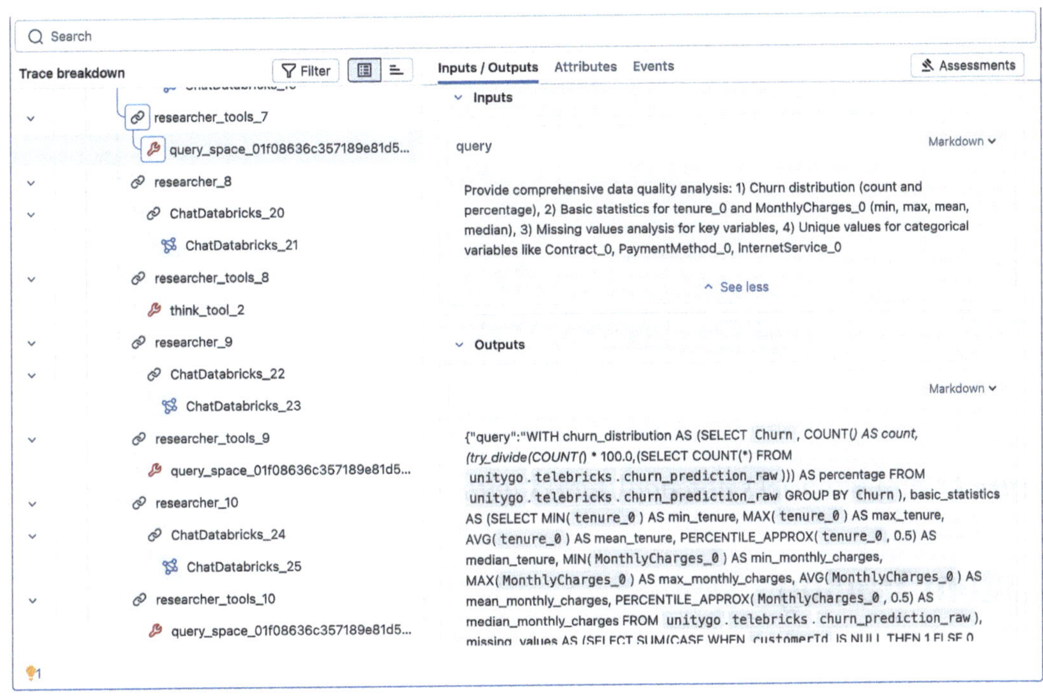

Figure 11-3. *Traces from deep research agents*

Scope of Auto-Trace

A single notebook is rarely sufficient for today's GenAI apps, as modern Generative AI applications are often too complex to be confined to a single notebook. You might wonder about the scope of the trace, as in whether it is confined to the notebook being executed or extends to the broader application. The good news is that `autolog()` captures everything within the execution environment. Take chess as an example; we can easily create a game in which an LLM plays against another LLM. Figure 11-4 shows a simple setup of a Python project along with a driver notebook. With this setup, we can download any non-Databricks GenAI projects from GitHub, then trace them with `autolog()`, and we will be able to see our traces in the Experiment tab.

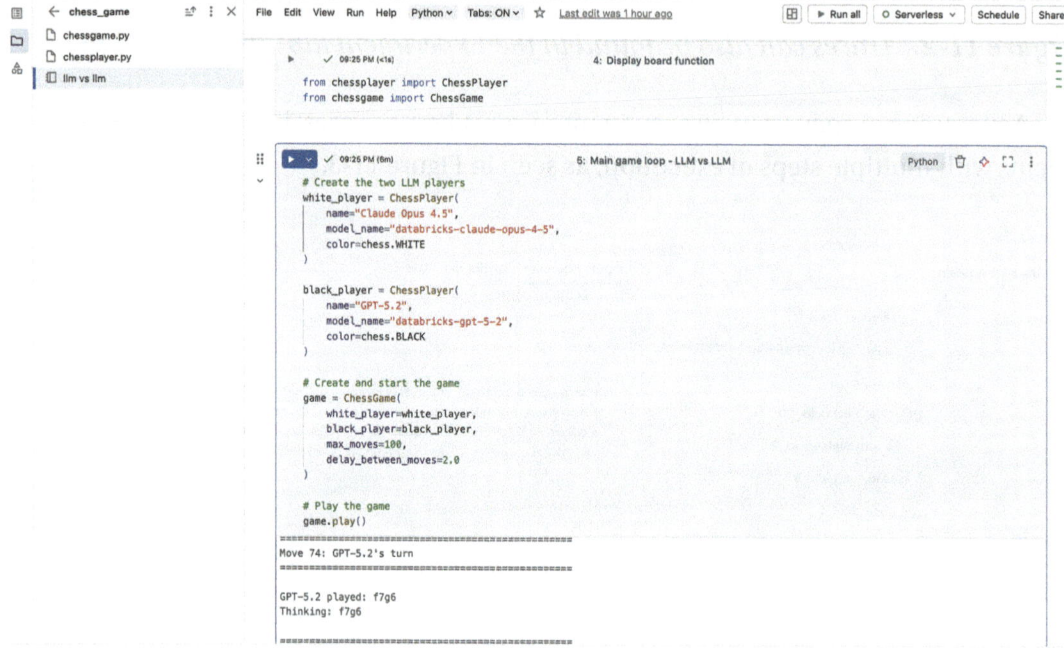

Figure 11-4. *Databricks notebook works with Python files*

Custom Tracing

While `autolog()` is very useful in understanding the structure of unfamiliar code, there are times when we want the tracing to be more predictable. For example, we may want to capture only the LLM's moves, or we may want to capture the reasoning behind the move. As shown in Figure 11-5, auto-logging lacks consistency, and the information isn't organized in a way that's easy to search.

Trace ID	Request	Response	Tokens
tr-47d59bd13...	You are playing chess as White. Current board p...	g2h3	392
tr-ac1370f09...	You are playing chess as Black. Current board p...	d7h3	390
tr-6cda46ecd...	You are playing chess as White. Current board p...	Looking at this position, I need to analyze the ke...	435
tr-20bd1902...	You are playing chess as Black. Current board p...	g4d7	396
tr-ae5737d81...	You are playing chess as White. Current board p...	f2f3	287
tr-b825e8b8...	You are playing chess as Black. Current board p...	d7g4	392
tr-c61127277...	You are playing chess as White. Current board p...	Looking at this position, I need to analyze the ke...	449

Figure 11-5. *Trace output from GPT 5.2*

There are three types of manual tracing: the first is the function decorator, the second is span tracing, and the third is a low-level API, which we don't recommend because it requires a lot of tuning. Table 11-1 summarizes the differences between the APIs.

Table 11-1. *Comparison between tracing APIs*

Feature	Function decorators	Span tracing	Low-Level APIs
Use Case	Trace entire functions with a one-line decorator. @mlflow.trace	Trace arbitrary code blocks within functions for fine-grained control.	Direct control over trace lifecycle for complex scenarios.
Automatic Parent-Child	Yes	Yes	No—manual management
Exception Handling	Automatic	Automatic	Manual
Works with Auto-Trace	Yes	Yes	No
Thread Safety	Automatic	Automatic	Manual

By this point, the limitations of the low-level APIs—such as manually creating, managing, and linking spans—should be clear. That's why this book focuses primarily on function decorators and span tracing, which provide higher-level abstractions and address most practical tracing needs.

Function Decorator

MLflow 3 includes a function decorator implemented with `@mlflow.trace`. It is recommended to always specify the decorator's parameters to give context to the traces. However, by default, the decorator will capture the inputs and outputs of the decorated function. The parameters are:

- name, the span name if different than the name of the decorated function

- span_type, the type of span. Set either one of built-in Span types, e.g. SpanType.LLM or a string

- attributes, add custom attributes to the span

If you have ever used telemetry in your project, you will already be familiar with the concept of spans—and the idea is the same here. A span represents a single unit of work within a larger operation. It captures what happened, when it happened, and how long it took. When multiple spans are connected together, they form a trace, which shows the full execution path of a request or workflow. MLflow's Span design maintains compatibility with OpenTelemetry specifications. MLflow provides predefined SpanType values for categorization (e.g., Agent or LLM) as well as different icons in the trace UI as shown in Figure 11-6. You can also use custom string values for specialized operations.

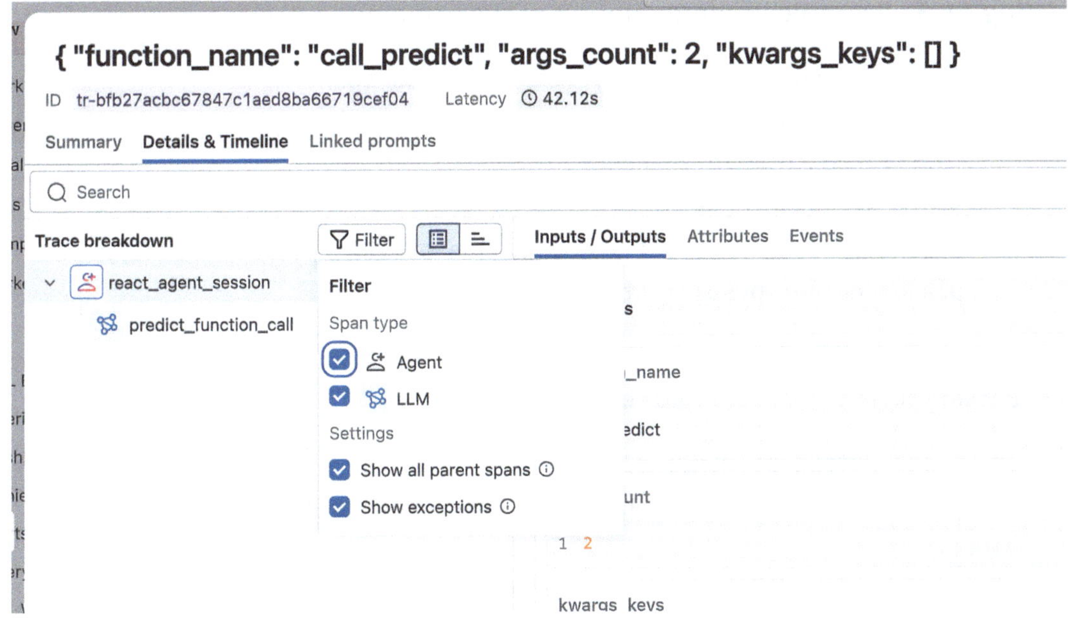

Figure 11-6. *Various span types captured in traces*

For details about the spans and their definitions, please refer to the Databricks documentation:

```
https://docs.databricks.com/aws/en/mlflow3/genai/tracing/span-concepts
```

Span Tracing

Span tracing is represented by `mlflow.start_span`. Note that this is not a function decorator, so we do not need to supply the @ sign. It is used inside a function, which is one level lower than a function decorator. The goal is to provide greater granularity in what we can capture beyond the function's input and output, allowing for deeper insights into its execution. Listing 11-5 shows an example of implementing span tracing in Google DeepMind's GameArena, a project that uses LLMs to play chess against each other to test their raw intelligence. The full source code can be found at the following GitHub repo:

```
https://github.com/rwforest/game_arena/tree/databricks-migration
```

Listing 11-5. Example implementation of span tracing

```python
with mlflow.start_span(name="parse_response") as span:
    parser_input = parsers.TextParserInput(
        text=response.main_response,
        state_str=pyspiel_state.to_string(),
        legal_moves=parsers.get_legal_action_strings(pyspiel_state),
        player_number=pyspiel_state.current_player(),
    )

    parser_output = parser.parse(parser_input)

    if parser_output is None:
        print(colored("Parser output is None, ending game.", "red"))
    else:
        print(colored(f"Parser output is {parser_output}.", "magenta"))

    span.set_inputs({"parser_input": parser_input})
    span.set_outputs({"parser_output": parser_output})
```

Evaluation

At the beginning of this book, we discussed how Agent Bricks, in particular, Information Extraction, and Custom LLM are evaluation-driven. Agent Bricks heavily utilized the features in MLflow, and this serves as a best practice for how we can use the evaluation functions to make our applications better.

Evaluating Traces with LLM Judges

When the traces are first captured using MLflow, only the input and output are recorded. However, it is often helpful to have assessments done to ensure the quality of the outputs. In Agent Bricks, they use LLM judges. We can also evaluate our custom traces with LLM judges using make_judge.

Taking the LLM vs. LLM game as an example (see Figure 11-5), we can define three different types of judges:

- Template-based judge

- Guidelines-based judge

- Pre-defined judge

Template-Based Judge

This type of judge is designed to be used with input/output values of the traces. The variables will be part of the prompt, allowing us to be very specific about what we are asking. They are represented by `{{ variable }}`. For example, Listing 11-6 is a judge that evaluates whether the chess move in a chess puzzle is the correct move.

Listing 11-6. A judge that evaluates if the predicted move matches the expected move

```
from mlflow.genai.judges import make_judge
from typing import Literal
from mlflow.genai import log_trace

move_correctness_judge = make_judge(
    name="move_correctness",
    instructions=(
        "Evaluate if the chess move predicted by the agent matches the "
        "expected correct move.\n\n"
        "Position (FEN): {{ inputs }}\n"
        "Agent's predicted move: {{ outputs }}\n"
        "Expected correct move: {{ expectations }}\n\n"
        "Consider that moves may be written slightly differently (e.g., "
        "'Qxf7#' vs 'Qxf7') but mean the same thing."
    ),
    feedback_value_type=Literal["correct", "incorrect", "equivalent_move"],
)
# example log trace
chess_position = "r1bqkbnr/pppp1ppp/2n5/4p3/8/5N2/PPPPPPPP/RNBQKB1R w
KQkq - 2 3"
predicted_move = "Qxf7"  # simulate LLM answer
```

```
correct_move = "Qxf7#"
log_trace(
    inputs=chess_position,        # input to the model
    outputs=predicted_move,       # model prediction
    expectations=correct_move     # ground truth
)
```

We can execute the judge with `mlflow.genai.evaluate()` as shown in Listing 11-7.

Listing 11-7. Executing the judge against a list of traces

```
import mlflow
traces = mlflow.search_traces(location='<experiment-id>')
results = mlflow.genai.evaluate(
    data=traces,
    scorers=[move_correctness_judge],
)
```

The allowed variables include inputs, outputs, expectations, and trace. As seen above, the values of these variables will be populated based on the traces. There is no need to set them explicitly, but at the same time, we cannot add additional variables ourselves. For the judge to work correctly, each trace must include the **expectations** value (the correct move, in this case). This value is added to the trace from the ground truth and serves as the reference the judge uses to evaluate the model's prediction, allowing it to determine whether the predicted move is correct, incorrect, or equivalent.

Key points for MLflow GenAI traces

- The ground truth label must be available at the time you log the trace.

- MLflow does not store expectations separately; they are logged directly in the trace metadata. When using the UI, humans can provide the expectations manually, which are then saved as part of the trace (see more in the next section).

- Where the truth comes from (database, file, vector store, human input) is determined in your pipeline—MLflow just needs the final value to include in the trace.

The result of `make_judge` is an assessment added to the individual trace, as shown in Figure 11-7.

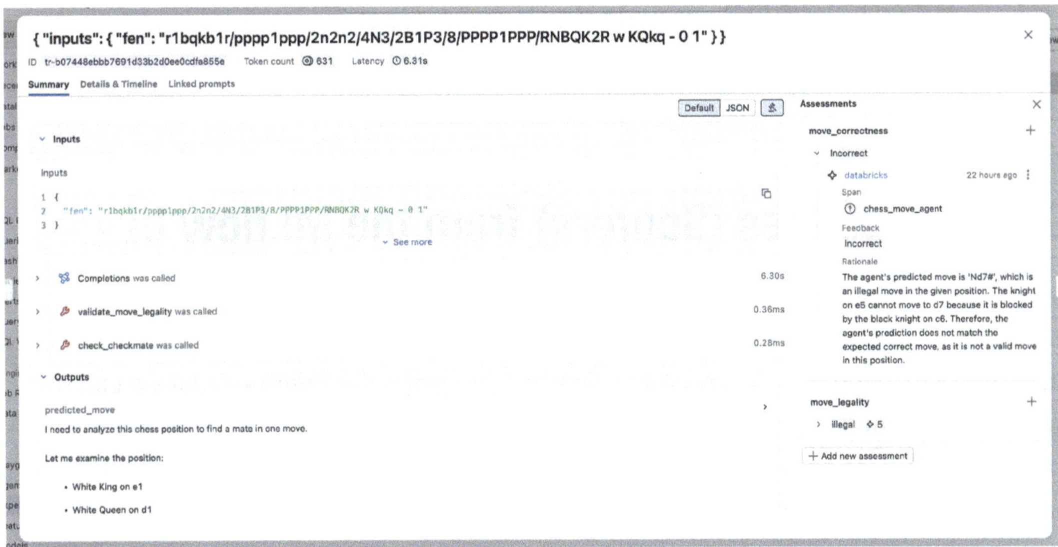

Figure 11-7. *Move correction judge shows up as an assessment*

Guidelines-Based Judges

The guideline-based judge is derived from the `Guidelines` class from `mlflow.genai.scorers`. It is ideal for checking compliance with rules, style guides, or information inclusion/exclusion. For example, we can check if a game follows all the standard chess rules as shown in Listing 11-8.

Listing 11-8. Evaluate using guidelines, which are just plain English

```
pgn_guideline = """
The output must be a valid PGN string.
1. Use Standard Algebraic Notation (SAN).
2. Ensure move numbers are sequential.
3. They cannot be illegal moves.
"""

from mlflow.genai.scorers import Guidelines
mlflow.genai.evaluate(
    data=eval_dataset,
    scorers=[
```

```
      Guidelines(name="pgn_validity", guidelines=pgn_guideline),
   ],
)
```

Creating Judges (Scorers) from the MLflow UI

For each of the "GenAI apps & agents experiment" (select from the dropdown next to the experiment name), there is a "Scorers" option, which we can use to define the judges and run them using the user interface. Figure 11-8 shows the interface that we can use to define these judges. They are called "scorers" in the MLflow interface, but they are, in fact, LLM judges because these scorers generally produce a score, but here they give pass/fail or qualitative assessments.

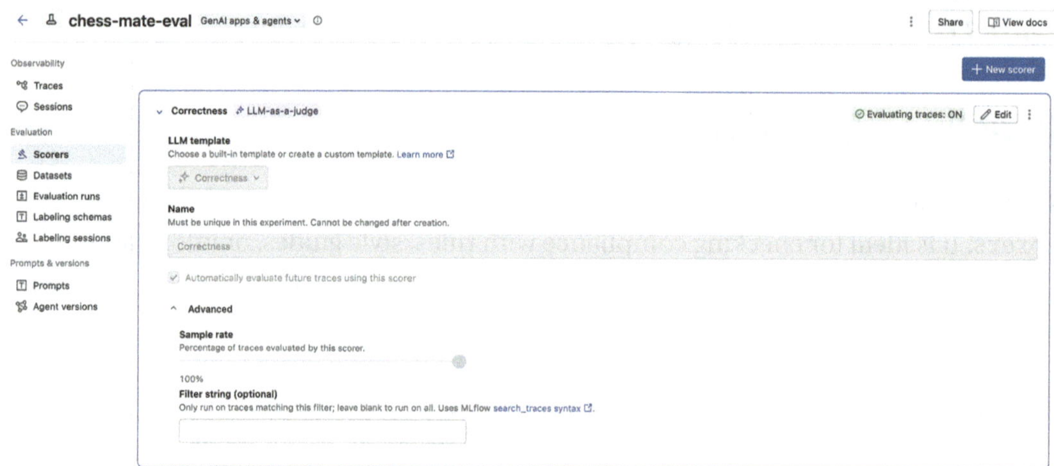

Figure 11-8. *Creating a score (LLM judge) from the MLflow UI*

Table 11-2 shows all the available pre-built judges. They are available on the UI to choose from, but there can be more judges available in the MLflow library because new judges may be added to the SDK before they appear in the UI.

Table 11-2. *Pre-built LLM judges in MLflow*

Template/Judge	Explanation	Ground Truth req?
Correctness	Evaluates if the app's response is correct when compared against a provided ground truth (expectations).	Yes
Retrieval Groundedness	Checks if the response is strictly based on the retrieved context. It identifies if the agent is "hallucinating" information not present in the sources.	No
Relevance to Query	Assesses whether the app's response directly and effectively addresses the specific user input or question.	No
Safety	Monitors the output for harmful, offensive, toxic, or otherwise inappropriate content.	No
Retrieval Relevance	Evaluates if the specific documents or context retrieved by the system are actually relevant to the user's request.	No
Retrieval Sufficiency	Determines if the retrieved context contains enough information to fully answer the question (usually compared against ground truth).	Yes
Guidelines	A flexible template where you provide specific natural language rules (e.g., "The response must be professional and under 100 words") for the LLM to verify.	No

As discussed above, we can create template-based judges, which is available on the interface as "Create custom LLM template." The instruction building process is more guided with the "Add variable" dropdown. But the usage is the same as what we discussed.

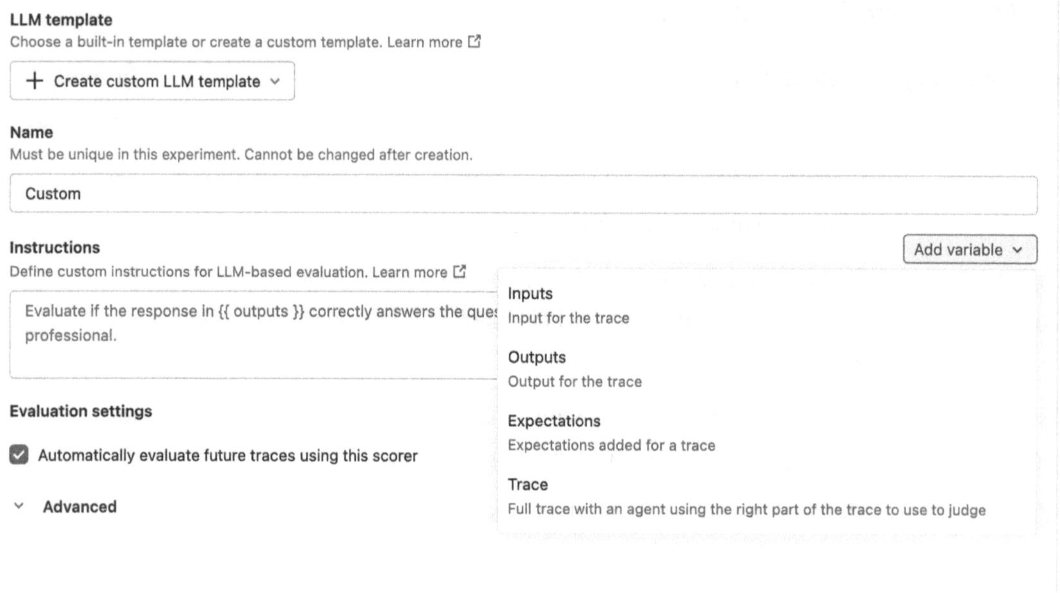

Figure 11-9. *Creating a custom LLM template*

Review App: Labeling Session

After we have gone through LLM-as-a-judge, it's time for human-as-a-judge. In general, it is generally easier to find human experts to set expectations than to give lengthy feedback (providing ground truth). That's why the labeling sessions are very simple, allowing feedback to be collected quickly. In certain cases, where an LLM is not performing, like playing chess, we can consider using code-based judges to generate labels as feedback. For example, we can use a chess engine to search for the best move as a ground truth label.

When we collect our traces, we can turn them into a dataset. The importance of a dataset isn't just another table for reference, but it can also be used for prompt optimization, a technique we will cover in the last section of this chapter.

MLflow offers a user interface for us to manage the review sessions without code, but it's a three-step process.

- Creating datasets from traces or manually via code. We will use this dataset for review later on.

    Figure 11-10 shows that we can select the traces to add to a dataset.

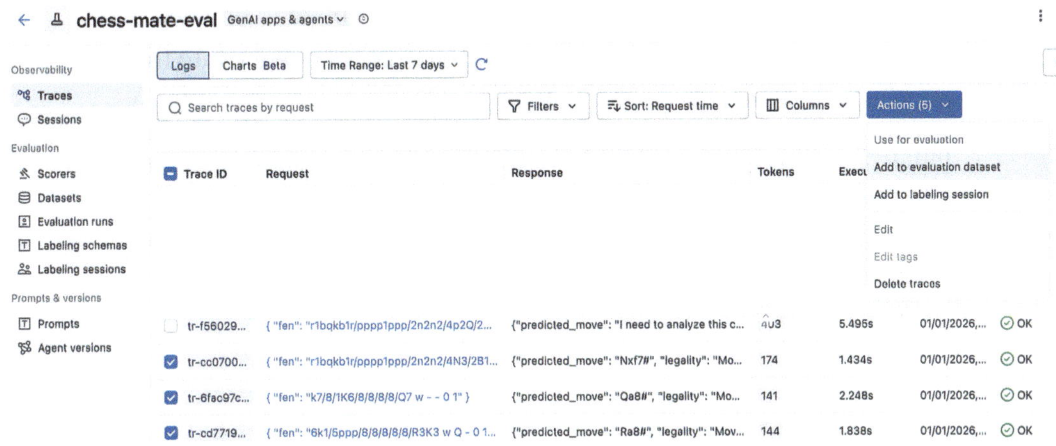

Figure 11-10. *Adding traces to a dataset*

- Set up a labeling schema, which is a reusable component throughout the entire review session

  Figure 11-11 shows the labeling schema page, which is an intuitive interface that will be populated on the right side of the review app, along with the traces.

Figure 11-11. *Label schema user interface*

- Create a review session, and the expectations will be merged back into the traces automatically. We can also merge them back into the dataset using code.

Finally, as seen in Figure 11-12, we need to provide a dataset and a label schema to create a review session.

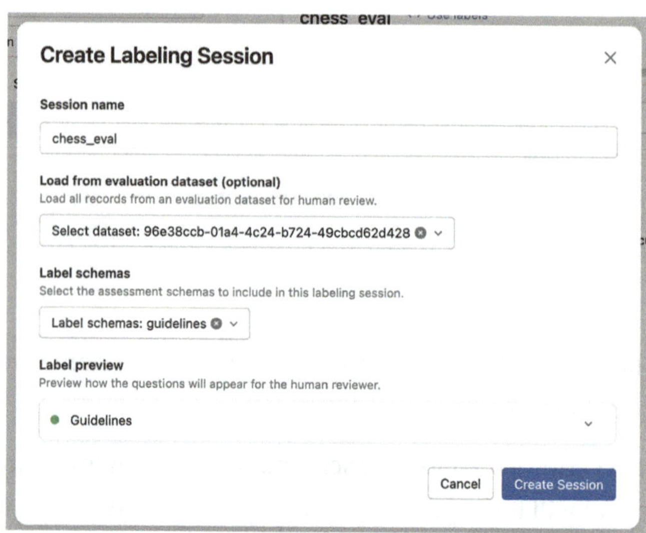

Figure 11-12. *Creating a review labeling (review) session*

The expectations/feedback entered in the review app will show up in the traces. This completes the full cycle: gathering the traces from the app, curating the dataset, setting up a review app, and allowing the results to flow back into the traces. Figure 11-13 shows the expectation in one of the traces.

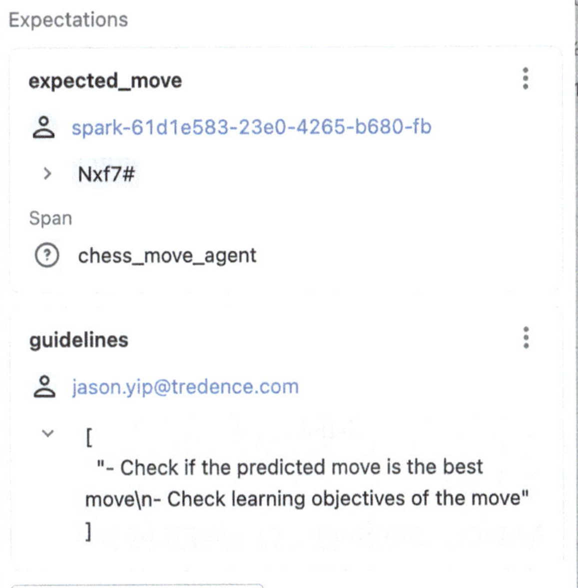

Figure 11-13. Expectations from one of the traces after going through the review

Hello Prompt Optimization, Goodbye, Prompt Management

The final feature we will introduce is prompt management. There isn't any magic in terms of managing prompts—all it takes is careful attention, and it gradually evolves over time. MLFlow provides an interface to manage these prompts. However, it's a daunting task to keep track of all prompts across models, since each model responds differently to a given prompt. In this final section, we will introduce a more advanced feature that's loosely integrated with MLflow, which is DSPy, aka declarative self-improving Python.

Introducing DSPy

DSPy is an open-source project from Stanford NLP, where Matei Zaharia, a co-founder of Databricks, is also a professor. That's why some of these features are being integrated back into MLflow.

A free course on DSPy is also available on DeepLearning.AI. With the details available in the source, we will be brief in this section. Feel free to go through the course to learn more.

```
https://www.deeplearning.ai/short-courses/dspy-build-optimize-
agentic-apps/
```

At a high level, DSPy has three different components:

- **Signatures**: Define the input/output structure and task description (the "contract") that tells the LLM (Large Language Model) what to do, like "`question -> answer`" or "`context, question -> reasoning, answer`".

- **Modules**: Wrap LLM calls with different execution strategies— Predict for simple generation, `ChainOfThought` for reasoning before answering, and `ReAct` for tool-using agents that iterate through observation-action cycles.

- **Optimizers**: Automatically improve prompts by testing instruction variations to find what works best and bootstrapping few-shot examples from your training data to teach the model your specific task.

The reason why this methodology is groundbreaking is that it reduces the manual work of structuring prompts for each call. We also don't need to build a complex regex to extract the output because the inputs and outputs are defined in a contract. We will once again illustrate the concept with chess. Figure 11-14 shows the three pillars.

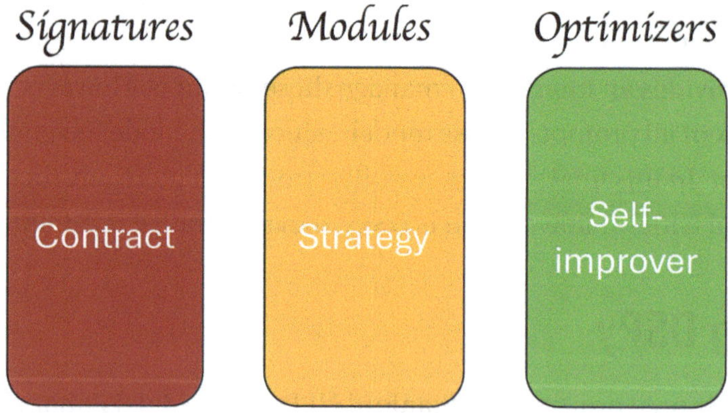

Figure 11-14. *The three pillars of DSPy*

Our objective here is to use LLM to solve checkmate-in-one-move puzzles. These puzzles are usually very straightforward with one correct answer and one objective, which has a very clear boundary, not like a game of chess, where one move would lead to another.

The Setup

Before we start, we need to configure our LLM, which is just two lines of code. For Databricks Foundation Model APIs, we can use databricks as llm_provider. This should be a string of the form "llm_provider/llm_name" that is supported by LiteLLM. For example, "openai/gpt-4o". As shown in Listing 11-9, we are using Claude Sonnet 4.5 hosted on Databricks.

Listing 11-9. Configuration of DSPy's Language Model

```
import dspy
lm = dspy.LM("databricks/databricks-claude-sonnet-4-5", max_tokens=1500)
dspy.configure(lm=lm)
```

The Contract

As stated above, the contract simply uses English to define the input and output. While we can be brief and just use "question -> answer", we would lose the only opportunity to describe the contract in a more descriptive way. For example, in chess, there are two notations, known as SAN (Standard Algebraic Notation) and UCI (Universal Chess Interface). While SAN is the official notation for recording chess moves for humans, computers often use UCI because it's more informative. Listing 11-10 shows a simple contract instructing the model to give the UCI format.

One thing to pay attention to here is that the docstring, the string literal right below the class definition, is also being considered as an input. Do not ignore this critical comment.

Listing 11-10. Sample contract for checkmate in one move

```
class ChessMateInOne(dspy.Signature):
    """Given a chess position in FEN notation, find the move that delivers
    checkmate in one move in UCI format.
    """
    fen: str = dspy.InputField(desc="Chess position in FEN notation (white
    to move)")
    move: str = dspy.OutputField(desc="ONLY the checkmate move in UCI
    notation (4-5 characters like 'h5f7')")
```

The Strategy

The most common modules are `Predict` and `ChainOfThought`. While `Predict` is just a simple chat completion, as its name suggests, `ChainOfThought` would show the reasoning behind it. Listing 11-11 is the definition of our chess solver.

Listing 11-11. Chess solver using ChainOfThought

```python
class ChessSolver(dspy.Module):
    """A module that uses Chain of Thought reasoning to find mate-in-one
    moves."""

    Def __init__(self):
        self.cot = dspy.ChainOfThought(ChessMateInOne)

    def forward(self, fen: str) -> dspy.Prediction:
        return self.cot(fen=fen)
```

Finally, without building out the prompt ourselves and constructing a ChatCompletion API call, we can easily use the model as a Python function, as shown in Listing 11-12. This is the beauty of DSPy—we no longer need to worry about keeping track of which prompt for which model and how the results stack up. We can focus on building our Python program, and DSPy will take care of the rest. What's more, to extract the move, we simply take it from our contract with `result.move`. We have defined it in the contract, and now we can use it!

Listing 11-12. Executing chess solver

```python
solver = ChessSolver()

# Quick test with one example
sample = trainset[0]
print(f"FEN: {sample.fen}")
print(f"Expected Move (UCI): {sample.move}")

result = solver(fen=sample.fen)
print(f"Predicted Move (raw): '{result.move}'")
```

The Self-Improver

Now we get the basic prompt and response without the need for lengthy prompt engineering or multiple manual attempts. DSPy also presented an interesting idea of using data to improve the prompt. These are called Optimizers. There are a few built-in optimizers in DSPy, which we can find in the documentation on dspy.ai, but we will choose `BootstrapFewShot` for our chess. This uses a few-shot learning strategy that automatically prepends a small set of example input-output pairs to a prompt, guiding the LLM to generate accurate predictions for new inputs.

There are a few advantages of `BootstrapFewShot` for chess:

- Automatic Example Selection

  `BootstrapFewShot` doesn't just randomly pick training examples—it bootstraps by running the model on training data, keeping only the examples where the model succeeded.

- Pattern Learning for Chess Reasoning

  For mate-in-one puzzles, this is powerful because the LLM learns reasoning patterns from successful examples (e.g., "look for back-rank threats," "check if queen can deliver check with no escape squares").

- `ChainOfThought` Reasoning

  The `ChainOfThought` reasoning traces show how to analyze FEN positions step-by-step. It teaches the model to output moves in the correct UCI format.

Listing 11-13 defines an evaluation metric that we can use for the training, which is simply comparing if the moves are the same in terms of UCI format.

Listing 11-13. Chess move metric comparing the UCI moves

```
def chess_move_metric(example: dspy.Example, pred: dspy.Prediction,
trace=None) -> bool:
    """Check if the predicted move matches the expected move.
    """
```

```
expected_uci = example.move.strip().lower()
predicted_uci = pred.move.strip().lower()

# Direct match
if expected_uci == predicted_uci:
    return True

# Handle promotion variations (e.g., 'a7a8q' vs 'a7a8')
if len(expected_uci) >= 4 and len(predicted_uci) >= 4:
    if expected_uci[:4] == predicted_uci[:4]:
        return True

return False
```

Now we can use `BootstrapFewShot` as shown in Listing 11-14.

Listing 11-14. BootstrapFewShot function definition

```
optimizer = dspy.BootstrapFewShot(
    metric=chess_move_metric,
    max_bootstrapped_demos=num_demos,
    max_labeled_demos=num_demos,
    max_rounds=1,
)

optimized_solver = optimizer.compile(solver, trainset=trainset)
```

The result of the optimization improved the accuracy of the solver by 10 percentage points, from 20% to 30%. While it's still low accuracy, we can see that LLMs are still not good at playing chess!

The Output

If you are worried about the black box operation in DSPy, remember that it's an open-source project. The second thing is to recall that at the beginning of this chapter, we have discussed `autolog()`. And yes, we can use `mlflow.dspy.autolog()`. As shown in Figure 11-15, all the call events, as well as the reasoning, are all shown in the trace.

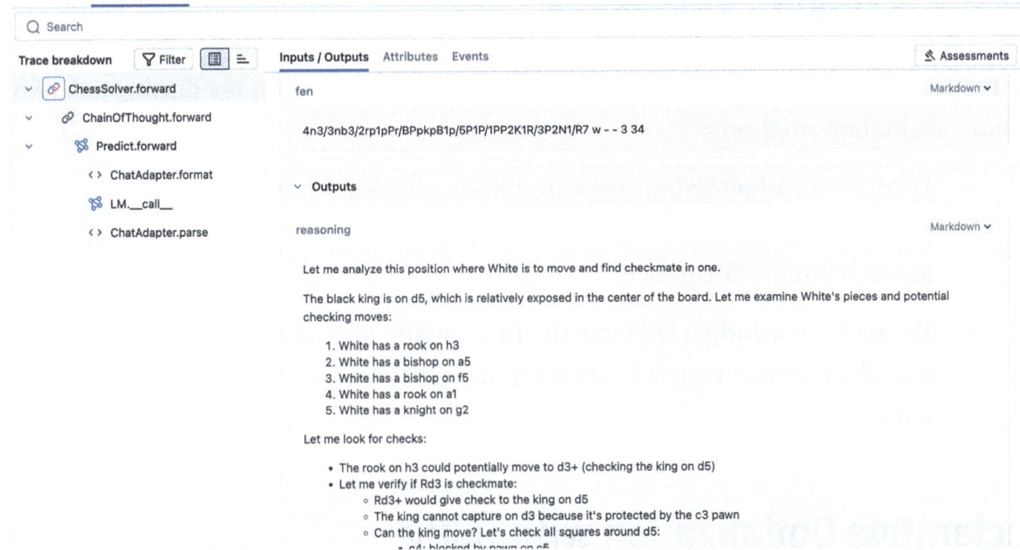

Figure 11-15. *Output of* `mlflow.dspy.autolog()` *on ChessSolver*

Conclusion

In conclusion, MLflow 3 represents a significant evolution from its 2018 origins as a machine learning lifecycle tool to its status as an industry-standard platform for Generative AI (GenAI) and agentic workflows. By 2025, the platform achieved massive scale, surpassing 30 million monthly downloads and serving over 5,000 organizations. (`https://pypistats.org/packages/mlflow?trk=public_post_comment-text`)

The advancements introduced in this chapter highlight critical pillars for modern AI development:

Enhanced Observability Through Tracing

MLflow has streamlined the developer experience by providing a single trace view that supports a wide array of libraries, including OpenAI, LangGraph, and AutoGen. Whether using auto-tracing to automatically capture a hierarchy of "spans" or manual tracing via function decorators for finer control, or even DSPy for auto-prompt optimization, developers can now easily visualize the "chain of thought" within complex agents.

Rigorous Quality Assurance

The platform bridges the gap between raw output and high-quality results through two primary evaluation methods:

- **LLM-As-a-Judge:** Using template-based, guideline-based, or predefined judges (such as correctness or safety) to automatically assess agent performance.

- **Human-As-a-Judge:** Utilizing the Review App and labeling sessions to collect human expert feedback and establish ground truth datasets.

Declarative Optimization with DSPy

The integration of DSPy (Declarative Self-Improving Python) marks a shift away from manual prompt management toward automated optimization. By defining clear signatures (contracts) and using optimizers like `BootstrapFewShot`, developers can use data to automatically improve prompt instructions and reasoning patterns.

MLflow 3 remains deeply integrated with the Databricks ecosystem, allowing developers to use tools like the Foundation Model APIs (FMAPI) while maintaining the flexibility of open-source frameworks. This synergy ensures that as GenAI features continue to evolve, the platform provides a stable foundation for building, tracking, and serving sophisticated AI applications.

## CHAPTER 12

Real-Time Intelligence with Spark Structured Streaming

Many people think of streaming as some very low-latency continuous real-time events like X, formerly Twitter, feeds or IoT devices. While that was the original use case, streaming has evolved over the years to allow integration with other non-real-time tables and as a useful technique to enable incremental processing for batch pipelines. In this chapter, we will first go back in time to visit Spark Streaming; then we will look at the latest Databricks Structured Streaming engine. They are largely the same, but Structured Streaming has better abstractions. Then we will look at how to use Lakeflow Declarative Pipelines to process streaming. Apache Spark offers two popular streaming processing engines: Spark Structured Streaming and real-time mode. While both engines are designed for real-time data processing, they have distinct advantages and use cases.

The Foundation of Structured Streaming

Micro-batching is the foundation of Spark Structured Streaming, which processes data in small increments, called batches, where each batch is processed before processing new batches. This approach allows for low-latency processing and high throughput. In micro-batching, illustrated in Figure 12-1, data is processed in small batches defined by time intervals, e.g., every 1–2 seconds. Spark Structured Streaming uses a write-ahead log, which means it logs received data to fault-tolerant storage before processing, tracking offsets to ensure disaster recovery. However, the write-ahead log introduces serialization overhead, making batch writing sequential and resulting in hundreds of milliseconds of latency between batches.

263

J. Yip et al., *Databricks Data Intelligence Platform*, https://doi.org/10.1007/979-8-8688-2524-8_12

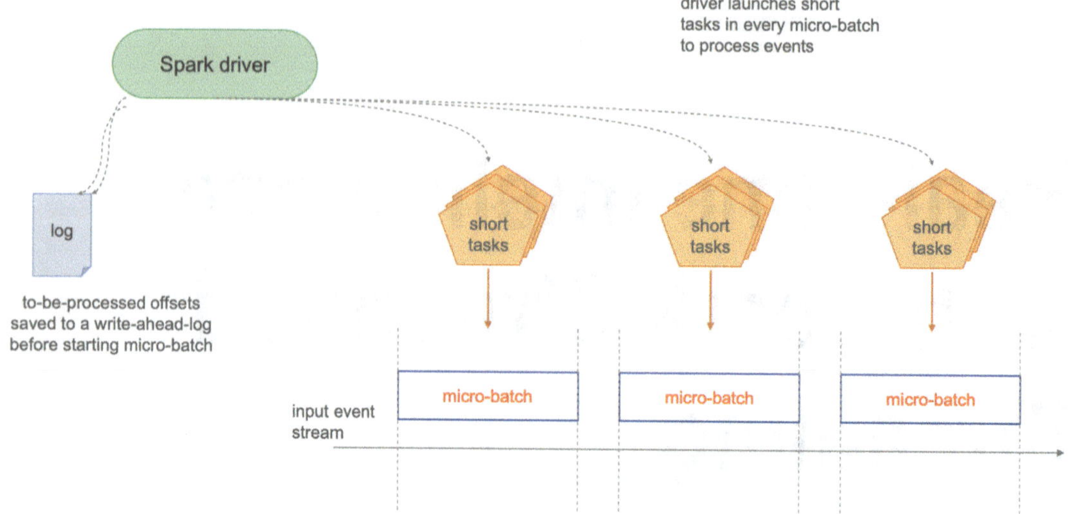

Figure 12-1. *Micro-batch processing*

The high-level architecture of Spark Structured Streaming consists of the following components:

- **Data Source:** The source of the data stream, such as Kafka or Amazon Kinesis

- **Receiver:** The component that receives the data from the data source and hands it over to Spark Streaming

- **Structured Streaming (Engine):** The core engine processes the data stream

- **Processing:** The component that performs various operations on the data stream, such as filtering, mapping, reducing, joining, and so on

- **State Management:** The component that manages the state of the streaming application, including checkpoint management

- **Output Sink:** The component that writes the processed data to a target system, such as a file, database, or messaging system

Figure 12-2 shows the workflow.

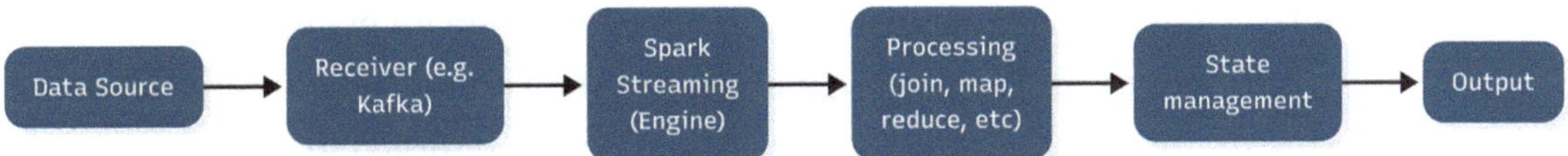

Figure 12-2. *Spark Structured Streaming workflow*

To meet the requirements of today's data engineering environments, Spark Structured Streaming is built to handle complexity that traditional batch-only systems cannot. When designing your pipeline, you should focus on the following patterns:

- **Arbitrary Stateful Processing**

  Because streaming applications run 24/7, keeping track of processing progress and state is important. There are multiple ways to manage state:

  a. **Checkpointing:** Spark Structured Streaming can checkpoint the application's state at regular intervals, allowing it to recover from failures and resume processing from the last checkpointed state.

  b. **Windowing:** Windowing operations (e.g., `window()`, `reduceByKeyAndWindow()`)allow you to manage state over a sliding window of data.

  c. **Arbitrary Grouped State (mapGroupsWithState):** These operators allow you to maintain and update custom state for specific groups of data (like user IDs or geographic regions) regardless of whether the data arrives in order.

  d. **Advanced Stateful Processing (transformWithState):** This is the most powerful operator, providing full, granular control over how state is stored, updated, and emitted for every incoming event, making it ideal for complex use cases like real-time stock tracking.

- **Event-Time Processing**

  Modern applications must distinguish between when an event
  occurred (event time) and when it arrived at the system (processing
  time). Structured Streaming uses watermarking to handle late-
  arriving or out-of-order data, allowing you to correctly assign delayed
  events to their historical windows.

Structured Streaming

Structured Streaming is the standard streaming engine built on the DataFrame/Dataset
API of Spark SQL. It processes data incrementally, supporting exactly-once guarantees in
micro-batch mode and robust state management.

Beyond putting a structure (dataframe) into the streaming data, Structured
Streaming is designed to address the following challenges:

- **Providing End-to-End Reliability and Correctness Guarantees:**
  When failure occurs, the engine automatically resumes from the
  last committed micro-batch using the configured checkpoint
  location and source offsets. With the increasing demand for
  streaming, pipelines must be continuously monitored, and failures
  automatically mitigated to ensure that highly available insights are
  delivered promptly.

- **Performing Complex Transformations:** In addition to streaming
  systems (e.g., Kafka), source data can often come in as flat file
  formats (CSV, JSON, Avro, etc.) that often must be restructured and
  transformed (e.g., parsing, metadata extraction, basic validation,
  schema capture) before being ingested into a bronze table.
  Structured Streaming is designed to process and transform this data
  with minimal latency.

- **Handling Late or Out-of-Order Data:** As discussed, there is a
  challenge in processing late arrival data because some events may
  arrive later than expected or out of order. We will discuss how
  Structured Streaming's watermarking and event-time processing
  capabilities address this issue.

Spark Real-Time Mode

Real-time mode is a new offering in Databricks Runtime 16.4 or above. Instead of performing micro-batches one after another, real-time mode launches a set of long-running tasks at the same time to keep reading, processing, and writing data continuously (see Figure 12-3). This means as soon as new data is available. This architecture drives very low latency and can reach single-digit milliseconds in best-case scenarios.

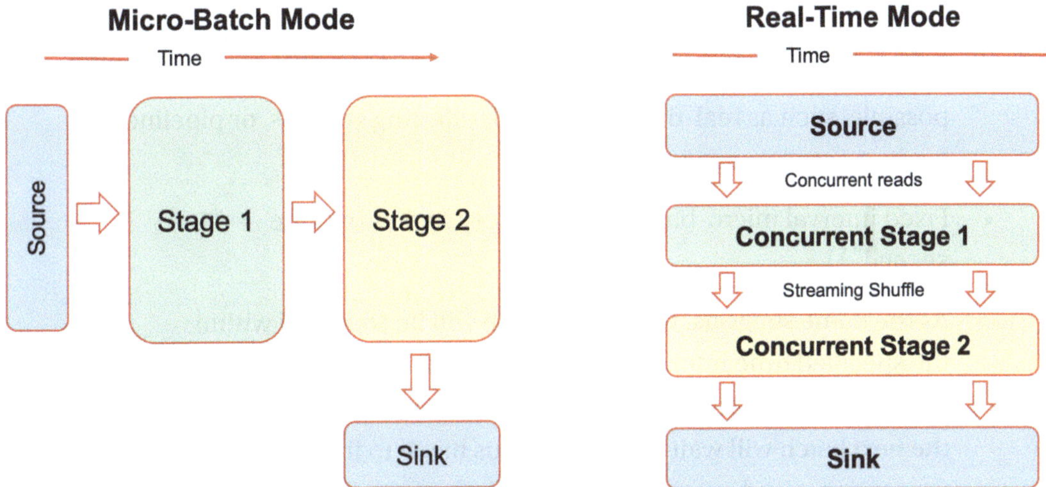

Figure 12-3. *Micro-batch mode vs. real-time mode*

In Spark Structured Streaming (default micro-batch), the system tracks progress by writing source offsets and commit metadata to the checkpoint location at each batch boundary, coordinated by the driver. In Databricks real-time mode, offsets and state are checkpointed periodically at configured intervals to preserve recovery semantics while supporting low-latency processing. With asynchronous progress tracking and asynchronous state checkpointing enabled, these durability operations can occur without blocking steady-state processing, further reducing the latency associated with maintaining offsets and state information.

Triggers

In Spark Structured Streaming, a trigger controls how often the engine checks for new data and advances the streaming query, shaping latency, cost, and operational behavior.

The following are the different trigger modes:

- Default mode (no trigger is specified)

  If the trigger option is not specified, then by default, the query will be executed in micro-batch mode, which schedules the next micro-batch immediately after the previous one completes. Default mode is useful when you want to process data continuously and as fast as possible, such as real-time dashboards, alerting systems, or pipelines feeding near-live analytics.

- Fixed interval micro-batches (`trigger(processingTime = "1 second")`)

  As the name suggests, the micro-batch will be triggered within the specified time interval. Since micro-batching is a sequential operation, if the previous batch cannot finish in the specified interval, the next batch will wait for the previous batch to finish before processing. Fixed-interval micro-batches are useful when you want to process streaming data at predictable, regular intervals to control latency, batch size, or downstream update schedules.

- Available-now micro-batch (`trigger(availableNow=True)`)

  Processes all available data when the streaming query starts and stops once it has been processed. You can control the amount of data processed per micro-batch using the maxBytesPerTrigger parameter. If the total data exceeds this limit, Spark will automatically split it into multiple micro-batches until all data is processed. This trigger is ideal if you need to enable incremental processing in batch jobs without continuously running a cluster.

- Real-Time mode (trigger(realTime="5 minutes"))

  Provides continuous processing of incoming data with minimal latency while maintaining a fault-tolerant state.

This trigger is useful for operational workloads requiring sub-second processing, such as fraud detection or real-time personalization. More about real-time mode in the next section.

Output Modes

In Spark Structured Streaming, output modes determine how the results of a streaming query are persisted (written to the sink). The modes can be configured using `writeStream.outputMode()`. These are brief descriptions of the modes:

- **Complete:** In complete mode, the result table is written to the sink on every trigger, rather than just the new or updated rows. This is particularly useful for aggregations because it ensures that late-arriving data is included without losing counts. However, be careful; if the sink does not replace the old data (e.g., does not support overwriting), this can lead to duplicate records.

- **Append (Default):** Only new rows since the last trigger are written to the sink. Note that for aggregations or stateful operations, late-arriving data may not be captured unless additional handling is applied.

- **Update:** This mode outputs only the rows in the result table that have been updated since the last trigger, rather than outputting the entire result table. This mode is mainly used for stateful operations or aggregations, where the results of previous rows may need to be updated as new data arrives.

Windowed Grouped Aggregation

Structured Streaming offers a way to group aggregations over time-based windows (sliding, tumbling/fixed, or session), for example, aggregating events every five minutes. This works similarly to a `groupBy` operation but is applied over a time dimension. Imagine we are converting a timestamp column to a time range, say 12:00 to 12:10, and for each trigger, we aggregate the events and save them to a table. In the table shown in Figure 12-4, we can notice that "cat" changed to 2 from 12:05 to 12:10 because another "cat" arrived at 12:07.

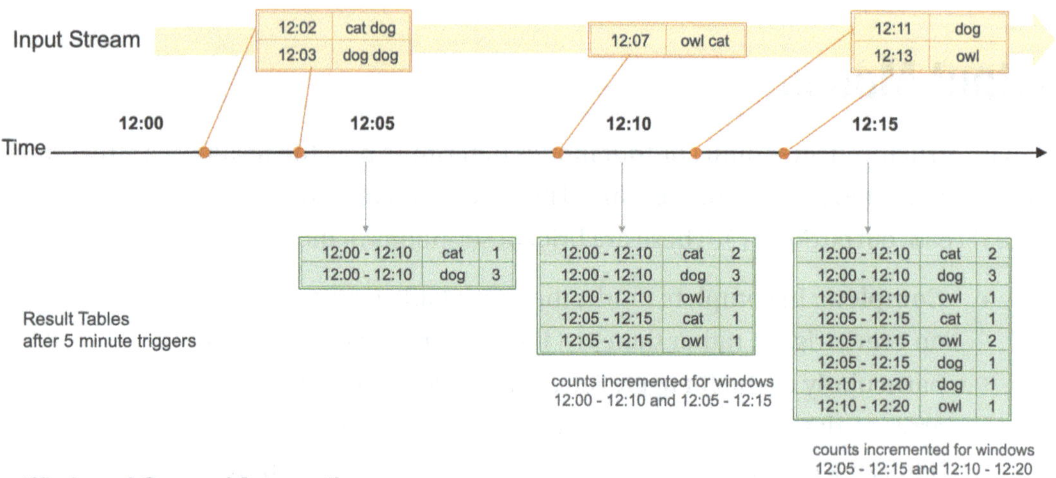

Figure 12-4. *Windowed grouped aggregation*

State Management

When you need to manage state in a streaming query, Spark Structured Streaming provides checkpointing. Checkpointing persists both the processing state and the source offsets, enabling the query to recover correctly after a failure. Databricks recommends always specifying a checkpoint location to ensure reliable recovery, as illustrated in Listing 12-1.

Listing 12-1. Spark Streaming Checkpoint

```
streamingDataFrame.writeStream
  .format("parquet")
  .option("path", "/path/to/table")
  .option("checkpointLocation", "/path/to/table/_checkpoint")
  .start()
```

However, there are times when more advanced stateful processing is required. That's where the new operators `mapGroupsWithState()` and `flatmapGroupsWithState()` come into the picture. These operators allow you to maintain state for arbitrary groups of data, where the grouping key does not need to follow a sequential or predefined pattern. For example, you might want to group users by a geographic region or spending threshold rather than by individual IDs. Incoming events may arrive at a finer granularity (e.g., city), but these operators allow you to maintain and update state at the desired grouping level, even when data is late or arrives out of order. These techniques are helpful to ensure that late or out-of-order items are correctly assigned to its intended group for accurate analysis (see Figure 12-5).

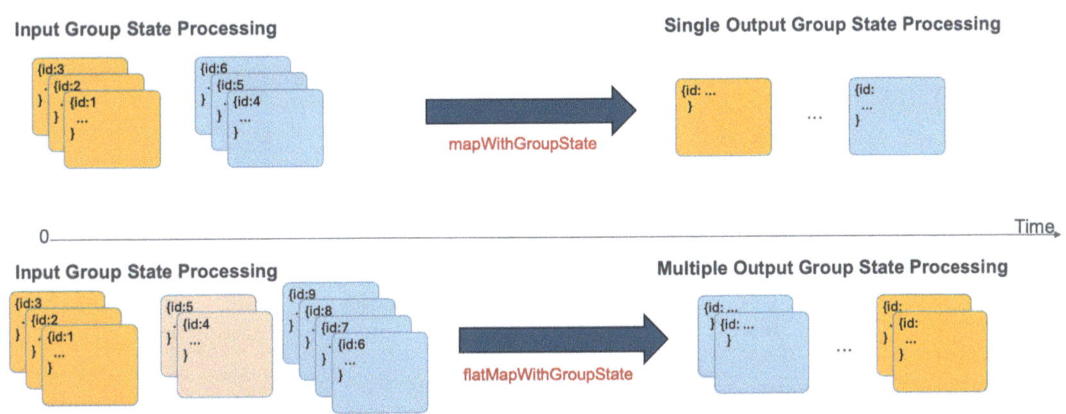

Figure 12-5. *Arbitrary stateful processing in structured streaming*

Late-Arrival Handling: Watermark

Structured Streaming uses a **watermark** to determine how long to continue processing updates for a given state entity before marking the window final. A watermark is a moving threshold computed as the maximum event time seen so far minus the specified lateness tolerance.

In other words, Watermark (W) at any point in processing time is:

$$W = max(ET_{seen}) - T_{lateness}$$

Where:

- $max(ET_{seen})$ is the maximum event time the engine has encountered so far.

- $T_{lateness}$ is your specified tolerance, e.g., 10 minutes

Any data arriving with an event time older than the watermark is considered late and can be either ignored or processed separately. Listing 12-2 is an example.

Listing 12-2.   Spark Streaming Watermark

```python
from pyspark.sql.functions import window
(df
  .withWatermark("event_time", "10 minutes")
  .groupBy(
    window("event_time", "5 minutes"),
    "id")
  .count()
)
```

Figure 12-6 is a visual walkthrough of the process.

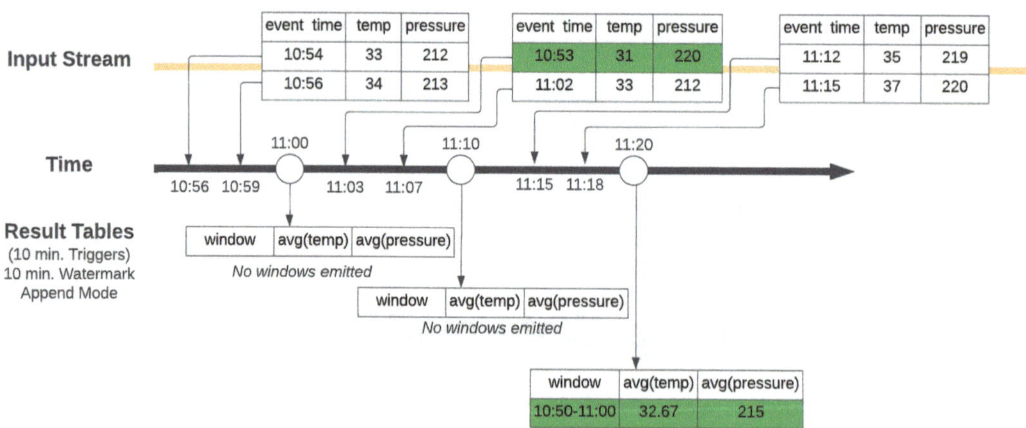

Figure 12-6.   *Visual walkthrough of the watermarking process*

- **The Tolerance Buffer:** By setting a 10-minute watermark, we tell Spark to keep the "state" of open windows even after their clock time has passed. For example, even though the **10:50–11:00** window technically "ended" at 11:00, Spark retains it in memory because the watermark (10:56 - 10 = 10:46) hasn't passed the window's end yet.

- **Late Arrival Acceptance**: Data is accepted if its event time is at least as late as the current watermark. For example, during the 11:10 trigger, the watermark is at 10:52, so an event with a timestamp of 10:53 is accepted for processing. Events earlier than 10:52 would be considered late and ignored.

- **The Final Emission (Append Mode)**: In Append Mode, a window is emitted only when Spark is certain no additional data can arrive. For instance, at the 11:20 trigger, the max event time reaches 11:15, moving the watermark to 11:05. Since 11:05 is beyond the 11:00 window end, Spark finalizes the window, computes the aggregation, and outputs the result.

Earlier, we discussed a new output mode, Update Mode, which is useful for aggregations with a watermark.

Unlike Append Mode, which waits for the watermark to "close" a window before showing any results, Update Mode continues updating the count on every trigger until the watermark threshold is reached, as shown in Figure 12-7.

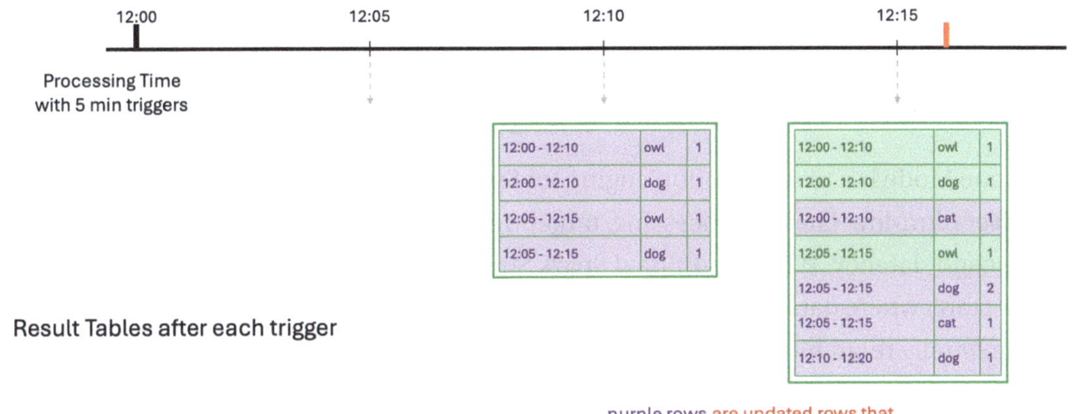

Figure 12-7. *Update mode in aggregation*

Auto Loader

Auto Loader incrementally and efficiently processes new data files as they arrive in cloud storage. It uses a Structured Streaming engine and a special source called "cloudFiles." The goal is to abstract the complexity of loading file arrivals in cloud storage. The cloudFiles protocol can automatically monitor ADLS, S3, GCS, and Azure Blob Storage files without requiring you to set up a file trigger in Databricks or an external system.

As Auto Loader discovers new files, it stores their metadata in a scalable key-value store (RocksDB) located in the pipeline's checkpoint directory. This mechanism ensures that each file is processed exactly once.

In the event of a failure, Auto Loader can resume processing from the checkpoint, continuing to provide exactly-once guarantees when writing data to Delta Lake.

Combined with Lakeflow Declarative Pipelines, Auto Loaders provide the following advantages:

- Autoscaling compute infrastructure for cost savings

- Data quality checks with expectations

- Automatic schema evolution handling

- Monitoring via metrics in the event log

Spark Real-Time Mode Deep Dive

Real-time Mode is a new execution engine for Spark Structured Streaming designed to deliver ultra-low-latency processing, targeting sub-second (often double-digit millisecond) latency. While traditional Spark Streaming relies on a micro-batch architecture, where data is processed in small, sequential batches, real-time mode fundamentally re-architects data flow through the system to minimize overhead and latency.

Architecture: Continuous Processing vs. Micro-Batching

In the traditional micro-batch model, latency is inherent because the system must plan, execute, and commit each batch sequentially. Real-time mode eliminates this bottleneck by using continuous processing. Figure 12-8 illustrates the differences.

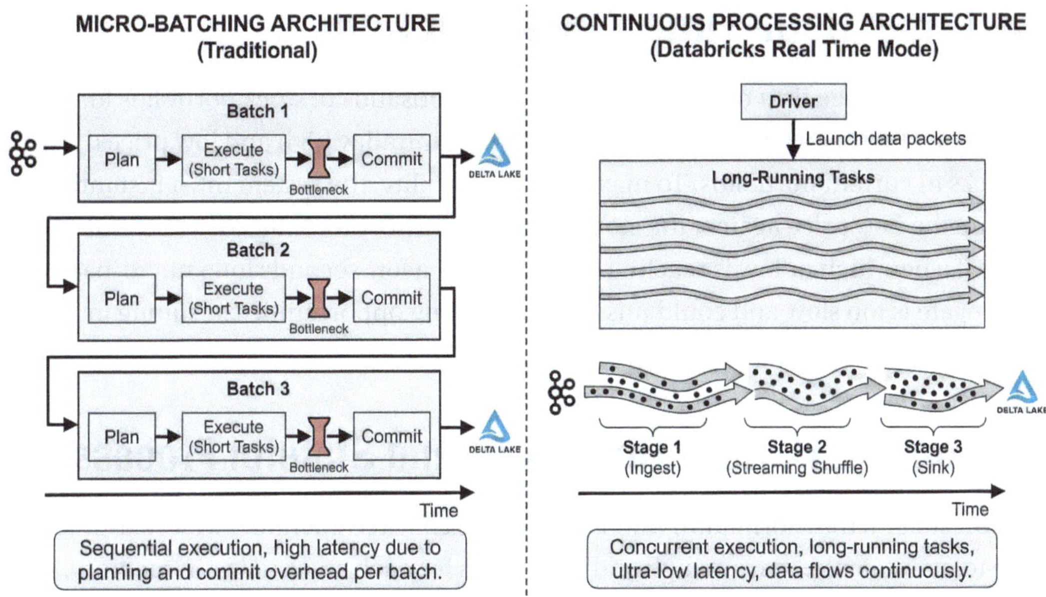

Figure 12-8. *Comparison between micro-batch and real-time mode*

- **Long-Running Tasks:** Instead of launching short tasks for every micro-batch, the driver launches long-running tasks that process data continuously for fixed, long durations (e.g., minutes).

- **Concurrent Execution:** Stages are scheduled continuously and concurrently, allowing data to flow from the source through transformations to the sink without waiting for previous stages to complete.

- **Streaming Shuffle:** Data is shuffled as soon as it arrives, rather than waiting for a map stage to finish before starting a reduce stage.

Advanced State Management

Spark Structured Streaming introduced the `transformWithState` operator to maintain and update arbitrary state for each key in a streaming pipeline; this is especially useful in real-time mode. Unlike standard aggregations or windowed operations, it provides full control over how state is updated, stored, and emitted for every incoming event. Below we will consider a real-life scenario.

Use Case: Real-Time Stock Tracking

Imagine a brokerage firm managing hundreds of thousands of stock portfolios for customers. They need a real-time feed showing the intraday high and low prices for stocks in current portfolios. To maximize profitability, the system must instantly compare the new price against the stored high/low history for that stock and emit an update if a new high or low is reached. Waiting even a few seconds for a micro-batch to aggregate is too slow and could miss critical trading opportunities, resulting in suboptimal performance.

Solution: Spark Real-Time Mode and Stateful Processing

We use a stateful transformation (the `transformWithState` operator) with real-time mode to maintain the current high and low values for each stock ticker in memory. Because real-time mode processes data continuously rather than in micro-batches, the state can be updated and the result emitted as soon as the tick arrives.

This code demonstrates grouping a stream of stock ticks by their ticker symbol and applying a stateful function to track highs and lows. The crucial element is combining this highly optimized stateful operation with the low-latency trigger.

Figure 12-9 highlights sample architecture for a brokerage firm. However, to simplify the process, we are only showing the price tracking logic in Listing 12-3.

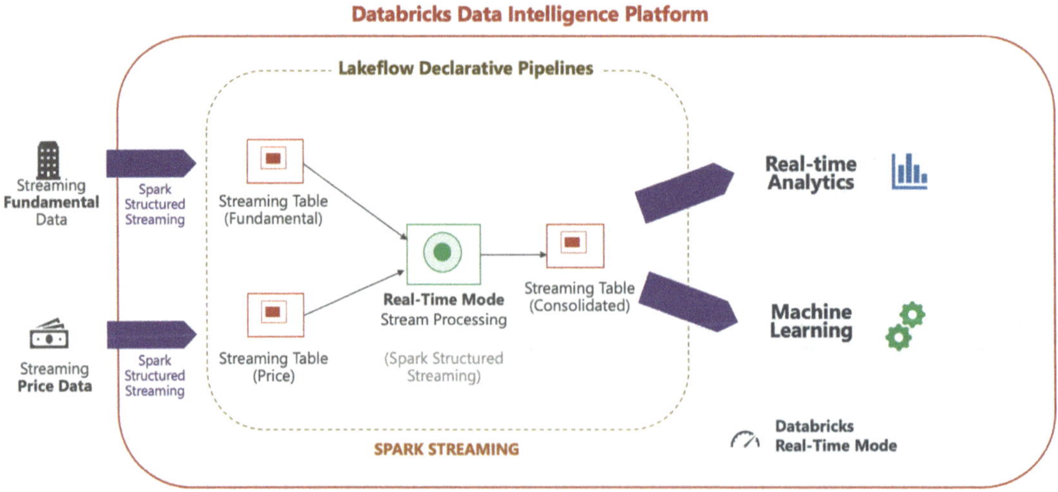

Figure 12-9. Sample stock brokerage firm streaming workflow

Listing 12-3. Stock price tracking using `transformWithState`

```python
from pyspark.sql.functions import col, current_timestamp, from_json
from pyspark.sql.streaming import StatefulProcessor,
StatefulProcessorHandle
from pyspark.sql.types import StructType, StructField, StringType,
DoubleType, TimestampType
import pandas as pd

# Schema for the source data (e.g., from Kafka)
input_schema = StructType([
    StructField("ticker", StringType()),
    StructField("price", DoubleType())
])

state_schema = StructType([
    StructField("current_high", DoubleType()),
    StructField("current_low", DoubleType())
])

output_schema = StructType([
    StructField("ticker", StringType()),
    StructField("latest_price", DoubleType()),
    StructField("day_high", DoubleType()),
    StructField("day_low", DoubleType()),
    StructField("updated_at", TimestampType())
])

class IntradayStatsProcessor(StatefulProcessor):
    def __init__(self, handle: StatefulProcessorHandle) -> None:
        """

        Initialize the state variable.
        This runs once per partition on the executor.
        """

        # We use ValueState for simple high/low tracking
        self.state_handle = handle.getValueState("intraday_stats",
        state_schema)
```

```python
    def handleInputRows(self, key, inputRows):
        """

        Process individual rows as they arrive.
        :param key: The grouping key (ticker).
        - inputRows: Iterator of rows for this key
        """

        # 1. Recover existing state from the state store (if any)
        if self.state_handle.exists():
            existing = self.state_handle.get()
            current_high = existing["current_high"]
            current_low = existing["current_low"]
        else:
            current_high = float('-inf')
            current_low = float('inf')

        # 2. Process the iterator of input rows for this specific ticker
        results = []
        for row in inputRows:
            price = row["price"]

            # Update local variables
            current_high = max(current_high, price)
            current_low = min(current_low, price)

            # Prepare the result rows for downstream
            results.append((
                key["ticker"],
                price,
                current_high,
                current_low,
                pd.Timestamp.utcnow()
            ))

        # 3. Persist the updated state back to RocksDB
        self.state_handle.update((current_high, current_low))

        # 4. Yield the results to the output stream
        yield pd.DataFrame(results, columns=output_schema.names)
```

```python
# Streaming pipeline
raw_stream = spark.readStream \
    .format("kafka") \
    .option("kafka.bootstrap.servers", "YOUR_KAFKA_BROKERS") \
    .option("subscribe", "stock_ticks") \
    .load()

stock_ticks_stream = raw_stream \
    .select(from_json(col("value").cast("string"), input_schema).
    alias("data")) \
    .select("data.*")

processed_stream = stock_ticks_stream \
    .groupBy("ticker") \
    .transformWithState(
        IntradayStatsProcessor(),
        outputStructType=output_schema,
        outputMode="Update",
        timeMode="None"
    )

query = processed_stream.writeStream \
    .format("delta") \
    .outputMode("Update") \
    .option("checkpointLocation", "/path/to/checkpoints/real_time_stock_
    stats") \
    .trigger(realTime="5 minutes") \
    .table("real_time_stock_analysis")
```

Structured Streaming Best Practices

Below are some best practices:

- **Use DataFrame Instead of Dataset:** Datasets offer compile-time
  type safety but may incur additional serialization overhead, while
  DataFrames are generally simpler and slightly more performant

because Spark can fully optimize untyped DataFrames using the Catalyst engine. Note that PySpark does not provide a typed Dataset API, so DataFrames are the standard choice in Python.

- **Specify Trigger Intervals:** Control the frequency of streaming data processing.

- **Specify Checkpoint Location:** Always specify a checkpoint location to ensure fault-tolerant and the ability to recover from failures.

- **Use Update Mode for Aggregation:** Efficiently update aggregates instead of recalculating.

- **Leverage Watermark for Event-Time Processing:** Handle late-arriving data while limiting the amount of state Spark needs to maintain, preventing unbounded growth.

- **Monitor and Adjust Resources:** For micro-batch streaming, use the Structured Streaming UI in Spark UI for detailed monitoring and troubleshooting. The UI provides real-time statistics, allowing you to quickly identify unusual behavior—such as spikes in processing rates—and take immediate action to determine whether it's due to a system issue, a surge in legitimate traffic, or an external event like trending news. Figure 12-10 illustrates this interface.

Streaming Query Statistics

Running batches for **26 minutes 56 seconds** since **2020/05/21 22:44:28** (**505** completed batches)

Name: myQuery
Id: 9d0dd219-995d-423e-b97d-36931110bb27
RunId: 04e101e6-8bf8-4165-9099-9104c951ba3b

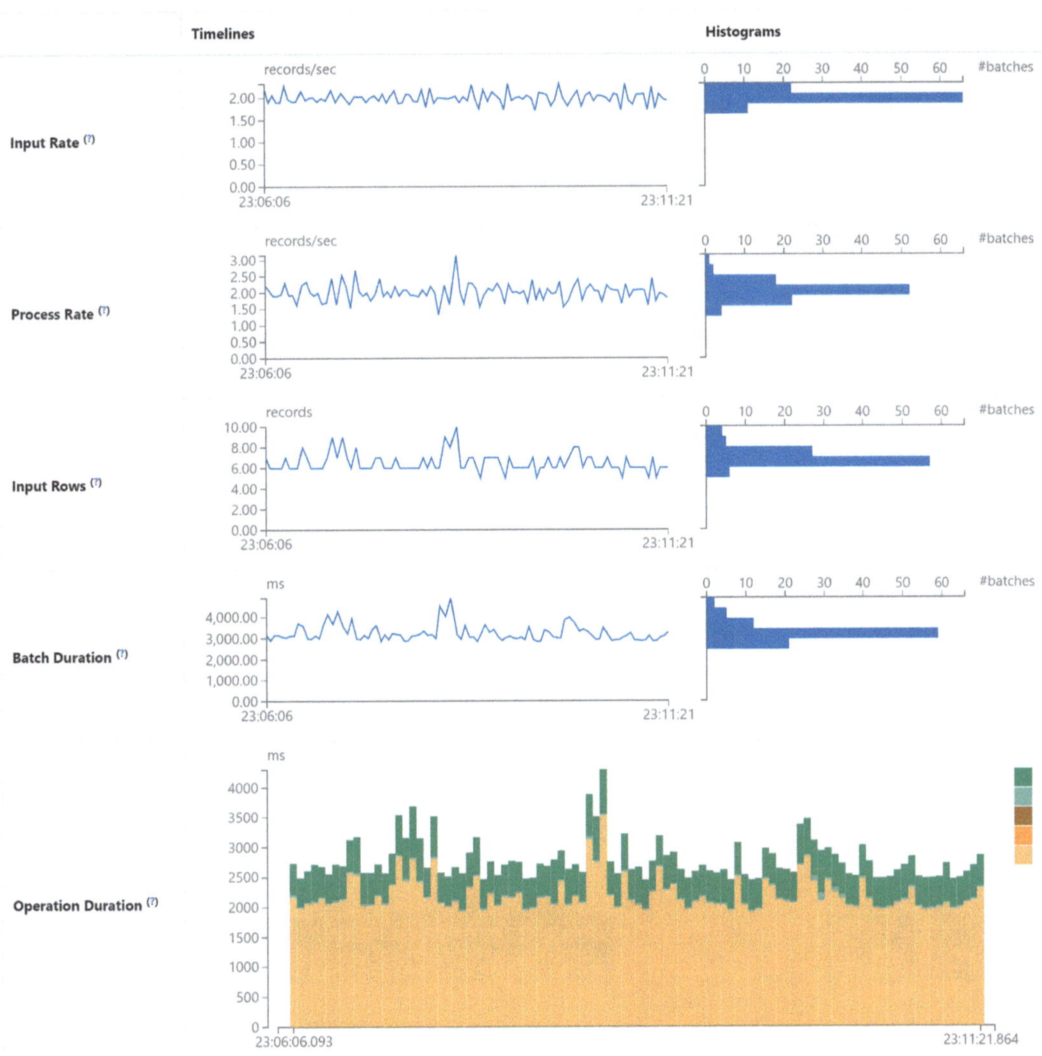

Figure 12-10. *Structured streaming monitor UI*

Real-Time Machine Learning

In the last edition of the book, we discussed machine learning operations in depth. If you are not familiar with the machine learning lifecycle, you are encouraged to explore it.

With the emergence of real-time mode and Databricks' acquisition of Tecton, a real-time feature store, we finally have a way to do real-time machine learning end-to-end within Databricks. The way we are managing real-time machine learning is very similar to traditional machine learning, also known as batch mode. There are a few critical infrastructure components that we can leverage to enable real-time inferencing with Databricks. Figure 12-11 shows this lifecycle.

- **Online Stores:** A specialized serving structure that enables real-time feature lookup and serving backed by a high-speed transactional database: Lakebase

- **Streaming Table:** Streaming tables are append-only Delta tables that support continuous ingestion of streaming data

- **Specialized Libraries:** Tecton is both a feature engineering client and a feature serving client, enabling low-latency feature retrieval and serving from online stores

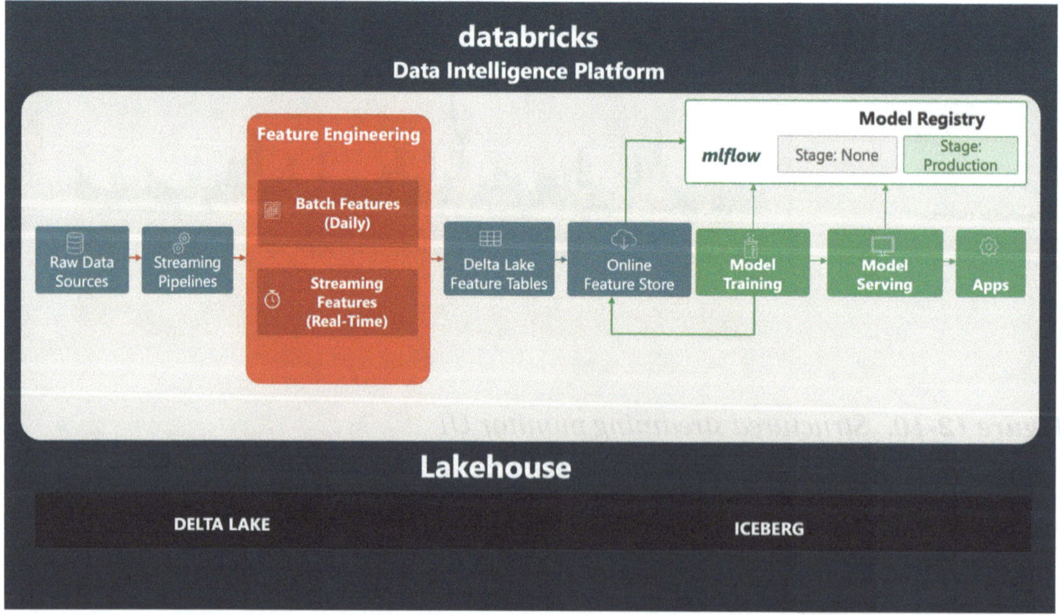

Figure 12-11. *Architecture diagram of real-time machine learning*

Online Store

The term "online" means high availability and low latency, often in the millisecond scale. Before Lakebase, the only option for the online store in Databricks was a third-party service, such as Amazon's DynamoDB or Microsoft's CosmosDB. Now with Lakebase, Databricks offers first-party solutions not only for application development but also for feature stores. Figure 12-12 shows that we can sync the offline features to an online feature store backed by Lakebase while remaining governed by Unity Catalog.

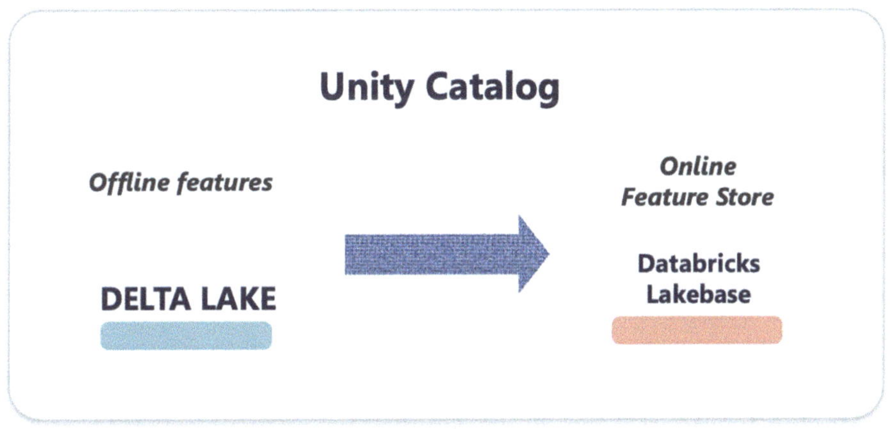

Figure 12-12. *Databricks online feature store*

Specialized Libraries: Tecton

Tecton's on-demand feature views (ODFVs) allow us to create our own real-time feature pipelines. These real-time features can be easily defined as a standard Python transformation. Instead of only transforming raw data sources, ODFVs can also transform request-time feature sources and pre-computed feature tables, enabling flexible, real-time feature pipelines. Using Batch/Stream Feature Views and ODFVs, the entire real-time feature pipeline can be defined in Python within Tecton's declarative feature engineering framework. Consider our stock trading feature store in Figure 12-13.

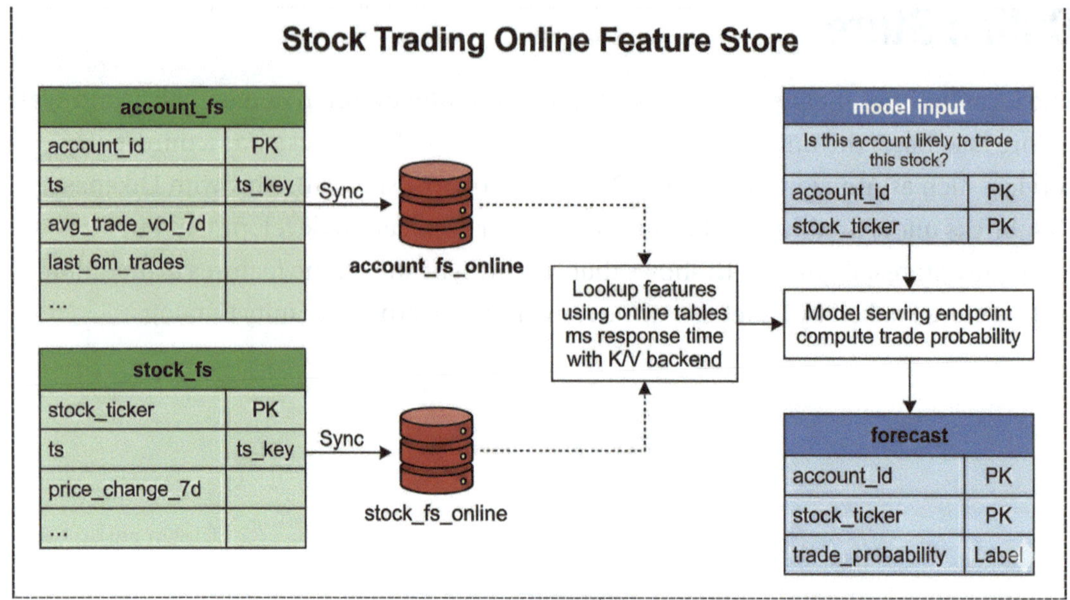

Figure 12-13. *Stock trading online feature store*

At a high level, we can define the following batch features:

- **Moving Averages**: 5-day, 20-day Simple Moving Average (SMA)

- **Volatility**: 20-day rolling standard deviation

- **Volume Metrics**: Average daily volume, volume ratio

- **Returns**: Daily returns, cumulative returns

We can then utilize on-demand feature views to derive real-time stock signals. Listing 12-4 is some simplified code. The full code will be available on this book's GitHub repo.

Listing 12-4. Code skeleton for real-time stock features

```
@on_demand_feature_view(
sources=[stock_features], # Pre-computed features
schema=[...])

def realtime_stock_signals(stock_features, request_data):
"""Compute signals at prediction time."""
```

```
price_vs_sma = request_data["current_price"] / stock_features["sma_20"]
# Intraday volatility from request
intraday_range = (request_data["intraday_high"] - request_data["intraday_
low"]) / request_data["intraday_low"]

# Real-time momentum signal
momentum = request_data["current_price"] - stock_features["prev_close"]
return {...}
```

Conclusion

More than 14 million Structured Streaming jobs run weekly on Databricks (source: https://www.databricks.com/blog/performance-improvements-stateful-pipelines-apache-spark-structured-streaming), and that number is growing at a rate of more than two times year over year (see Figure 12-14).

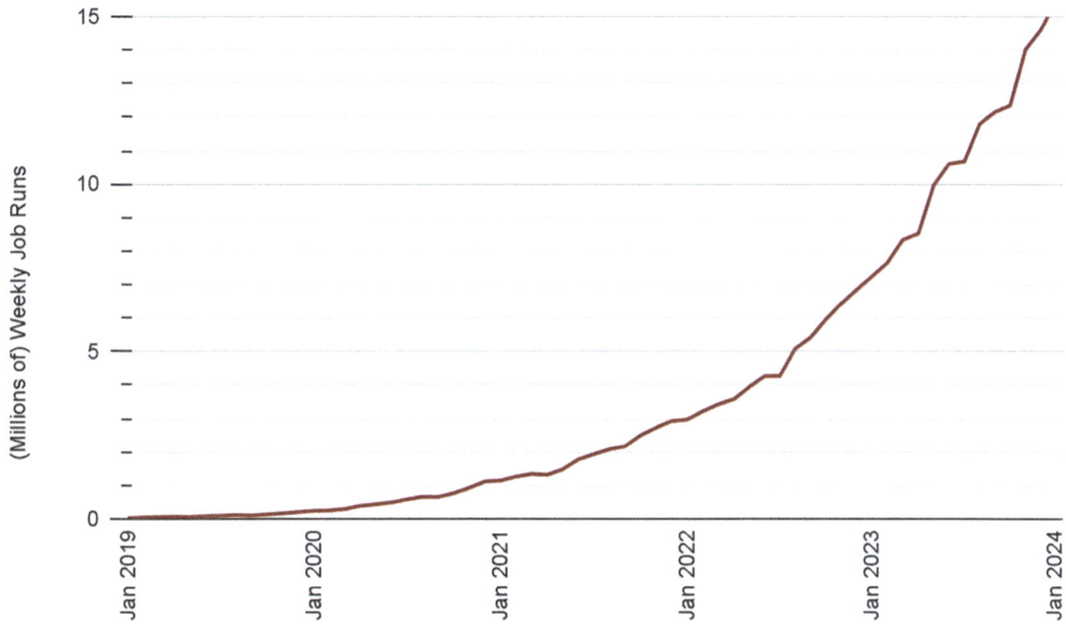

Figure 12-14. *Streaming job runs on Databricks since 2019*

Streaming in Spark has evolved from simple micro-batch processing to a sophisticated, low-latency architecture capable of handling real-time workloads at scale. Structured Streaming, combined with innovations like continuous processing, advanced state management, and Lakeflow declarative pipelines, delivers reliability, flexibility, and performance for modern data pipelines.

The Auto Loader, particularly when combined with Lakeflow Declarative Pipelines, provides groundbreaking resilience support when processing source files from cloud storage. This combination also provides cost savings, comprehensive monitoring, and lineage support.

Looking ahead, the integration of real-time machine learning and feature stores like Tecton positions Databricks as a leader in enabling end-to-end real-time analytics and AI. By adopting these best practices and leveraging the latest enhancements, teams can build streaming systems that are not only fast and resilient but also future-ready.

## CHAPTER 13

Lakeflow Connect: Data Ingestion for the Lakehouse

Organizations have a wealth of information siloed in various data sources. It could be relational databases, on-premises data warehouses, big data storage systems like Hadoop, ERP/CRM systems, or real-time streaming sources. Many analytics use cases require not only efficient processing of this data but also a unified approach to produce meaningful reports and predictions. To start this journey, organizations need to ingest data from different sources into a single location. In this chapter, we will look at how to ingest data from various sources incrementally and efficiently, using Lakeflow Connect and other techniques, along with failover, into your Delta Lake.

In a Databricks Lakehouse, organizations can ingest data from a variety of sources to create a "single source of truth," enabling comprehensive analytics and data science across all their data. To break down the ingestion process, especially for batch data, it can be a single-step or two-step process.

In the single-step process, source data is ingested directly into the target Delta tables. This approach simplifies data ingestion by reducing the number of steps. Lakeflow Connect supports this direct-to-Delta ingestion method for SaaS connectors (see Figure 13-1). In contrast, database connectors in Lakeflow Connect use a gateway plus a landing storage area (a Unity Catalog volume) before applying changes to the Delta destination.

J. Yip et al., *Databricks Data Intelligence Platform*, https://doi.org/10.1007/979-8-8688-2524-8_13

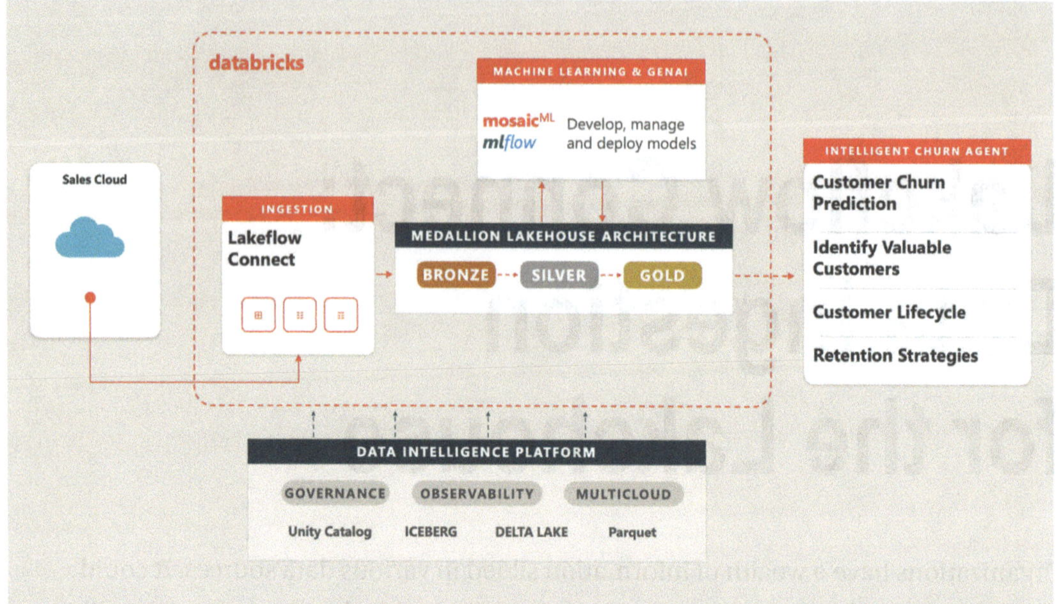

Figure 13-1. *Databricks reference architecture: ingestion*

In a more general two-step process, the first step is to upload (replicate) raw data from a variety of sources, be it on-prem or other systems, into your cloud storage (S3, ADLS, or Google Cloud Storage). This is normally referred to as *cloud ingestion*. Once it lands in your cloud storage, the second step is to move it into your Delta Lake layer. This is referred to as *delta ingestion*. Now, for Delta ingestion, there are two popular and efficient techniques: the Auto Loader and the `COPY INTO` (`legacy`) command. Later in the chapter, we will discuss both in detail.

We will discuss Delta Lake at length in Chapter 15, but we'll touch on it here. Databricks' integration with Delta Lake ensures reliability and performance at scale, providing ACID transactions and a unified process for batch and streaming data. This unification of data not only simplifies data management but also empowers organizations to derive more valuable insights, make data-driven decisions, and, ultimately, drive business growth.

Now, let's move in and learn the various methods used for both cloud and Delta ingestion.

Cloud Ingestion

As a first step, we need to move data to the cloud and, more specifically, to your cloud data storage. Usually, we call this layer the *landing zone,* where data from various sources is stored in any format, such as CSV, Parquet, or JSON. This layer is a source for Delta ingestion into the Delta bronze layer.

There are several alternatives for bringing data to the cloud. The first method is via the built-in Databricks Lakeflow connectors, which ingest data from sources such as Workday, MySQL, and Salesforce. Moreover, the Databricks UI provides an intuitive way to move the data directly to Delta Lake. Next are native cloud tools like Azure Data Factory for Azure Cloud. Finally, ingestion can occur through third-party tools such as Fivetran via Partner Connect (in the Marketplace).

Next, we will investigate these options in much more detail:

- **Databricks Lakeflow Connectors, Add Data, and File Upload:** The File Upload UI and Add Data UI (see Figure 13-2) allow you to easily move data for ingestion into Delta tables in Unity Catalog. It enables you to ingest data from a wide range of data sources in a secure manner via notebook templates or drag-and-drop functionality.

- **Add Data UI:** The Add Data UI acts as a central location for all your ingestion needs from various data sources into the Databricks Lakehouse.

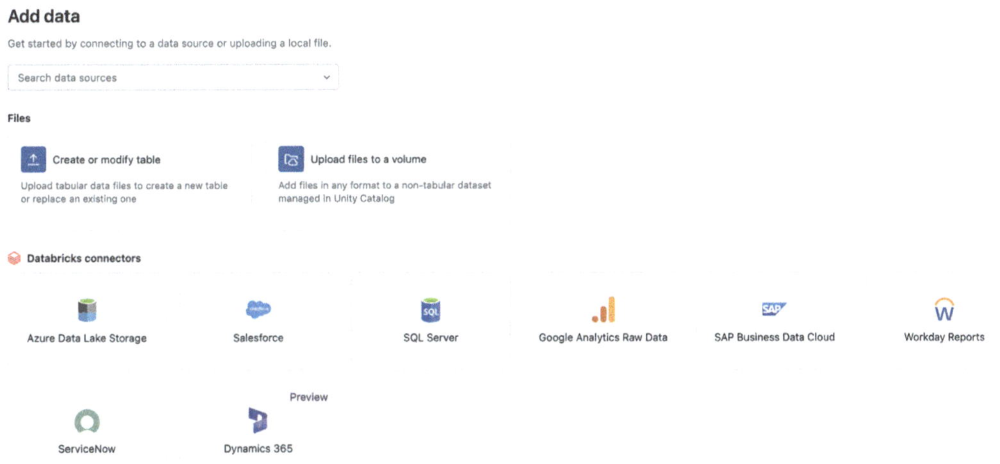

Figure 13-2. The Add Data UI Interface

Now developers can click any data source they want to ingest data from and then follow the UI flow and instructions to complete data ingestion step by step directly into Delta Lake.

Databricks supports several integrations, such as Azure Data Lake Storage or Amazon S3 as the destination. Furthermore, it supports both transactional databases, such as SQL Server, and SaaS sources, such as SAP, Workday, and Google Analytics.

When we click on an icon, a wizard is created to walk us through the steps to ingest data. Figure 13-3 shows the SQL Server wizard.

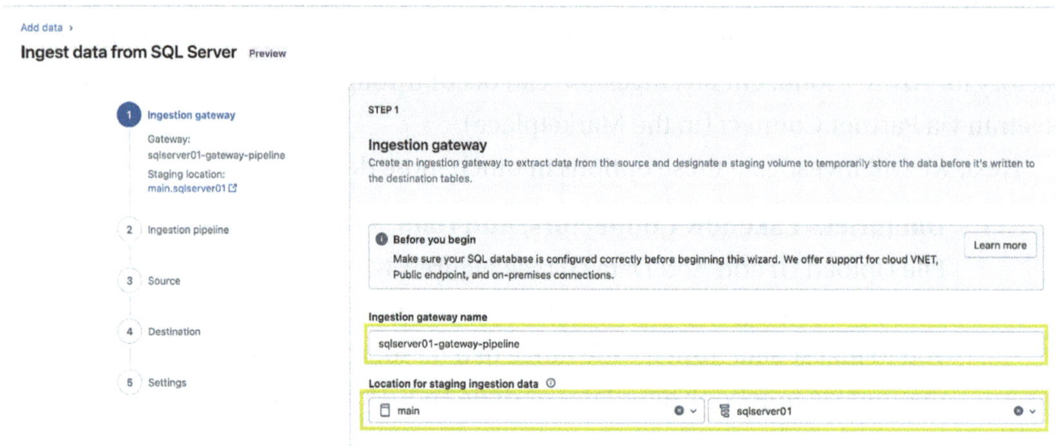

Figure 13-3. *Wizard for SQL Server to Delta table ingestion*

Behind the scenes, the wizard will create a Lakeflow job, and everything is backed by serverless compute—no cluster is required.

Furthermore, you can leverage more than 150 additional connectors in the UI that Fivetran supports.

- **File Upload UI:** The File Upload UI allows you to drag and drop local files seamlessly and enables the secure uploading of these files to a Unity Catalog Volume or the creation of a Delta table. The UI is accessible across all personas through the navigation bar (Figure 13-4) or from the Catalog UI by clicking the + Add icon. The File Upload UI offers the option to create a new table or overwrite an existing table.

Figure 13-4. *Data ingestion, File Upload UI*

You can use the File Upload UI to ingest via the following features:

- Select or drag and drop one or multiple files (CSV, JSON, etc.)

- Preview and configure the resulting table, and then create the Delta table

- Auto-select default settings such as column types

- Modify various formats and table options

Both the Add UI and File Upload UI provide user-friendly interfaces to ingest data, which could be local or in other data storage platforms, into the Databricks Lakehouse platform. Next, we move on to cloud data ingestion using cloud-native tools.

- **Ingestion via Cloud-Native Tools:** Another popular way to ingest data into the cloud is via cloud-native technologies. For example, for batch ingestion, we can use ADF (Azure), Glue Connectors (AWS), or Data Fusion (GCP). For stream ingestion, Event Hubs (Azure), Kinesis (AWS), Google Pub/Sub, or Kafka are popular choices.

- **Ingestion via Third-Party Tools:** The next ingestion method is to leverage the extensive Databricks partner ecosystem, particularly for ingestion partners such as Fivetran, Hevo, and Rivery. To make this a seamless process, Databricks has worked closely with them to not

only validate their technology but also build integrations that enable
you to load data into cloud storage. These integrations enable low-
code, scalable data ingestion from various sources into a Databricks
Lakehouse. These partners are featured in Databricks Partner
Connect (Figure 13-5), which provides a UI interface that simplifies
connecting third-party tools to your Lakehouse for data ingestion.

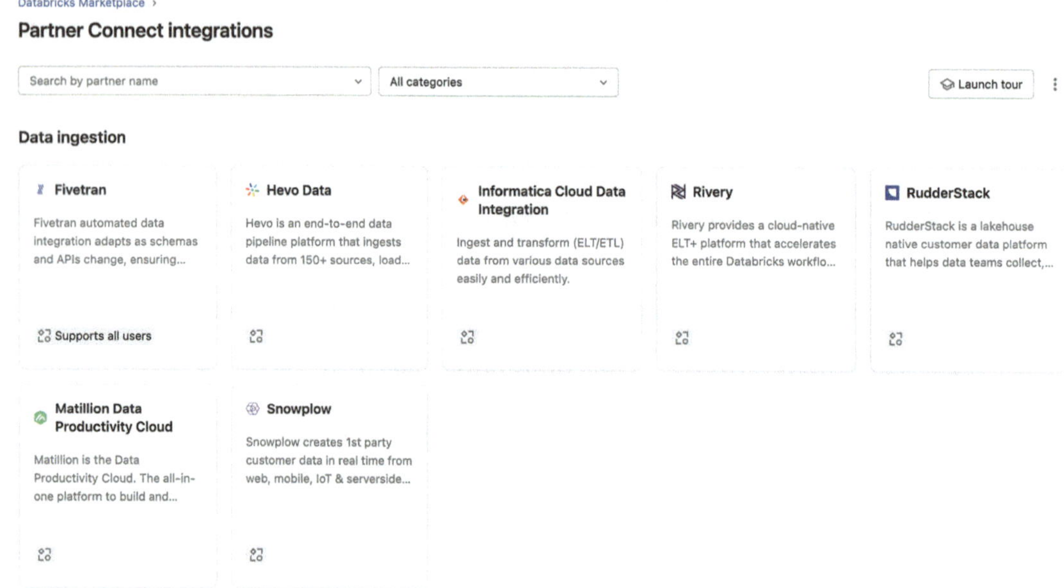

Figure 13-5. *Databricks Partner Connect*

- **Partner Connect** lets you create trial accounts with selected
  Databricks technology partners and connect your Databricks
  workspace to partner solutions in the Databricks UI. With just a few
  clicks, Partner Connect will automatically configure resources such
  as clusters, tokens, and connection files for customers to connect
  with data ingestion, prep, and transformation and BI and ML tools.

  Fivetran is a popular third-party data ingestion partner for
  Databricks that offers simple, no-code connectors to ingest data
  from more than 150 sources (e.g., MySQL, DynamoDB, SFTP)
  into destination data stores such as Databricks Delta Lake.
  Fivetran's ingestion solution helps customers avoid the overhead

of setting up, optimizing, and maintaining manual or open-source connectors, simplifying ongoing management and reducing operational risk. The connector for Fivetran works as follows:

- Set up a Databricks connection with a SQL Warehouse (recommended), an all-purpose compute, or a serverless compute.

- Specify the data source in the connection and the schedule (takes just five minutes).

- Once complete, Fivetran will run SQL commands on the warehouse to load/update Delta tables, which will contain the data from the source as scheduled.

Databricks, with its vast partner ecosystem, allows you to utilize third-party technology to move data from a variety of sources into the Lakehouse.

Files Ingestion

The data has now landed in your cloud storage, or the landing zone. Here, the data could be in any format, such as CSV, JSON, Parquet, etc. For connectors that ingest to cloud storage (rather than directly into Delta tables), the next step is to move that data into Delta Lake (the bronze layer) to complete your second-step data ingestion process (Figure 13-6).

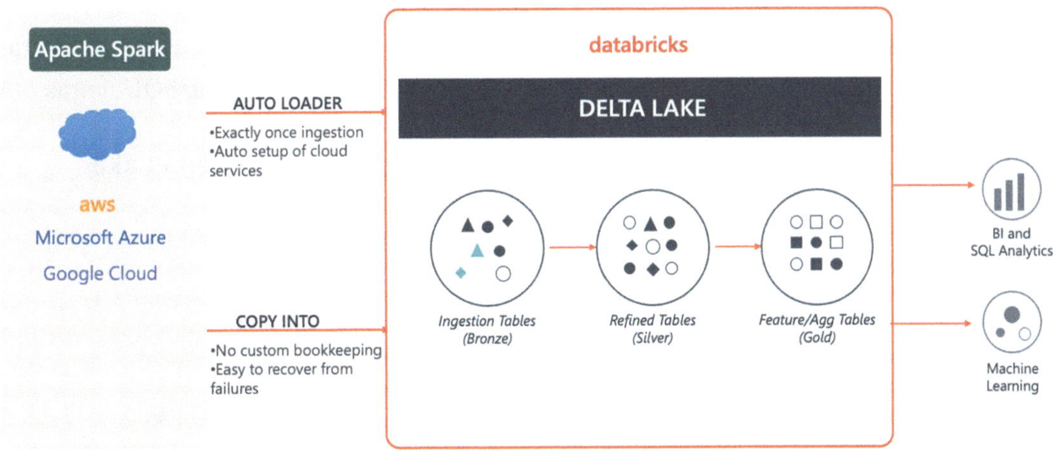

Figure 13-6. *Delta ingestion via the Auto Loader and COPY INTO*

This might sound simple, but there are a couple of ways in which things could go wrong. For example, you could accidentally miss some files to process, which leads to missing data, or you could ingest previously ingested files, leading to duplicates, and reverting or deleting those files would be even more complicated. Further, if there is a schema change in the source system, it could lead to failed jobs or even lost or corrupted fields in your data files.

To overcome these challenges, Databricks has developed Auto Loader and COPY INTO. Now, let's investigate both in detail.

Auto Loader

The Auto Loader provides a highly efficient way to incrementally and efficiently process large amounts of data as it arrives in cloud storage. It also guarantees that each data file is processed exactly once. This is important because processing only new files incrementally prevents missing or duplicate data. Incremental processing also helps save time and reduces the cost of data ingestion. This assumes the files have been uploaded in full and without duplicates to the landing zone, though in practice, deduplication may still be needed later.

The Auto Loader is designed for structured, semi-structured, and unstructured data. The Auto Loader can ingest JSON, CSV, XML, Parquet, Avro, ORC, text, and BINARYFILE file formats into Delta Lake.

Under the hood, the Auto Loader provides a structured streaming source called "cloudFiles." Given an input directory path in cloud file storage, the cloudFiles source automatically processes new files as they arrive and can also process all existing unprocessed files in that directory. The Auto Loader can be set up easily using the syntax shown in Listing 13-1. Note that this functionality is also accessible using SQL syntax.

Listing 13-1. Auto Loader setup with streaming ingestion into Delta Lake

```
df = spark.
    readStream.
    format("cloudFiles") \
    .option("cloudFiles.format", "json") \
    .load("<path-to-source-data>") \
    .writeStream \
    .option("maxFilesPerTrigger", "2000") \
```

```
.trigger(availableNow=True) \
.option("mergeSchema", "true") \
.option("cloudFiles.inferColumnTypes", "true") \
.option("checkpointLocation", "<path-to-checkpoint>") \
.start("<path-to-target>")
```

Let's investigate the previous code and discuss a few important parameters. In the first part, we are creating a readStream to read in input JSON files that have landed in the raw folder. In the second part, we do a writeStream and ingest the data into Delta Lake. The following are some noteworthy options in the previous syntax:

- **Checkpoint:** In the event of failures, Checkpoint helps the Auto Loader resume processing from where it left off by using information stored at the checkpoint location while continuing to provide exactly-once guarantees when writing data to Delta Lake. You don't need to maintain or manage any state yourself to achieve fault tolerance or exactly-once semantics.

- **Trigger.AvailableNow:** The Auto Loader can be scheduled to run in Databricks Jobs as a batch job by using Trigger.AvailableNow. The AvailableNow trigger instructs the Auto Loader to process all files that arrived before the query start time. New files that are uploaded after the stream has started are ignored until the next trigger. Let's assume that the incoming data is spiky, and instead of processing continuously, you want to process the data nightly as a batch job. Trigger.AvailableNow allows you to do that without changing your code/architecture. Alternatively, you can use other triggers (e.g., default) to run the process continuously (see Chapter 12 for more details).

- **mergeSchema:** The mergeSchema option tells the Auto Loader to dynamically detect schema evolution, such as new fields added to the data. This prevents users from having to track and handle these changes manually. However, to achieve maximum performance in production, you can still explicitly monitor table schema changes and handle them as needed.

- **manually.inferColumnTypes:** The schema inference has always been expensive and slow at scale, especially with dynamic JSON. The Auto Loader efficiently samples data to infer the schema and stores it under `cloudFiles.schemaLocation` in your bucket.

- **Rescued Data:** The source system often sends data that might be malformed and not fit in the table structure. The Auto Loader automatically adds the `_rescued_data` column, which stores data that does not conform to the expected schema, allowing it to be processed later.

Now we will look under the hood as to how the Auto Loader discovers files. When you begin to scan hundreds of files and millions of rows, it becomes an expensive operation, leading to ingestion challenges and higher storage costs.

Scanning folders with many files to detect new data is expensive, leading to ingestion challenges and higher cloud storage costs. To solve this issue and support an efficient listing, Databricks Auto Loader offers two modes: Directory Listing and File Notification (Figure 13-7).

- **Directory Listing (Default):** This is the default mode in which the Auto Loader identifies new files by periodically listing the contents of the input directory on the cloud storage. This mode allows you to quickly start without any additional permission configurations if you have access to the data on cloud storage. To ensure eventual completeness of data, the Auto Loader automatically triggers a full directory listing after completing a configured number of consecutive incremental listings. Directory Listing mode is suitable for small to medium-sized directories or for moderate volumes of incoming files.

- **File Notification (Recommended):** In this mode, the Auto Loader sets up a managed cloud notification and queue service that subscribes to file events from the input cloud storage directory. This requires additional cloud permissions to set up. File notification is more performant and scalable for very large input directories or for a high volume of files, such as millions/hr.

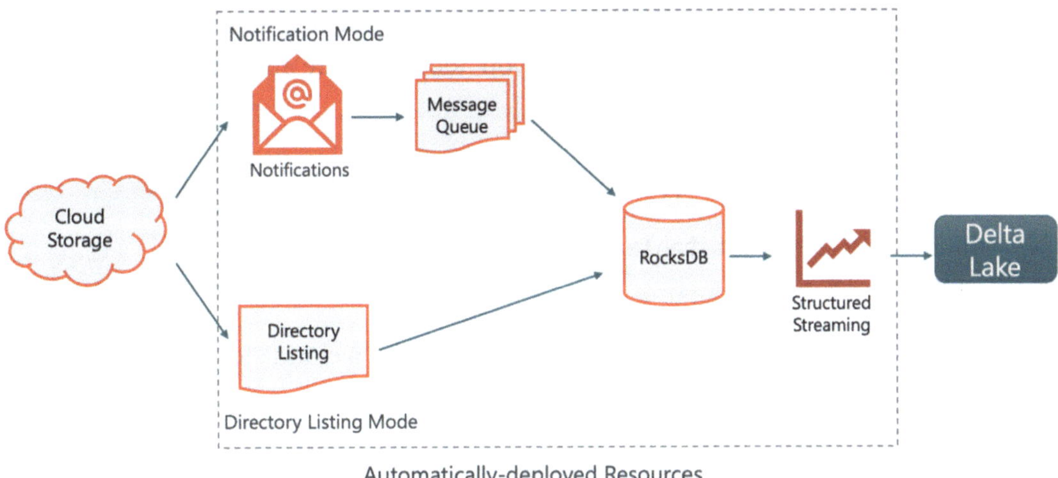

Figure 13-7. *Auto Loader modes: direct listing and file notification*

To conclude, the Auto Loader is a scalable solution that handles the incremental ingestion of billions of files and guarantees exactly-once processing. Further, it comes with features like schema inference and schema evolution and rescues data that would have been otherwise ignored or lost. Next, let's investigate the second option, COPY INTO command.

COPY INTO (Legacy)

COPY INTO is a SQL command that lets you load data from cloud storage into a Delta table. It is considered legacy, and Databricks' recommendation is to use Auto Loader instead when authoring new code. COPY INTO supports many common file formats, including JSON, CSV, Parquet, Avro, and text files. COPY INTO is idempotent by default. It maintains an internal state (often using a key-value store like RocksDB) to record and track the metadata of processed files, so files are processed only once. This saves time and reduces costs, as your ETL pipeline processes each file only once rather than performing a full load each time. Now, the COPY INTO command is perfect for scheduled or ad hoc ingestion use cases in which the data source has a small number of files, typically in the thousands. It is recommended that, for a larger number of files, the Auto Loader be used. COPY INTO supports target schema evolution, merging, mapping, and inference.

COPY INTO requires a target table to exist, as it ingests the data into a target Delta table. If the ingestion is run for the first time, you need to create an empty Delta table beforehand, as shown in Listing 13-2.

Listing 13-2. Creating an empty Delta table for COPY INTO ingestion

```
DROP TABLE IF EXISTS test_table;
CREATE TABLE test_table;
```

Once the table is created, you can ingest the data from a cloud storage location to the Delta table using the COPY INTO command (Listing 13-3).

Listing 13-3. COPY INTO command with VALIDATE mode for data validation

```
COPY INTO test_table
FROM 's3://my-bucket/exampleData'
FILEFORMAT = CSV
VALIDATE
FORMAT_OPTIONS ('header' = 'true', 'inferSchema' = 'true', 'mergeSchema'
= 'true')
COPY_OPTIONS ('mergeSchema' = 'true')
```

VALIDATE: The COPY INTO validate mode (available in runtime 10.3 and above) lets you preview and validate your source data before writing or ingesting files. Some of the validations check whether the schema matches the target table, whether it needs to change, whether all null constraints are met, and whether the data can be parsed. The validation mode produces a sample table you can view.

If you find inconsistencies, such as mismatched column names or formatting issues, you can go back and fix them in the code.

Once you are satisfied with the preview table, you can remove the VALIDATE keyword and rerun the COPY INTO command as shown in Listing 13-4.

Listing 13-4. COPY INTO command for actual data ingestion

```
COPY INTO test_table
FROM 's3://my-bucket/exampleData'
FILEFORMAT = CSV
```

```
FORMAT_OPTIONS ('header' = 'true', 'inferSchema' = 'true', 'mergeSchema'
= 'true')
COPY_OPTIONS ('mergeSchema' = 'true')
```

To conclude, the COPY INTO SQL command lets you load data from a file location into a Delta table. This is a retriable and idempotent operation. Files in the source location that have already been loaded are skipped. It's worth noting that with the file-level tracking, modified files with the same name will also be skipped. It is generally recommended to use these ingestion tools for immutable files only.

Beyond Ingestion

Lakeflow Connect is the latest offering from Databricks. At first glance, it is no more than another source-to-destination wizard, but there's much more to it than that.

Change Data Capture (CDC)

Data teams often need to develop pipelines to keep the ingested data in sync with the source. It will involve determining the primary key and upsert timestamp on each table; the effort can be highly inefficient. One major differentiation of Lakeflow Connect is its support for Change Data Capture (CDC) out of the box. In other words, all changes made to the source will be automatically refreshed in the destination delta table.

However, there are two categories of the CDC capability. For transactional databases like SQL Server, manual setup is required because Lakeflow directly pulls from the transaction log, and such administrative tasks should not be automated for security reasons. Figure 13-8 shows the architecture of Lakeflow Connect for transactional databases. For an end-to-end tutorial, please refer to the following Databricks blog post:

```
https://www.databricks.com/blog/lakeflow-connect-efficient-and-easy-
data-ingestion-using-sql-server-connector
```

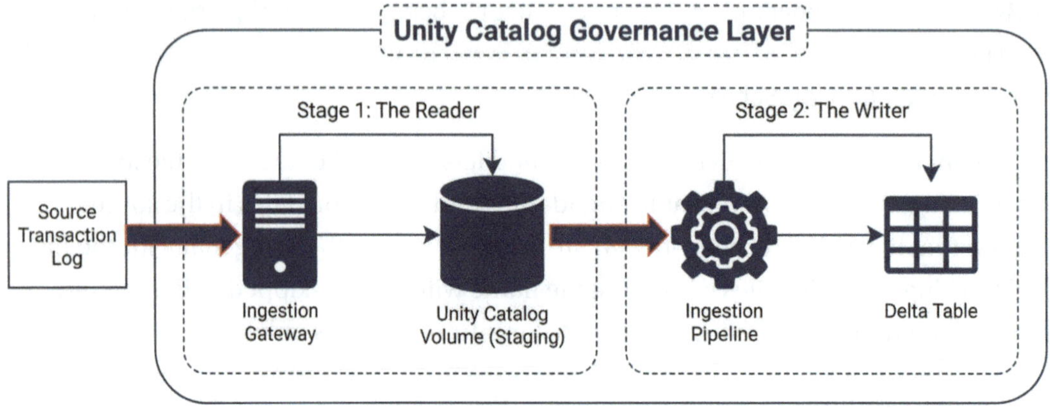

Figure 13-8. *Architecture of Lakeflow Connect for transactional databases*

For Software as a Service (SaaS) like Salesforce, it gets easier because Databricks manages the snapshot by leveraging the APIs provided by the source. The process is the same as what we discussed in the beginning, but it's fully managed by Lakeflow. Figure 13-9 is an example workflow for Salesforce data ingestion.

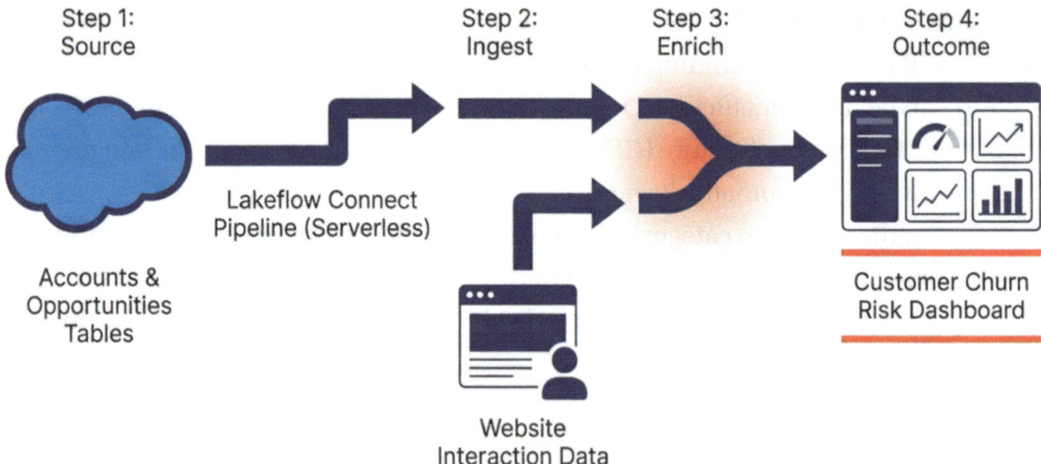

Figure 13-9. *Data ingestion for SaaS sources*

Automated Schema Evolution

Automated schema evolution is an extension of the CDC capability. Database schemas can change over time. And when that happens, the old pipeline will either break, or if we only limit the specific fields while pulling, we'll miss out on the additional details.

Lakeflow Connect does not just move data; it adapts to changes in the source structure without breaking the pipeline. Figure 13-10 illustrates this evolution process.

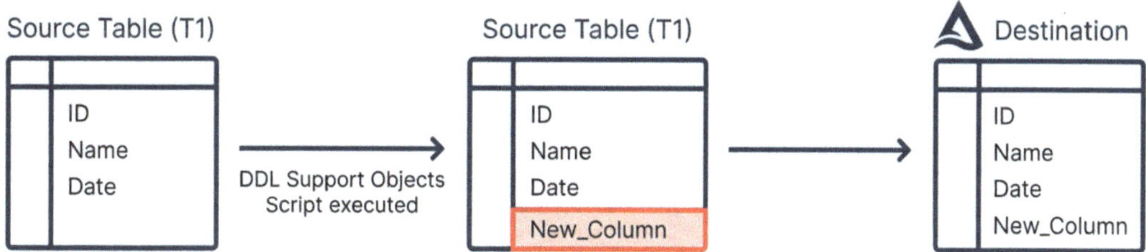

Figure 13-10. Handling schema evolution automatically

Unified Governance with Unity Catalog

As shown in Figure 13-8, all Lakeflow Connect pipelines are integrated directly into Databricks' Unity Catalog, providing security and lineage that standalone connectors often lack.

- **Managed Connections:** Connections to sources (like Salesforce or SQL Server) are stored as securable objects within Unity Catalog, allowing for centralized access control and auditing.

- **Lineage and Discovery:** Ingested tables (bronze layer) are immediately visible in the Catalog Explorer, complete with lineage data that tracks the flow from the source system to the final Delta table.

Zerobus: Direct to Delta streaming

The latest member of the Lakeflow Connect family is Zerobus. It is built for near-real-time use cases such as IoT or telemetry that require high volume but are not mission-critical. While Zerobus simplifies the ingestion process by providing an always-on server, it serves a single purpose, which is to ingest data into a Delta table. Zerobus is not a replacement for streaming but rather a niche product that fits a specific purpose. Figure 13-11 shows the differences between Zerobus and streaming. Caution comes from the fact that Zerobus is not a message bus—it does not retain events or provide backfill. If a producer or network failure occurs before an event is acknowledged, and no retries are performed, the event can be lost. Zerobus does not automatically replay missed messages, so delivery guarantees must be handled by the producer or downstream systems.

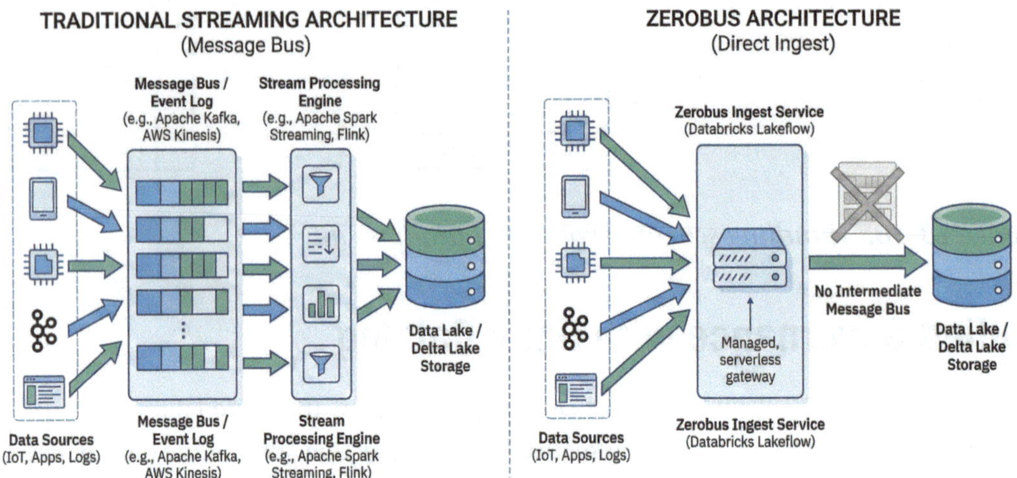

Figure 13-11. *Comparison between streaming and Zerobus*

The Zerobus service is a multi-tenant, regional service managed by Databricks. Clients do not connect to their specific workspace driver; rather, they connect to a regional Zerobus endpoint, which routes traffic to the appropriate storage containers based on the workspace ID.

The endpoint format varies by cloud provider:

- AWS: `<workspace-id>.zerobus.<region>.cloud.databricks.com`

- Azure: `<workspace-id>.zerobus.<region>.azuredatabricks.net`

It is crucial to note that we must manually construct the Zerobus endpoint with the workspace ID and region of the Databricks subscription. If you are not sure about the region of your subscription, please refer to Listing 13-5 for the usage of the `nslookup` command.

Listing 13-5. nslookup command

```
nslookup adb-xxxxxxxxxx.13.azuredatabricks.net
```

You will get an output similar to Listing 13-6.

Listing 13-6. Partial output from nslookup

```
Azure: canonical name = eastus-c3.azuredatabricks.net.
AWS: public-ingress-xxxxxx.elb.us-west-2.amazonaws.com
```

The last limitation is that we must create a target table ourselves; otherwise, Zerobus will throw an error. Once we have everything ready, we can write directly into the target delta table as seen in Listing 13-7.

Listing 13-7. Sample Zerobus ingestion

```
sdk = ZerobusSdk(server_endpoint, workspace_url)
table_properties = TableProperties(table_name)
options = StreamConfigurationOptions(record_type=RecordType.JSON)

stream = sdk.create_stream(client_id, client_secret, table_properties,
options)

try:
    for i in range(100):
        record = {"device_name": f"sensor-{i}", "temp": 22, "humidity": 55}
        ack = stream.ingest_record(record)
        ack.wait_for_ack()
finally:
    stream.close()
```

Conclusion

In this chapter, we learned the methods and best practices for ingesting data into a Databricks Lakehouse.

Data ingestion is defined as a two-step process:

- **Cloud Ingestion:** Moving data from sources (SaaS, on-prem DBs) to a cloud landing zone using Databricks connectors, cloud-native tools (ADF, Glue), or Partner Connect (Fivetran).

- **Delta Ingestion:** Moving data from the landing zone to Delta Lake (Bronze layer) using "exactly-once" processing methods.

- **Direct-to-Delta Ingestion:** SaaS connectors in Lakeflow Connect write directly into destination Delta tables without a landing zone, while still preserving transactional consistency.

Ingesting data into a Databricks Lakehouse can be a single-stage (direct-to-delta ingestion) or a two-stage workflow (cloud ingestion followed by delta ingestion). In the single-stage process, the source data is directly ingested into delta tables. In the two-stage process, the data first lands in the cloud storage ("Cloud Ingestion") and then is processed into Delta Lake ("Delta Ingestion").

While cloud ingestion can be achieved through various Databricks-native and third-party tools, the preferred approach for moving data into Delta Lake is through the optimized methods. **Lakeflow Connect** is a recommended solution, providing scalable ingestion across a wide range of sources, including direct-to-delta ingestion for SaaS sources. When Lakeflow Connect cannot be used, **Auto Loader offers scalable** incremental file ingestions, and Zerobus supports incremental streaming ingestion.

The introduction of *Lakeflow Connect* and *Zerobus* further modernizes this architecture by automating complex tasks like Change Data Capture (CDC) and eliminating the need for complex message bus infrastructure in real-time use cases. By leveraging these tools, data engineers can ensure data reliability, handle schema changes gracefully, and maintain a unified, governed ingestion pipeline.

## CHAPTER 14

Open Data Governance with Unity Catalog

Data is one of an organization's most significant assets. An important determinant of a company's performance and growth is how well its data is handled regarding quality, management, and ownership. Organizations today, especially with ever-expanding use cases for GenAI, face increasingly stringent data privacy regulations. Nonetheless, the reliance on data is increasing as organizations look to help optimize operations and drive business decision-making. Therefore, they are looking for data governance on their data platforms to ensure that not only their data assets but, more importantly, their AI products are consistently developed and maintained and their precise guidelines and standards are adhered to.

In this chapter, we will look at Unity Catalog—Databricks' data governance solution. We will introduce the concept of Unity Catalog and how it differs from traditional Databricks' Hive metastore. Further, we will look at how you can enable Unity Catalog in your workspace and architect your data estate. Finally, we will deep-dive into some key features of Unity Catalog, such as centralized management, data lineage, and Delta Sharing.

What Is Databricks Unity Catalog?

Unity Catalog is Databricks' governance solution, a unified system for managing its data and AI assets (see Figure 14-1). It is a central metadata repository for all data assets, with tools for data governance, model management, access control, auditing, and lineage. Unity Catalog has been open-sourced since 2024.

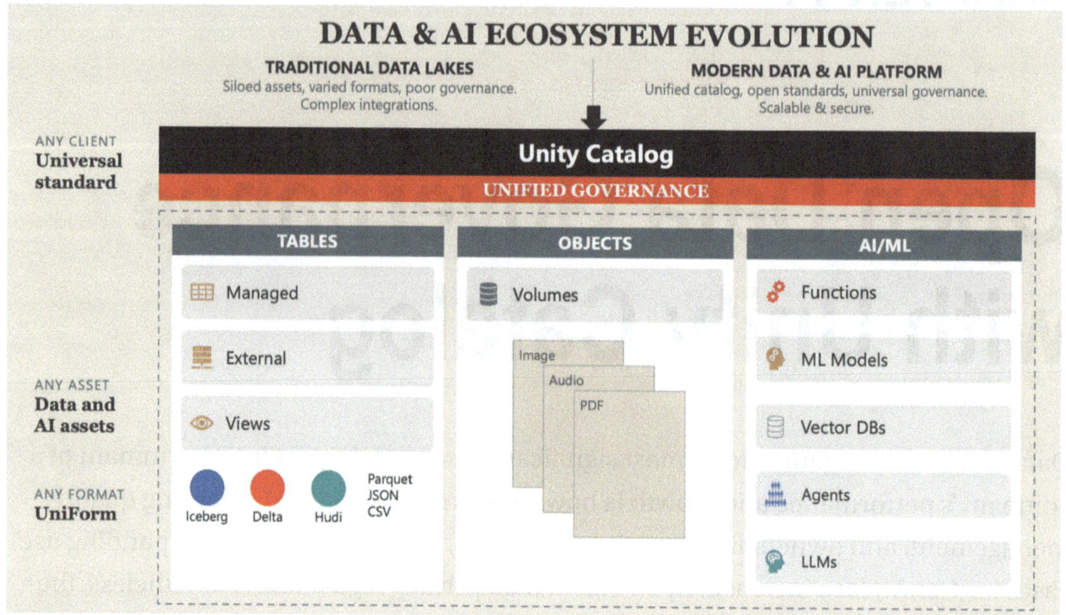

Figure 14-1. *Databricks Unity Catalog*

It maintains an extensive audit log of actions performed on data across all Databricks workspaces in your account. It provides capabilities such as effective data discovery, centralized metadata and user management, data lineage, and much more. It offers views and controls across all structured, semi-structured, and unstructured streaming data, AI models, notebooks, workspaces, files, tables, and dashboards.

In short, it offers fine-grained management of data and AI assets across all Databricks workspaces. This streamlines operations by reducing maintenance overheads, accelerating processes, and increasing efficiency and productivity.

Unity Catalog is the foundation of the Databricks Data Intelligence Platform, which understands the uniqueness of your data. If you are building your next GenAI application, it is essential to enable Unity Catalog in your Databricks environment.

Unity Catalog: Before and After

Before Unity Catalog, Databricks workspaces were separate and independent units (see Figure 14-2). Each workspace had its own metastore, user management (syncing users from an Identity Provider to Databricks), and Table access control list (ACL). Before Unity Catalog, there was no built-in mechanism to make a table in one Databricks workspace available in another workspace. This led to data and governance isolation boundaries between workspaces, and if you wanted to bring consistency between your workspaces, it would mean duplication of effort. Some users handled this by developing pipelines or code to synchronize their metastores and ACLs, while others set up their self-managed external metastores to use across workspaces. However, these solutions added more complexity and maintenance.

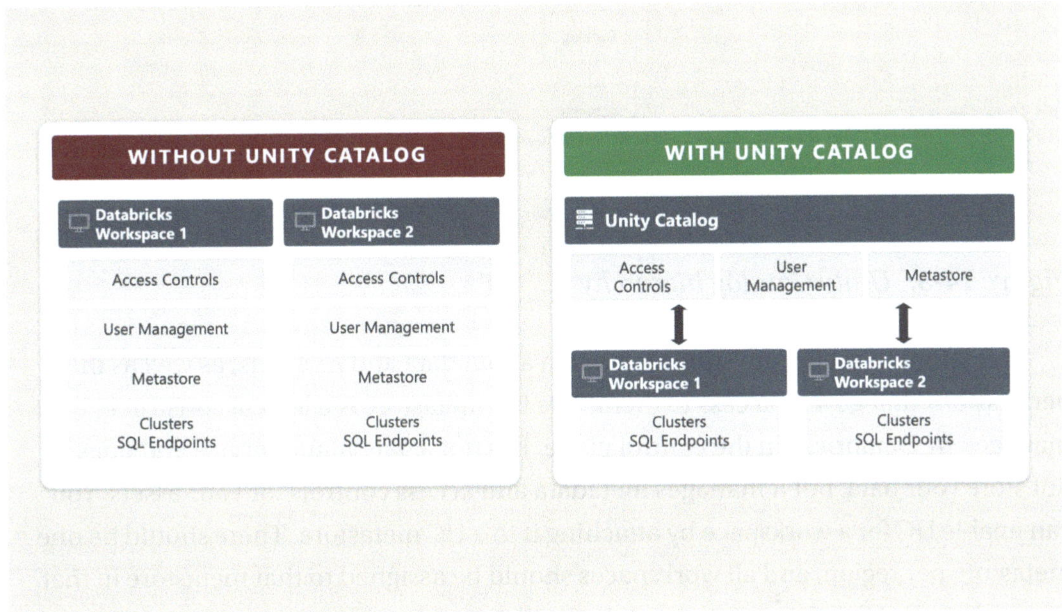

Figure 14-2. *Before and after Unity Catalog*

With Unity Catalog, Databricks has moved all three (User Management, Metastore, and Access Controls) out of workspaces to Databricks *account* that works across all workspaces. As a best practice, there should be a single Databricks account per organization (i.e., your entire company) per cloud provider. A Databricks account lets you set up a Unity Catalog metastore, global controls, and user management in one place and use them across multiple Databricks workspaces. For more information, please refer to the Databricks documentation: `https://docs.databricks.com/aws/en/data-governance/unity-catalog/best-practices`

Unity Catalog Hierarchy

Now let's move on and understand some key concepts with Unity Catalog (UC), such as the metastore, catalog, etc. (see Figure 14-3).

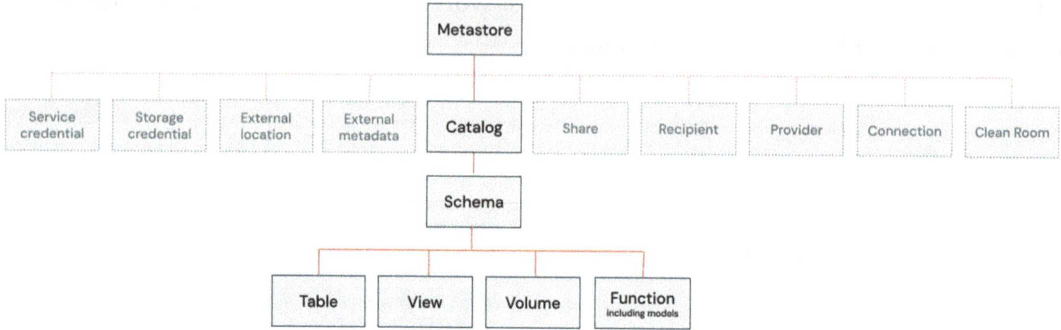

Figure 14-3. *Unity Catalog hierarchy*

Metastore: A metastore stores metadata about data and AI assets, as well as the permissions that govern access to them. The UC metastore is a logical construct managed by Databricks in the control plane. It is not a data plane service and does not store your data, but it manages metadata and access controls for your assets. You can enable UC for a workspace by attaching it to a UC metastore. There should be one metastore per region, and all workspaces should be assigned to that metastore in that region. New Databricks accounts automatically create a metastore.

The metastore has a three-level hierarchy: catalog, schema, and tables.

1. **Catalog:** A catalog serves as the top-level container in the three-level namespace hierarchy. It organizes the data assets and contains schemas (databases), tables, views, volumes, models, and functions.

2. **Schema (Database):** The second logical level in the three-level namespace, containing tables, views, volumes, functions, and models.

3. **Tables:** Tables are defined within a schema and govern tabular data. There are two types of tables: External and Managed.

- **External Tables:** In external tables, data is stored outside the
  UC-managed storage location(s) in a user-defined cloud storage
  location. It is used when direct access to data outside Databricks
  is required or when data needs to be shared across multiple
  systems. Unity Catalog governs access to external tables but does
  not manage the underlying data. This means that when you drop
  a table, it deletes only the metadata, not the underlying data.
  You can use Delta and other file formats (CSV, JSON, etc.) while
  creating an external table.

- **Managed Tables:** The default way to create tables in Unity
  Catalog. They are stored in a UC-managed storage location
  (at the schema, catalog, or metastore-level storage location).
  Unity Catalog manages the data lifecycle as well as file and
  folder layouts for these tables. The underlying data format is
  Delta. When a table is dropped, the underlying data is deleted
  from cloud storage by default within 30 days. This should be
  your default choice, as it simplifies table lifecycle management
  (e.g., no need to specify a table location) and maintenance
  (automatically optimized for best performance).

Volumes: Volumes are defined within a schema and provide governance for
non-tabular data (e.g., image files, etc.). They can store and access files in any format
(unstructured, semi-structured, structured) but cannot store tables. Volumes can be
"managed" by defaulting to the schema's managed storage location or "external" by
specifying an external storage location.

Unity Catalog Admin Roles

It is important to understand various admin roles associated with Unity Catalog.

1. **Account Admin:** Account admins administer and control
   everything at the account level, including SCIM, SSO, metastore
   creation/deletion, assigning metastores to workspaces, and
   creating low-level UC objects such as storage credentials for
   external location access. Account admins do not, by default, have
   access to all data unless granted explicitly.

2. **Metastore Admin:** Metastore admins can create catalogs and assign their ownership (via grants) to groups or individuals. They can also create external locations. Metastore admins have visibility into all securable objects within the metastore that they are an admin of.

3. **Data Owners:** Data owners can perform grants on data objects they own and create new nested objects. For example, a catalog owner can create a schema and then a table within that schema.

4. **Workspace Admin:** Workspace admins are similar to cloud administrators and, as the name suggests, manage the workspaces. They can define cluster policies on workspaces, add/remove user assignments, elevate user permissions within a workspace of various objects like notebooks, etc., and change job ownership.

Getting Started with Unity Catalog

In this section, we will quickly review how to get started with Unity Catalog by creating a metastore and assigning users and groups to workspaces via the account console. Please note that on the new Databricks Accounts, Unity Catalog is enabled by default.

Create a Metastore

The steps to create a metastore are detailed in Databricks documentation (`https://docs.databricks.com/en/data-governance/unity-catalog/create-metastore.html`). As a quick overview, account admins can log into Databricks Account Console and create a metastore (see Figure 14-4).

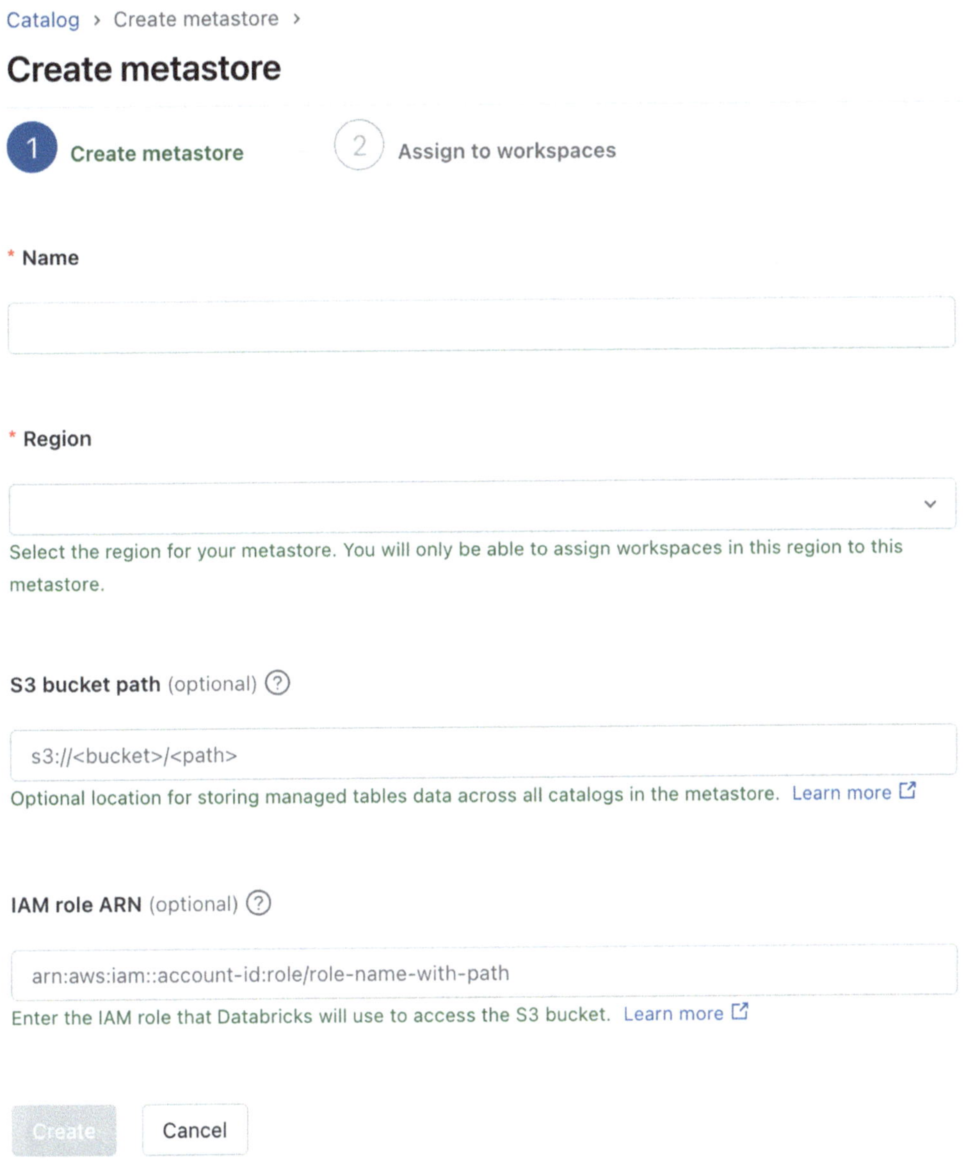

Catalog › Create metastore ›

Create metastore

1 Create metastore   **2 Assign to workspaces**

* **Name**

* **Region**

Select the region for your metastore. You will only be able to assign workspaces in this region to this metastore.

S3 bucket path (optional) ⑦

s3://<bucket>/<path>

Optional location for storing managed tables data across all catalogs in the metastore. Learn more ⌁

IAM role ARN (optional) ⑦

arn:aws:iam::account-id:role/role-name-with-path

Enter the IAM role that Databricks will use to access the S3 bucket. Learn more ⌁

Create   Cancel

Figure 14-4. *Unity Catalog metastore setup interface*

The key inputs required are as follows:

1. Name of the metastore.

2. The region in which the metastore is created. It is important to note that you can only have one metastore per region.

3. ADLS Gen 2 Path or the S3 bucket or GCS, which will be the root bucket for the metastore.

4. Access Connector ID (Azure): An Access connector in Azure allows you to use Managed Identity to access storage containers on behalf of Unity Catalog users.

5. IAM role ARN (AWS): Amazon Resource Name for the bucket that was set up in step 2 (`https://docs.aws.amazon.com/IAM/latest/UserGuide/reference_identifiers.html#identifiers-arns`).

Once the metastore is created, you can assign it to a workspace and thereby enable Unity Catalog (Figure 14-5) for that workspace.

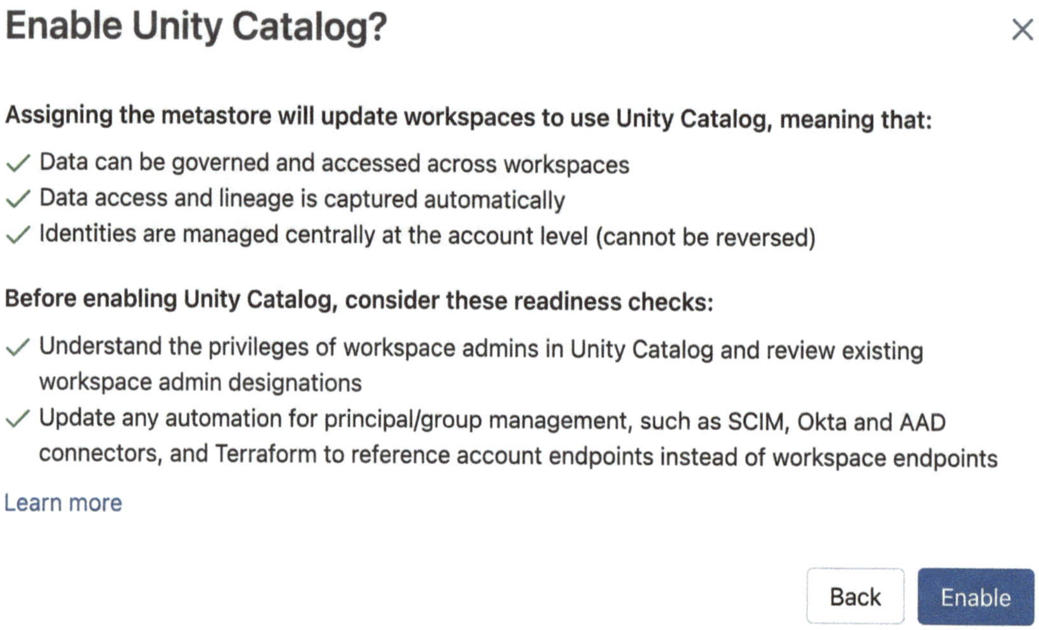

Enable Unity Catalog?   ✕

Assigning the metastore will update workspaces to use Unity Catalog, meaning that:

✓ Data can be governed and accessed across workspaces
✓ Data access and lineage is captured automatically
✓ Identities are managed centrally at the account level (cannot be reversed)

Before enabling Unity Catalog, consider these readiness checks:

✓ Understand the privileges of workspace admins in Unity Catalog and review existing workspace admin designations
✓ Update any automation for principal/group management, such as SCIM, Okta and AAD connectors, and Terraform to reference account endpoints instead of workspace endpoints

Learn more

Back   Enable

Figure 14-5. Final screen before enabling Unity Catalog

Organizing Data in Unity Catalog

As discussed earlier, the catalog is the top-level container in the three-level namespace. As a best practice, you should use catalogs to segregate your organization's data assets. This simply means catalogs typically correspond to a department, team, business unit, or development environment scope (Dev, UAT, Prod), as shown in Figure 14-6.

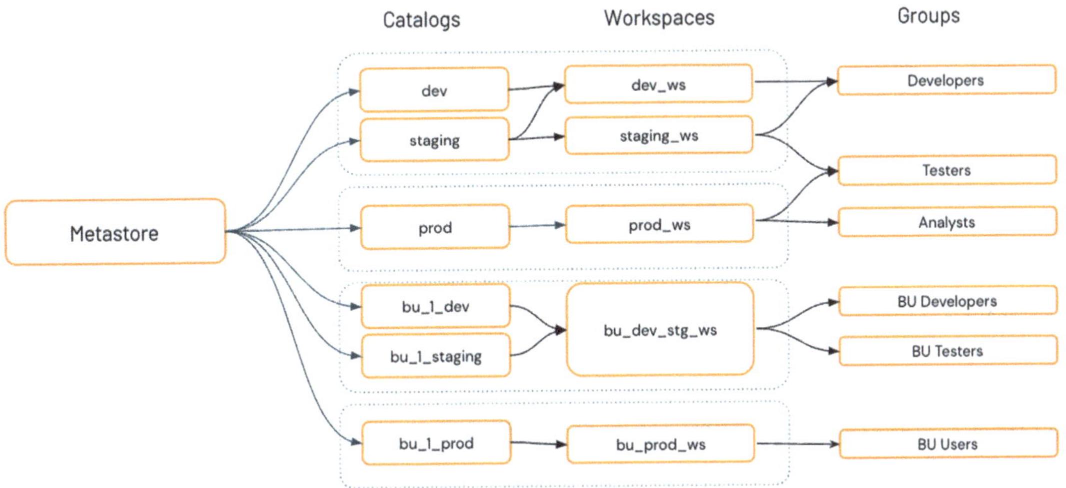

Figure 14-6. *Sample Unity Catalog structure*

Another common pattern is for developers to use workspaces as a data-processing isolation tool—for example, using different workspaces for prod and dev environments, or a dedicated workspace for processing sensitive data.

Therefore, when developers work in a particular workspace, such as the dev workspace or the sensitive-data workspace, they want to see only the catalogs associated with that environment. For example, while working in a dev environment, you want only the dev catalog visible, not prod. Unity Catalog has a feature that allows you to bind a catalog to specific workspaces. This ensures that all specified data processing is handled in the appropriate workspace. These environment-aware ACLs allow you to ensure that only specific catalogs are available within a workspace, regardless of a user's individual ACLs. This means the metastore admin or the catalog owner can define the workspaces where a data catalog can be accessed.

To learn more, please go to the following website:

```
https://docs.databricks.com/en/data-governance/unity-catalog/create-
catalogs.html#optional-assign-a-catalog-to-specific-workspaces
```

Figure 14-7 illustrates the workspace setup architecture.

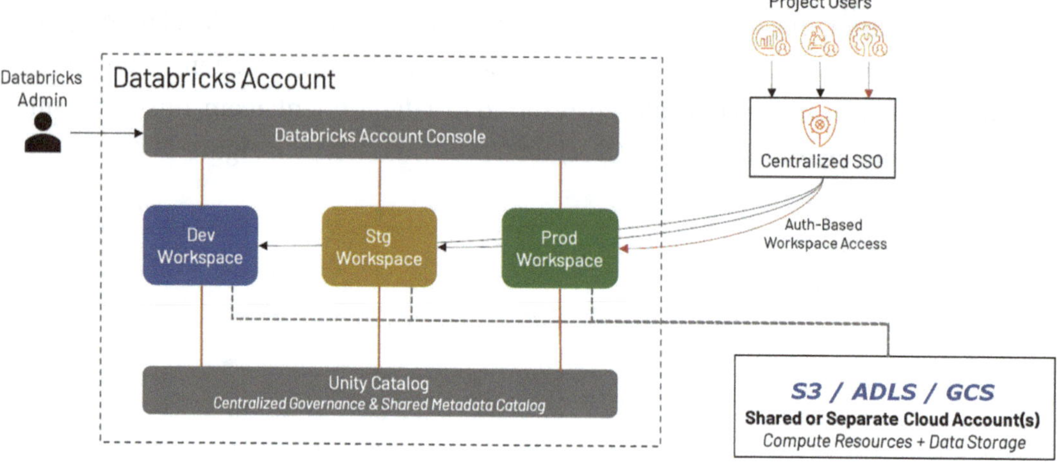

Figure 14-7. *With Unity Catalog, we can attach a catalog to different workspaces*

Key Features of Unity Catalog

Centralized Metadata and User Management

As explained earlier, Unity Catalog provides a single metastore across all workspaces in an account. This enables users to create and access tables, views, etc., across workspaces. Now, you can create multiple catalogs; set up schemas, tables, and views in one place; and access them across workspaces.

You can only create a single metastore per cloud region, but in cross-region metastore scenarios, we can use multiple metastores. It is important to note that when multiple metastores are set up in an organization, the catalogs cannot be attached to the workspaces in other metastores. The solution is to use Delta Sharing, which will be discussed later in this chapter (see Figure 14-8).

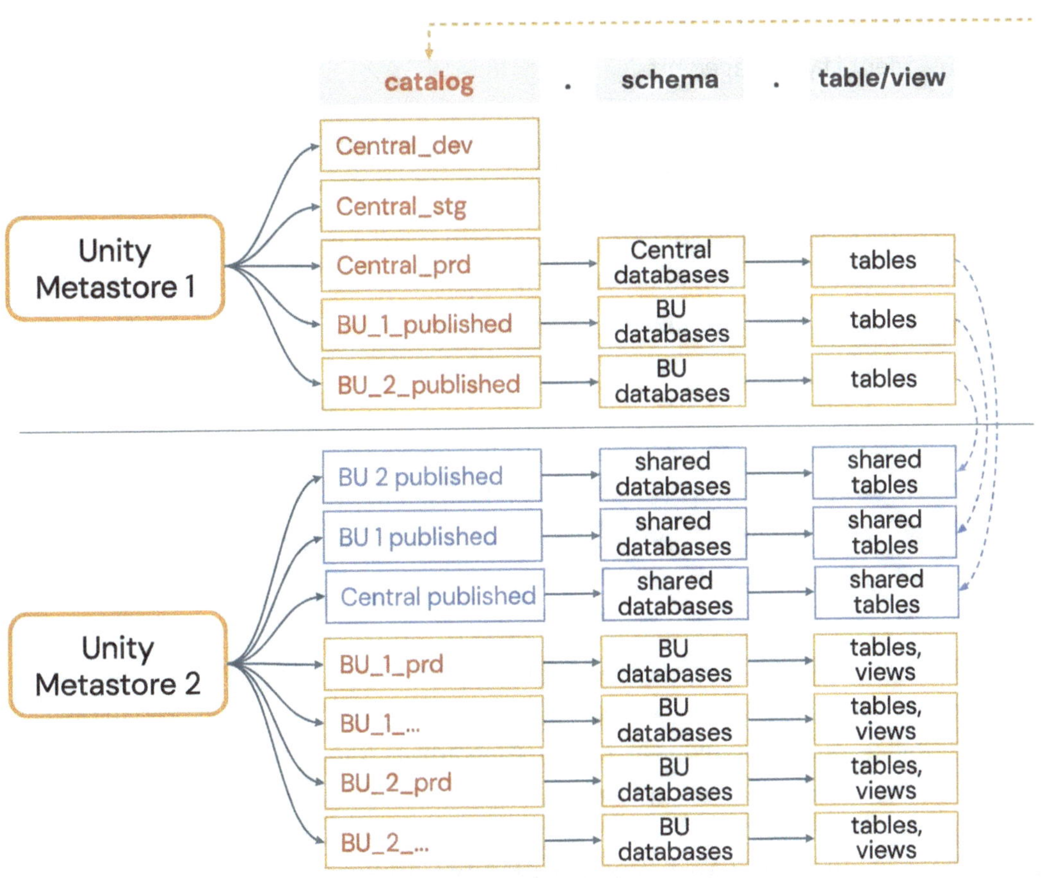

Figure 14-8. *Delta Sharing strategy with multiple metastores*

Another important feature of Unity Catalog is centralized user management. Before Unity Catalog, admins had to add users from an Identity Provider (IdP) to each new workspace, either manually or through SCIM synchronization, and maintain user assignments on a workspace-by-workspace basis. With Unity Catalog, once you have synced users and groups from your identity provider of choice, for example, Microsoft Entra ID (formerly Azure Active Directory), via SCIM to Databricks Account Console, you can assign users/groups to all different workspaces via the account console, hence centrally managing users across workspaces. As a best practice, you should enable SCIM integration at the account level and sync users to workspaces with Identity Federation (only available once UC is enabled). Do not use SCIM at the workspace level at all. This is to avoid user/group conflicts at the account level. For more information, please refer to the Databricks documentation:

```
https://learn.microsoft.com/en-us/azure/databricks/admin/users-groups/
automatic-identity-management
```

Centralized Access Controls

One of the main requirements of any data platform is strict control over data access to safeguard it and ensure compliance with your organization's data protection policies. Unity Catalog provides a centralized management approach for applying data access policies across all relevant workspaces and data assets.

The access control mechanisms use identity federation, where principals can be service principals, individual users, or groups. In addition, SQL-based syntax, the Databricks UI, or even Terraform and APIs can be used to provide and control fine-grained access across a wide range of resources, including schemas, tables, views, clusters, notebooks, and dashboards.

Let's investigate how you can use ANSI SQL to grant permission scopes on *securable objects* like tables or locations to *principals* like groups, users, or service principals. As a best practice, use groups to secure access to UC objects, such as tables, and to own securable objects. If a group owns an object, then any user in that group is an owner. The generic GRANT syntax is shown in Listing 14-1.

Listing 14-1. Generic GRANT syntax for Unity Catalog access control

```
GRANT <privilege> ON <securable_type> <securable_name> TO <principal>
```

For example, to grant SELECT permission on the *iot.events* table to the *engineers* group, you would use the syntax shown in Listing 14-2.

Listing 14-2. Example: Granting SELECT permission to the engineers group

```
GRANT SELECT ON iot.events TO engineers
```

The same functionality is also available via the Databricks UI in an easy-to-use point-and-click manner, which helps with on-the-spot access and auditing, as shown in Figure 14-9.

Grant on unitygo.iot.events   ✕

Principals

Type to add multiple principals

Privileges

APPLY TAG   gives ability to apply tags to an object

MODIFY   gives ability to add, delete, and modify data to or from an object

SELECT   gives read access to an object

ALL PRIVILEGES   gives all privileges ⓘ

MANAGE   gives ownership-like ability for the object, such as managing permissions, dropping, or renaming

Cancel   Confirm

Figure 14-9. *Granting permissions on a UC table*

In addition, Databricks offers the ability to set these ACLs on objects via the Databricks REST API, CLI, SDK, and Terraform, enabling fully automated Unity Catalog deployment and management.

Data Lineage

Data lineage is the process of tracking data at the metadata level (e.g., which columns in table A are used to compute columns in table B) as it flows from its source to its destination. It has gained significance due to the large volume of data processed through complex transformations and serves various purposes, including auditing and debugging. Thus, data lineage has become vital in understanding data movement, tracking, monitoring jobs, debugging failures, and tracing transformation rules.

Unity Catalog has end-to-end data lineage for all workloads, giving visibility into how data flows are consumed. Data lineage is automatically aggregated across all workspaces connected to a Unity Catalog metastore, so lineage captured in one workspace can be seen in any other workspace that shares the same metastore.

Unity Catalog provides users with both table- and column-level lineage in a single lineage graph, giving them a better understanding of what a particular table or column is composed of and where the data comes from. Users can easily follow the data flow across different stages, gaining insight into the relationships between tables and fields. Access permissions in UC are enforced, so users can only see and interact with assets (e.g., catalogs, schemas, tables) for which they have been granted access, including lineage information.

Further, the Unity Catalog tracks lineage for notebooks, workflows, ML models, and dashboards. This improves end-to-end visibility into how data is used in your organization and allows you to understand the impact of any data changes on downstream consumers (see Figure 14-10).

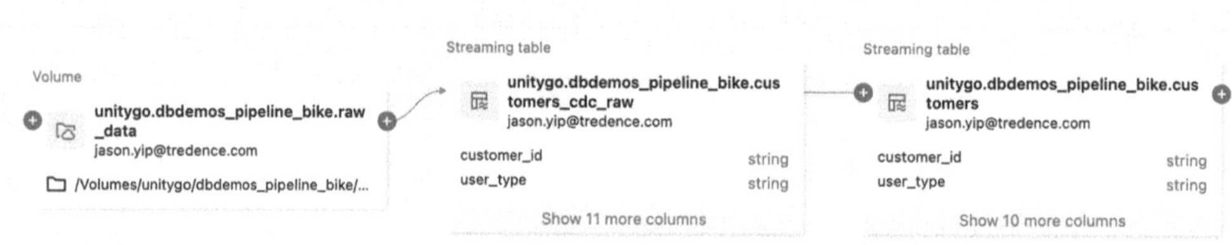

Figure 14-10. *Unity Catalog lineage*

Data lineage holds critical information about the data flow and uses Unity Catalog's common permission model. This means that users with appropriate permissions can view the lineage data flow diagram, thus adding an extra layer of security.

Finally, Unity Catalog also offers rich integration with various data governance partners, such as Collibra or Atlan via Unity Catalog REST APIs, enabling easy export of lineage information to these partner catalogs.

Data Access Auditing

Unity Catalog automatically captures user-level audit logs and records all data access activities. These logs capture various catalog events, such as creating, deleting, and altering multiple components within the metastore, including the metastore itself. Additionally, they cover actions related to storing and retrieving credentials, managing access control lists, handling data-sharing requests, and more.

The built-in system tables let you easily access and query the account's operational data, including audit logs, billable usage details, and lineage information.

Data Search and Discovery

Unity Catalog offers a unified UI across the platform with enhanced search capabilities. Furthermore, it uses a common permission model to ensure security, enabling users to access only the assets they are authorized to use. It allows tagging and documenting data assets, offers a comprehensive search interface, and utilizes lineage metadata to represent relationships within the data.

What's more, Databricks has greatly enhanced the platform's search and discovery capabilities through GenAI.

Row-Level Security and Column-Level Masking

Organizations are continually striving to protect and secure their data, and one important way they are doing so is through row-level security and column-level masking. This feature is now available in the Unity Catalog–enabled Databricks workspaces.

Row Filters

Row filters, used for row-level security (RLS), allow you to apply a filter to a table so that subsequent queries only return rows for which the filter predicate evaluates to true. A row filter is implemented as a SQL user-defined function (UDF). A row filter accepts zero or more input parameters, where each input parameter binds to one column of the corresponding table.

Create a Row Filter

Listing 14-3. Syntax for creating a row filter function

```
CREATE FUNCTION <function_name> (<parameter_name> <parameter_type>, ...)
RETURN {filter clause whose output must be a boolean};
```

Apply the Row Filter to a Table

Listing 14-4. Syntax for applying a row filter to a table with column specification

```
ALTER TABLE <table_name>
SET ROW FILTER <function_name> ON (<column_name>, ...);
```

Let's look at an example. We want to create a function to filter data for the U.S. region. If the function is called by a user in the `admin` group, the condition will be passed, and all the data will be returned. Otherwise, only the rows with `region='US' will be` returned.

Listing 14-5. Creating a region-based row filter function

```
CREATE OR REPLACE FUNCTION main.pii.us_filter(region STRING)
RETURNS BOOLEAN
RETURN IF(IS_ACCOUNT_GROUP_MEMBER('admin'), true, region = 'US');
```

Listing 14-6. Applying the 'US'_filter to the sales table

```
ALTER TABLE main.pii.sales SET ROW FILTER main.pii.us_filter ON (region);
```

Column Masks

Column masks are used for Column-Level Security (CLS) and let you apply a masking function to a table column. The masking function is evaluated at query runtime, substituting each reference to the target column with its result. In most use cases, column masks determine whether to return the original column value or redact it based on the identity of the user who invoked it. Column masks, like row filters, are expressions written as SQL UDFs as shown in Listings 14-7 and 14-8.

Listing 14-7. Syntax for creating a column mask function

```
CREATE FUNCTION <function_name> (<parameter_name> <parameter_type>, ...)
RETURN {expression with the same type as the first parameter};
```

Listing 14-8. Syntax for applying a column mask to a table

```
ALTER TABLE <table_name> ALTER COLUMN <col_name> SET MASK <mask_func_name>
[USING COLUMNS <additional_columns>];
```

In this example, if the user is not part of the `admin` group, the query results will mask the SSN values for non-admin users, as shown in Listing 14-9.

And we can apply the mask like Listing 14-10.

Listing 14-9. Creating an SSN masking function

```
CREATE OR REPLACE FUNCTION main.pii.ssn_mask(ssn STRING) RETURN IF(IS_
ACCOUNT_GROUP_MEMBER('admin'), ssn, '****');
```

Listing 14-10. Applying the ssn_mask to the users table

```
ALTER TABLE main.pii.users ALTER COLUMN table_ssn SET MASK ssn_mask;
```

Dynamic Views vs. Row Filters and Column Masks

Now, an important question is why row-level filters are needed when Databricks already offers dynamic views that let users create abstracted, read-only views of one or more source tables. After all, dynamic views, row filters, and column masks all let you apply complex logic to tables and process their filtering decisions at query runtime.

Let's discuss an important distinction between them. Creating a dynamic view creates a new view name that must not match any source table names. This abstraction layer ensures data integrity and prevents unintentional alterations to the core data. However, it requires a new view object (a separate name from the source tables). As a best practice, use dynamic views if you need to apply transformation logic, such as filters and masks; to read-only tables, you cannot modify the source table, and it is acceptable for users to refer to the dynamic views using different names than the source tables.

On the other hand, row-level filters and column masks apply logic directly to the table itself, and users don't have to deal with new or different table names or aliases. They are applied to tables and can be managed either per table or centrally via ABAC policies to ensure scale and consistency. Use row filters and column masks if you want to filter or compute expressions over specific data but still provide users access to the tables using their original names.

Delta Sharing

Organizations seek to securely exchange data with their customers, suppliers, or partners to unlock further business value. However, a key requirement is that data sharing should happen securely to establish trust in data security and privacy. Some of the most common use cases for data sharing include data monetization with customers, B2B sharing with partners and suppliers, and intra-company data sharing among various departments.

Data sharing is not a new concept, and organizations have traditionally used two main methods to share data. The first is through self-built solutions or tools that use APIs, JDBC/ODBC, or file transfers via SFTP. The second is via commercial software vendors. The problem with the first is scalability and infrastructure maintenance, while the problem with the second is cost and limited flexibility in data access.

An Open Standard for Data Sharing

Databricks Unity Catalog comes with Delta Sharing, an open protocol for securely sharing data internally and across organizations in real time. Delta Sharing is fully integrated with Unity Catalog and allows you to centrally manage and audit the shared data across organizations.

Some of the differentiators or benefits of Delta Sharing include accessing data where it resides without creating any copies or moving to other platforms. Further, you can integrate Delta Sharing with either open-source clients (e.g., Pandas, Spark) or commercial clients (e.g., Power BI, Databricks, and others) that support the protocol.

Delta Sharing is a transformative solution to access and share data. Let's see how Delta Sharing works.

How Delta Sharing Works

There are three main ways to share data with Delta Sharing:

1. **The Databricks-to-Databricks sharing** in which both the provider and the recipient are on Unity Catalog–enabled workspaces. It has some advanced features like notebook sharing, AI model sharing, data governance, auditing, and usage tracking

for both providers and recipients. Most importantly, it does not require the user to handle any tokens or credentials. The platform handles identity verification and secure end-to-end connections.

2. **The Databricks open sharing protocol** allows a provider with a Unity Catalog–enabled workspace to share data with a recipient on any computing platform outside Databricks. Access is handled using either a bearer-token credential file (downloaded via an activation link) or OIDC federation (tokenless for users).

3. **A user-managed implementation of the open-source Delta Sharing server,** which lets you share from any platform to any platform, whether Databricks or not. This is open-sourced with instructions at `https://github.com/delta-io/delta-sharing` (See Figure 14-11.)

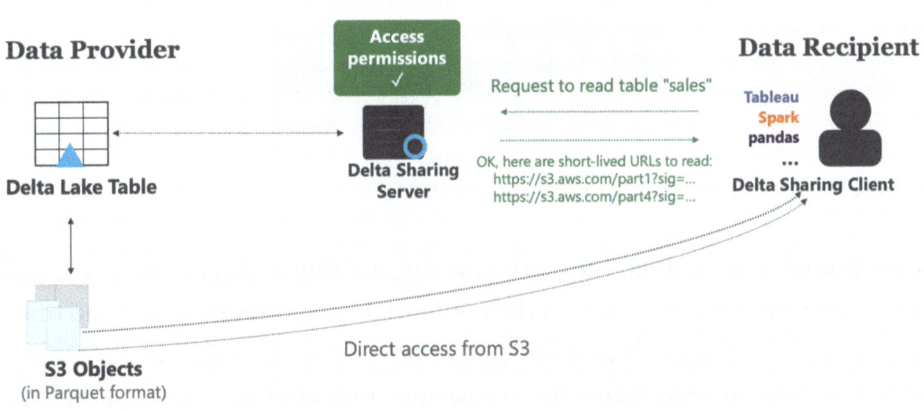

Figure 14-11. *Delta Sharing in action*

Let's examine how to set up Delta Sharing. The first step is for the data provider to register a Delta Lake table with the Delta Sharing server. This is done by creating a share (a read-only collection of data objects, such as tables and views) in a UC-enabled workspace. As shown in Listing 14-11.

Listing 14-11. Creating a share for Delta Sharing

```
CREATE SHARE IF NOT EXISTS test_share
ALTER SHARE test_share
ADD TABLE test_table
```

The next step is to create a recipient, an individual or organization, gaining access to a share. As shown in Listing 14-12.

Listing 14-12. Creating a recipient for Delta Sharing

```
CREATE RECIPIENT IF NOT EXISTS recipient;
```

Once the recipient is created, each recipient gets an activation link that they can use to download their credentials. The Delta server identifies and authorizes the recipient/consumer based on these credentials (see Figure 14-12).

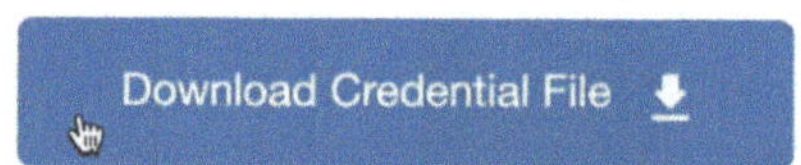

Figure 14-12. *Confirmation of Delta Sharing*

The recipient can download the credential file and then use it to access the tables in the share. One key point to note is that the credential file is a single download only. The recipient can now use the file to authenticate and access data using various methods, such as Pandas, Java, or even Power BI. In case the credential file is lost, you must contact the provider for a new URL.

Many open-source and commercial partners trust Delta Sharing, and Databricks also works with data providers to share data across the ecosystem.

One notable example of Delta Sharing is the Databricks Product called **Clean Rooms**, which is an open and secure marketplace for data and AI assets. It enables organizations to collaborate on data and AI assets without sharing or exposing the underlying raw, PII, or proprietary data. Participants can run analyses, combine datasets, or train ML/AI models in a controlled environment where privacy and governance rules are strictly enforced.

To summarize, Unity Catalog unlocks Delta Sharing, which allows you to securely share Delta Lake data with any tool that supports Delta Lake.

Delta Sharing: Iceberg Version

Databricks open-sourced Unity Catalog in 2024. Its goal is to build an open ecosystem for data governance. We have discussed that Delta Sharing offers an Open Access mode for external consumers, with a long list of partners. Under the hood, not only can providers set up Unity Catalog on their servers, but Databricks also hosts the Catalog API in each workspace. The API documentation can be found at the below link:

```
https://docs.unitycatalog.io/swagger-docs/
```

Previously, the Delta tables could be written in Iceberg format, and then we could set up external catalog integration in Iceberg clients like Snowflake. Figure 14-13 outlines the 3-step process. Database developers know it well: if we were to share data between Microsoft SQL Server and Oracle, data must be copied. Despite being configuration-heavy, an external catalog is the only way to be vendor-neutral.

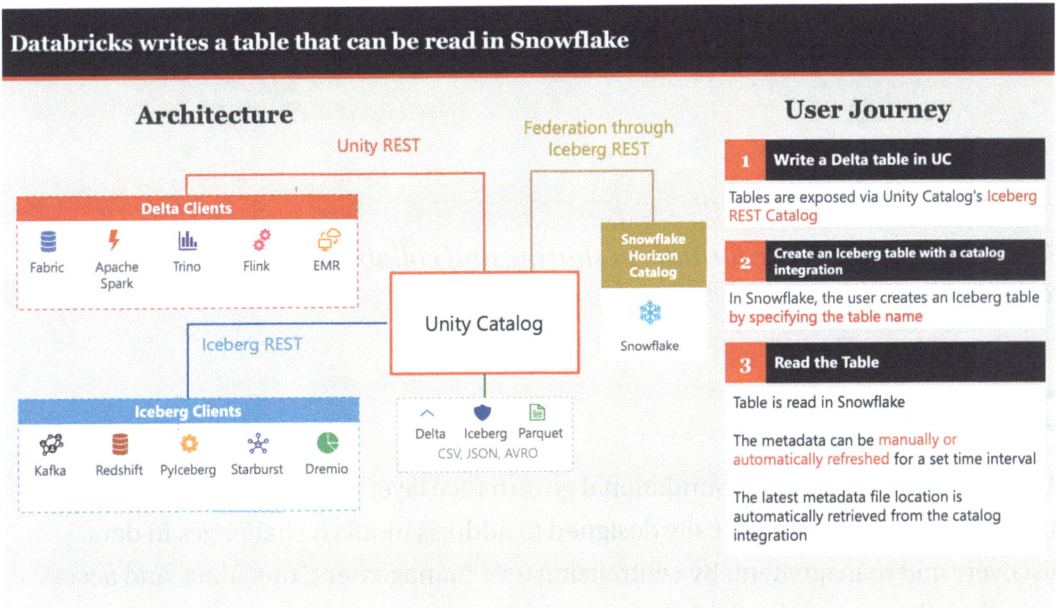

Figure 14-13. *Catalog integration between Iceberg clients via Unity Catalog API*

The Catalog War

In the chapter Delta Lake, we discussed that Databricks created the Uniform format to end the format war, but when Databricks created Unity Catalog, other vendors could also create their own catalogs, which created comparisons and a catalog war.

Iceberg Sharing is built on top of Delta Sharing and follows the Iceberg Catalog API format. Similar to the Unity Catalog API, the Iceberg Catalog API can also be found here:

```
https://iceberg.apache.org/rest-catalog-spec/
```

With this new addition, it supports both the source and the destination. In other words, we can share foreign Iceberg tables that reside in external catalogs, such as AWS Glue or Snowflake Horizon, thereby eliminating catalog lock-in. Figure 14-14 illustrates the new way of interaction.

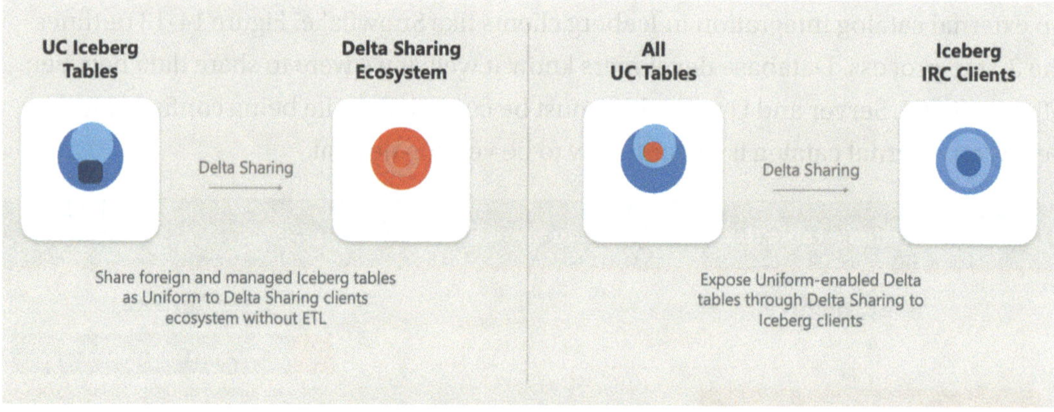

Figure 14-14. Unify the formats in sharing and collaboration for an open ecosystem

Conclusion

Unity Catalog serves as the foundational governance layer for the Databricks Data Intelligence Platform, specifically designed to address modern challenges in data discovery and management. By centralizing user management, metadata, and access controls at the account level, it eliminates the boundaries that previously existed between independent workspaces.

Key Governance Capabilities

1. **Unified Metadata Management:** Acts as a central repository for diverse assets, including tables, files, volumes, AI models, notebooks, and dashboards.

2. **Automated Data Lineage:** Provides end-to-end visibility through table- and column-level lineage graphs, tracking data movement from source to destination across all workloads.

3. **Granular Security:** Features advanced security tools such as SQL-based access controls, row-level filters, and column-level masking to protect sensitive information at query runtime.

4. **Comprehensive Auditing:** Automatically captures detailed user-level audit logs for all data access and management activities performed across the organization's workspaces.

Open Data Sharing

A critical component of Unity Catalog is **Delta Sharing**, an open protocol that facilitates secure, real-time data exchange without the need for data replication. This system supports:

1. **Internal and External Collaboration:** Securely sharing data with partners, customers, or across internal departments.

2. **Vendor Interoperability:** Utilizing both **Unity Catalog REST API** and **Iceberg REST Catalog API** to enable seamless integration with external ecosystems like Snowflake or AWS Glue, thereby reducing vendor lock-in.

3. **Cross-Platform Access:** Allowing recipients to access shared assets using a variety of open-source and commercial tools, including Spark, Pandas, and Power BI.

Unity Catalog streamlines operations and enhances productivity by providing a consistent, secure, yet open framework for managing the data and AI assets essential for modern GenAI applications. Figure 14-15 illustrates the latest evolution of Unity Catalog as of 2026.

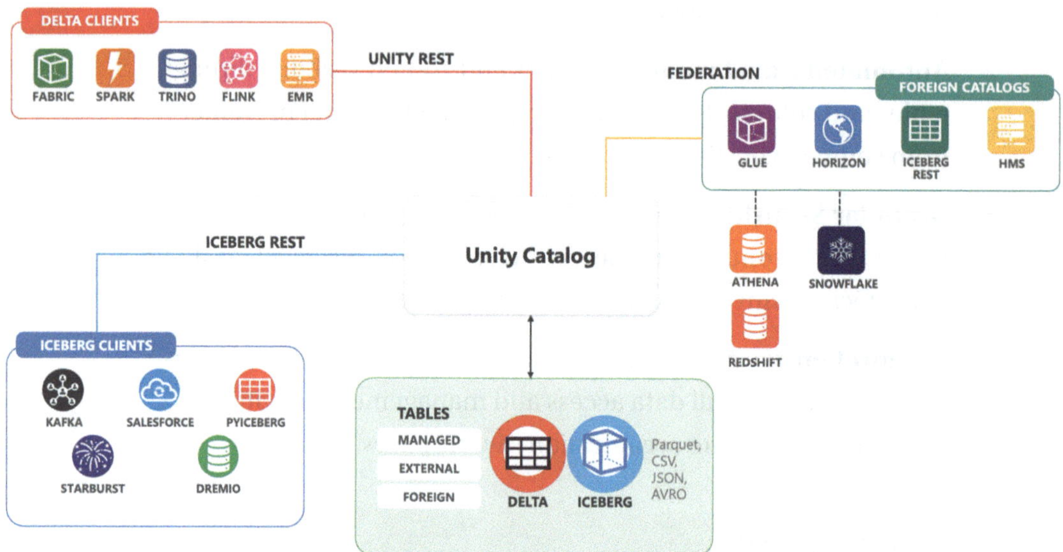

Figure 14-15. *The latest evolution of Unity Catalog*

Delta Lake: The Foundation of Reliable Data and AI

In this chapter, we will examine a crucial aspect of the lakehouse paradigm: the data storage format. As discussed in Chapter 1, the ideal storage format for a lakehouse is one that provides similar data management and performance features of a data warehouse but is an open format and built on top of cloud data storage. Delta Lake is a storage protocol that exactly fits the requirements. Delta Lake is an open, performant storage format that enables organizations to build data lakehouses, allowing data warehousing, machine learning and GenAI directly on the data lake.

We will focus on understanding why we need Delta Lake as the storage protocol in the lakehouse architecture. Thereafter, we will discuss the medallion architecture and some key features of Delta Lake, including merge capabilities, liquid clustering, and optimizations. We will end with some best practices when working with Delta Lake.

The Challenges of Other Formats

Before we start looking into Delta Lake, let's first understand some of the challenges of data lakes and other storage formats. Data lakes and standard storage formats, such as CSV, Parquet, JSON, etc., have been around for quite some time. However, there have been inherent challenges with reliability and performance when storing data in these traditional storage formats. Let's discuss some of these in a bit of detail.

First, when you use formats like Parquet and CSV, it is extremely difficult to roll back the data to its original state if your ETL job fails, leaving data corrupt. Not only that, but it is also hard to apply inserts, updates, and deletes to data stored in traditional storage formats. Next, a lack of schema enforcement leads to lower data quality, which is one of the main reasons for the low adoption of data lakes. Another important reason was that the performance of data in traditional data lakes was significantly behind that of warehouses due to issues such as small-file problems (a large number of very small files slowing processing) and the inability to cache queries or input data.

All these issues prevented the large-scale adoption of data lakes using common file formats, such as CSV and Parquet, as the de facto storage layer. However, the introduction of Delta Lake is a game-changer. Let's understand why.

What Is Delta Lake?

Delta Lake is an open-source storage layer that sits on top of storage services like Azure Data Lake Storage or Amazon S3 and provides reliability, data governance, and performance. At its core, Delta Lake, like transactional databases, provides ACID compliance to data files with schema enforcement, called a Delta table. Figure 15-1 shows the core components of a Delta table.

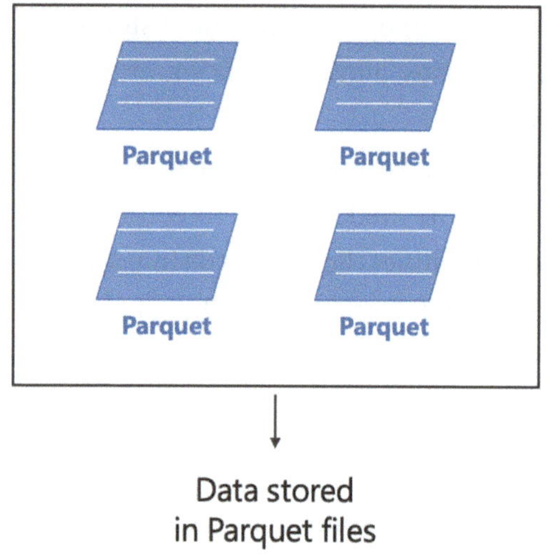

Figure 15-1. *Contents of a Delta table*

The components of a Delta table are as follows:

1. **Parquet Files:** Parquet, which organizes data in a highly efficient columnar format, has been the de facto format for storing big data for quite some time. Delta is built on top of Parquet, where the actual data is stored in Parquet format, enabling compression and encoding optimizations.

2. **Delta Log:** The Delta log is an ordered, immutable record of every transaction performed on a Delta Lake in the form of json files. It acts as the single source of truth for the Delta table. The Delta log is found in the _delta_log subdirectory within the Delta folder, which contains the Parquet data files for the table. The Delta log enables key features such as ACID transactions, time travel, scalable metadata handling, etc.

3. **Cloud Object Storage Layer:** Note that data is always stored in your cloud object storage layer (S3 for AWS, ADLS for Azure, and GCS for GCP). This storage layer ensures the durability and scalability of data in Delta Lake, enabling users to store and process large datasets without managing the complexities of the underlying infrastructure.

After reviewing what Delta Lake comprises, here are some key features.

- **Schema Enforcement:** Delta enforces the schema by default and blocks bad writes to the data. However, it provides the flexibility to evolve the schema as needed.

- **ACID Transactions:** ACID transactions ensure reliability and consistency, even during failures.

- **Version Control:** As discussed earlier, the Delta log serves as a ledger that tracks all changes to tables. If required, say when your job fails midway, the older version can be easily restored.

- **Unified Batch and Streaming:** Delta provides a unique capability: a unified source and sink for streaming and batch processing. For example, you can stream and add batch data to the same Delta table.

- **Time Travel:** The transaction log gives Delta the ability to time travel, enabling users to revert or access any version of the table as it was at a specific point in time.

- **Compliance:** Delta logs help improve data governance, security, and regulatory compliance needs.

Transaction Log: Single Source of Truth

The transaction log, also known as the Delta log, is an ordered record of every transaction performed on a Delta Lake table, stored as JSON files in the _delta_log directory. Each transaction includes metadata about the operation performed, the files added or removed, and the table schema at the time of the transaction.

Understanding the Transaction Log Protocol

The Delta transaction log protocol defines how clients interact with tables consistently. Key principles:

- All interactions begin by reading the transaction log to know what files to read

- Modifications create new data files and insert new metadata into the log

- Different clients (Rust, Spark, Trino) can work on the same table without conflict

Traditional DELETE Operation (Copy-on-Write)

Traditional deletes in Delta Lake use optimistic concurrency control to allow multiple readers and writers to access the table simultaneously while minimizing conflicts. This means that if two writes conflict, only one will succeed, while the other will fail to commit. When modifications occur, Delta Lake uses a *copy-on-write* approach for delete operations on object stores.

Every time you make a change, the system rewrites the entire affected data file with the changes applied before saving it.

- **The Write Process:** Parquet is an immutable file format, therefore in-place updates are not possible. Any change requires reading the full Parquet file, applying the modification, and rewriting a new version of the file.

- **The Penalty:** This causes high write latency. Rewriting large files for minor updates is computationally intensive and time-consuming.

- **The Benefit:** Since the data is already merged and "clean," the read process is very fast.

Deletion Vectors (Merge-on-Read)

Deletion Vectors introduce a new paradigm to the Delta Lake protocol, Merge-on-Read (MoR), in which existing data files remain untouched during operations such as DELETE, UPDATE, and MERGE. Changes are instead written separately for readers to merge when the data is read. Hence, "merge-on-read."

- **The Write Process:** Instead of rewriting the big data file, the changes are recorded in a separate, compact deletion vector file (a bitmap stored alongside the data files) that marks which rows are logically deleted or modified.

- **The Benefit:** This results in low write latency. It's incredibly fast because you're just "tagging on" the new information.

- **The Penalty:** When it's time to read the data, the system must perform a "Merge On-The-Fly." It reads the original data file and the associated deletion vector, then filters the marked rows in memory to produce the correct result. This makes for slower reads.

Figure 15-2 presents the comparison between the two.

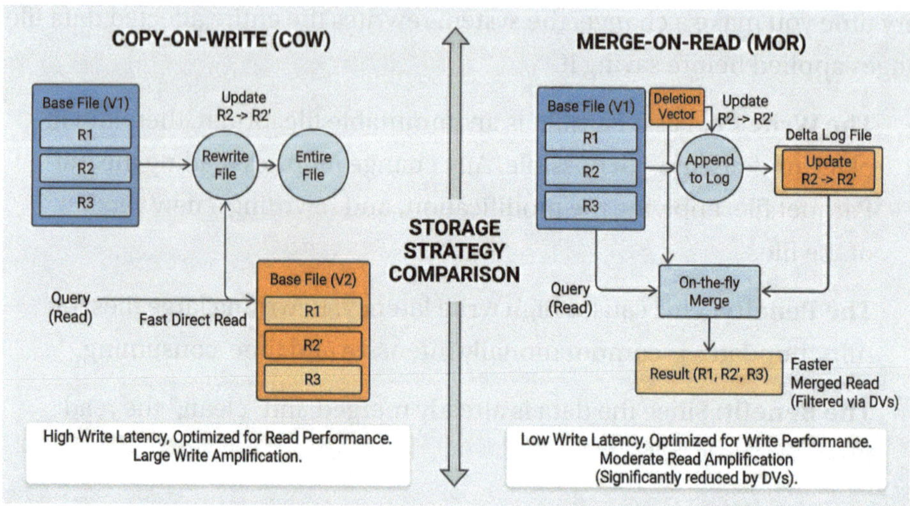

Figure 15-2. *Comparison between CoW and MoR*

With every new technology, there's always a trade-off. Copy-on-Write is best for
analytical workloads where data is written once (or infrequently) but read many,
many times by BI tools or data scientists, whereas Merge-on-Read is best for real-time
streaming or high-frequency update scenarios where getting data into the system quickly
is the top priority. Figure 15-3 illustrates this trade-off.

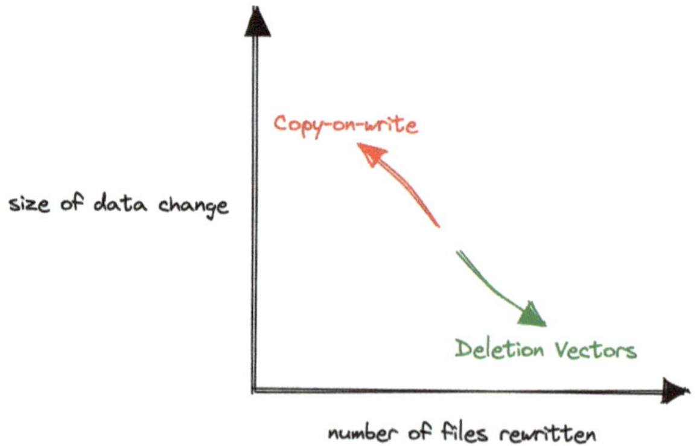

Figure 15-3. *The trade-off between CoW and MoR*

Delta Lake: Medallion Architecture

The Medallion architecture, sometimes referred to as the *multihop* architecture, is the concept of logically separating the data in a lakehouse into multiple layers, each with specific properties. A standard medallion architecture consists of three main layers: Bronze, Silver, and Gold. It is best practice to curate your data using a layered architecture, as it allows data teams to structure data according to quality levels and define roles and responsibilities for each layer. It also avoids reprocessing by persisting intermediate data states, allowing pipelines to process only new or changed data and reuse previously computed results instead of recomputing everything from raw data (see Figure 15-4).

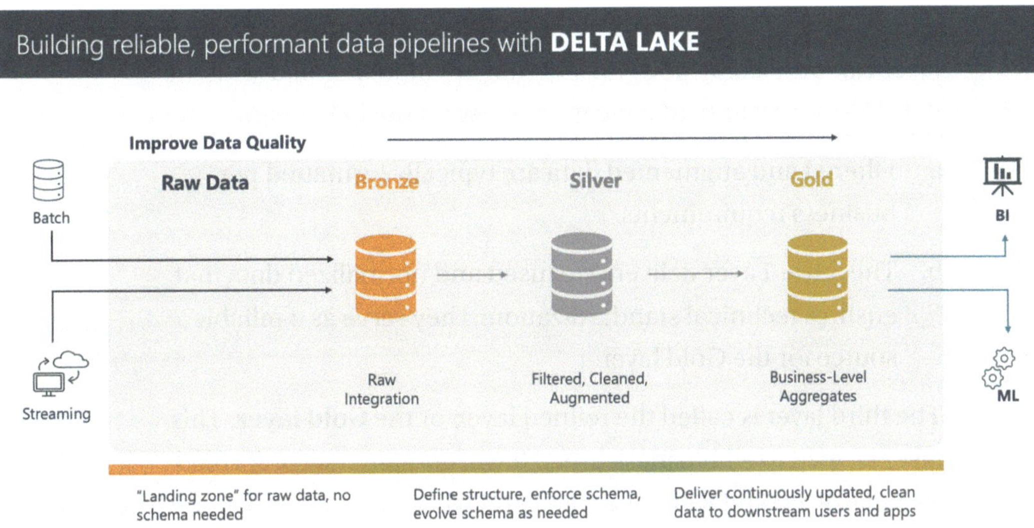

Figure 15-4. *Delta Lake medallion architecture*

Below we will examine the three layers and see what they mean.

- The first layer is the raw layer, often called the **Bronze layer**. This layer will preserve the data as close as possible to the original form. It is always best practice to maintain a copy of your source system data for the following reasons:

    a.   The source system copy helps to back out a production workload in case of any error.

    b.   The Bronze layer helps to reprocess a data if needed.

    c.   The Bronze layer preserves historical data for analytical processing and enables insights and trend analysis.

A source system copy helps hydrate a data lake, enabling new use cases and often required by data scientists to provide access to non-transformed, unbiased data.

- The second layer is the staging layer, also called the **Silver layer**. This layer can include multiple stages to help troubleshoot and process data in various forms and at different levels of conformity. The Silver layer can be used by power users who are more familiar with the data. A silver layer load typically consists of the following:

    a.   Filtered and augmented data are typically formatted per business requirements.

    b.   The Silver Layer delivers cleansed and normalized data that ensures technical standardization. They serve as a reliable source for the Gold layer.

- The third layer is called the refined layer, or the **Gold layer**. This layer is open to all business users. It will contain the conformed (standardized and validated) data and will be treated as the one true version for the business. This layer can contain smaller subsets of the data for a specific purpose (sometimes called data marts). The Gold layer often does the following:

    a.   Answers very specific business questions

    b.   Most likely contains fully aggregated data

    c.   Is the data that is ready for the presentation layer for BI tools to slice and dice this information (e.g., OLAP-style analysis)

    d.   Contains summary data that is high quality data (dimensions serve as a single source of truth)

Now that we understand the inner workings of Delta and the medallion architecture, we will look at some key features of Delta Lake.

Delta Lake Key Features

The following sections cover the key features of Delta Lake.

Update, Delete, and Upsert in Delta Tables

Delta supports both UPDATE and DELETE commands, neither of which is natively supported by the traditional Parquet format. Furthermore, it supports upserts using the MERGE Command.

Let's examine how the MERGE operation can be used in SQL to upsert data into a Delta table from a source table, view, or DataFrame as shown in Listing 15-1.

Listing 15-1. MERGE statement

```
MERGE INTO target
USING source
ON source.key = target.key
WHEN MATCHED THEN
  UPDATE SET *
WHEN NOT MATCHED THEN
  INSERT *
WHEN NOT MATCHED BY SOURCE THEN
  DELETE
```

These are important operations that are supported by relational databases, and now you can also perform them within your Delta Lake layer.

Schema Evolution

It is important to note that Delta enforces the schema by default. This prevents users from adding data that does not conform to the existing schema, avoiding unwanted data additions to your table and maintaining data quality. Any new write to a table is checked for compatibility with the table's schema, and the target table's schema, before it is committed. If the data is incompatible, Delta Lake cancels the transaction (no data is written) and raises an exception to notify the user of the mismatch.

However, data sources evolve over time due to changing requirements, which might involve adding new fields to or dropping existing fields from tables. So, to fulfill this use case, although Delta, by default, enforces schema, it also supports schema evolution using the `mergeSchema` option or `ALTER TABLE ... ADD COLUMNS`.

Schema evolution allows users to easily modify a table's schema to accommodate changing data, such as adding one or more new columns, during append or overwrite operations. Schema evolution can be used when you intend to change the schema of your table automatically without altering the schema explicitly by either setting the option mergeSchema to true or setting the property spark.databricks.delta.schema. autoMerge.

enabled *to true. However, enabling schema evolution will add overhead to the pipeline due to metadata updates, schema validation, and potential file rewrites. For large scale production workloads, it is recommended to manage the change explicitly to avoid excessive overhead.*

By including the mergeSchema option in your query, any columns present in the DataFrame but not in the target table are automatically added to the end of the schema as part of a write transaction. Nested fields can also be added, and these fields will be added to the end of their respective struct columns as well.

From Spark 3.0 onward, explicit DDL (using ALTER TABLE) is fully supported. The following code snippets provide some examples of how this can be utilized:

- Adding new columns (at arbitrary positions)

```
ALTER TABLE table_name ADD COLUMNS (col_name data_type
[COMMENT col_comment] [FIRST|AFTER colA_name], ...)
```

- Reordering existing columns

```
ALTER TABLE table_name ALTER [COLUMN] col_name (COMMENT
col_comment | FIRST | AFTER colA_name)
```

- Renaming existing columns

```
ALTER TABLE table_name RENAME COLUMN old_col_name TO
new_col_name
```

To conclude, Delta supports both schema enforcement, which prevents adding data that does not conform to the existing schema, and schema evolution, which gives users the flexibility to make intended changes to the table.

Time Travel

Delta Lake's time travel feature allows users to access and query historical versions of data stored in Delta tables. This is important because it eliminates the need to maintain point-in-time copies of data, which is cumbersome and costly. The Delta Log acts as a transaction log that maintains a granular view of changes made to data over time.

Some of the most common use cases where you might need to access previous versions of data are auditing data as it changes over time, reproducing ML experiments or reports, or rolling back to the earlier version in case of job failures.

Let's see this in action. As explained earlier, every operation that executes on a Delta table is automatically versioned in the Delta log (Figure 15-5).

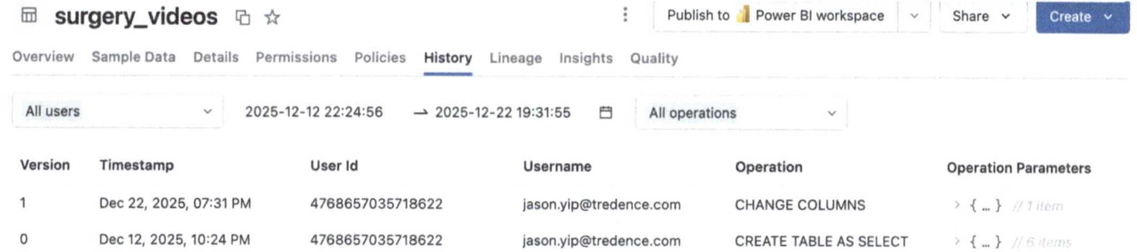

Figure 15-5. *Delta Log snapshot*

You can query the previous versions of the Delta table by doing the following, as shown in Listing 15-2:

Listing 15-2. Time travel queries

```
Using a timestamp
SELECT count(*) FROM my_table TIMESTAMP AS OF '2023-01-01'
Using a version number
SELECT * FROM employee_delta VERSION AS OF 2
```

A key question is how far back one can go to query previous versions of the Delta table. By default, you can query historical versions of the table for 30 days, whereas data files used for time travel are only guaranteed to be retained for 7 days, unless you modify the `delta.deletedFileRetentionDuration` setting or avoid running VACUUM with default parameter. Now, depending on the use case, one can increase or decrease the time using the `delta.logRetentionDuration` command. This gives users the flexibility to manage storage costs versus the need to go back and access historical data.

Clone Delta Tables

When you clone a table, you create a replica of it at a specific point in time. As the name suggests, clones share the same metadata as the source table but behave as separate tables with their own lineage or history. Therefore, any changes made to a clone affect only the clone, not the source. Furthermore, if the source data changes after the clone is created, those changes are not automatically reflected in the cloned table.

You can create a copy of an existing Delta Lake table on Databricks at a specific version using the `clone` command. Also, clones have a separate, independent log history from the source table. Time-travel queries on your source table and clone may not return the same results.

There are two types of clones: deep and shallow. Let's examine both:

Deep clone: A deep clone makes a full copy of the source table's metadata and data files. This is similar to copying a table using a CTAS (`CREATE TABLE ... AS SELECT ...`). Since the metadata is being copied from the source table, you do not need to re-specify partitioning, constraints, or other information, as you would with CTAS.

Deep clones are helpful when creating a completely independent copy of a Delta table for use cases like archiving specific tables or doing transformations on a new copy to test some transformations

Deep clones can be quickly created using the following syntax (Listing 15-3):

Listing 15-3. Deep clone syntax

```
CREATE OR REPLACE TABLE db.target_table CLONE db.source_table
```

Shallow Clone: A shallow (*also known as Zero-Copy*) clone duplicates only the metadata of the source table. The table's data files are not copied, so no additional physical copy is created, which helps reduce storage costs. These clones are not self-contained and depend on the source from which they were cloned for data.

Shallow clones are useful when you want to perform experiments on a new table, such as testing new code on production data, without affecting the production tables.

A shallow clone can be created using the following syntax (Listing 15-4):

Listing 15-4. Shallow clone syntax

```
CREATE OR REPLACE TABLE my_test SHALLOW CLONE my_prod_table;
```

One point to remember is that shallow clones are not self-contained tables like deep clones. If the data in the source table is deleted for any reason, your shallow clone may no longer be usable.

Generated Column

Generated columns are a special type of column whose values are automatically generated based on user-specified functions over the columns in the Delta table.

When you write to a table with generated columns, and you do not explicitly provide values for them, Delta Lake automatically computes the values. If you explicitly provide values for them, the value must satisfy the constraint (`<value>` `<=>` `<generation expression>`) IS TRUE or the write will fail.

Change Data Feed

Change Data Feed (CDF) is a feature in Delta Lake on Databricks that tracks row-level changes between table versions. When enabled, it records "change events" that describe not only the final state of the data but also how it got there.

This capability is the backbone for efficient Change Data Capture (CDC) architectures.

How CDF Works

When you enable CDF on a Delta table (via `delta.enableChangeDataFeed = true`), the system starts capturing metadata for every transaction. Unlike standard Delta tables that only show the latest snapshot, a CDF-enabled table includes:

- **_change_type**: Identifies the action—`insert`, `delete`, `update_preimage` (the row before the change), or `update_postimage` (the row after the change).

- `_commit_version`: The specific table version of the change.

- `_commit_timestamp`: When the change occurred.

Figure 15-6 illustrates the process of a MERGE operation and the Change Data Feed output.

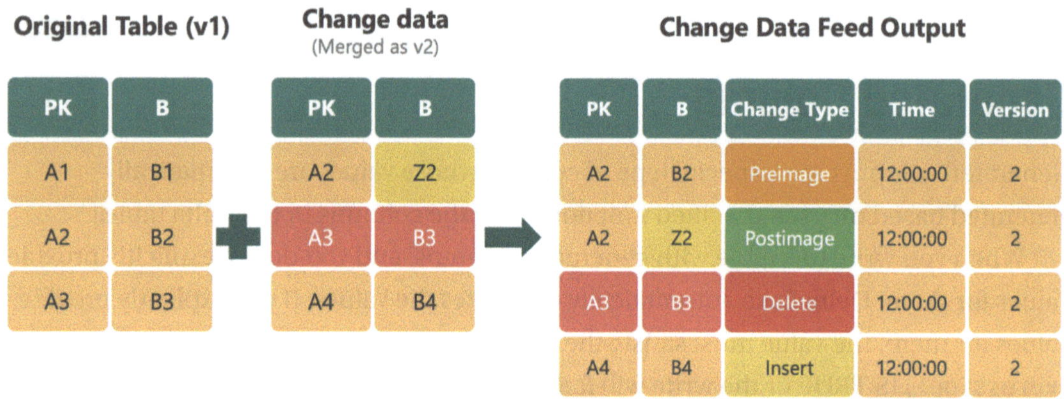

Figure 15-6. *Change Data Feed in action*

Application in CDC (Change Data Capture)

CDC is the process of identifying and capturing changes to a source database so they can be applied to a downstream target. CDF simplifies by tracking changes to records, thereby reducing computing costs and speeding up processing windows.

It is good to understand one way to solve this problem. However, in Chapter 16 Lakeflow Declarative Pipelines, we will discuss how to use the AUTO CDC API to perform Change Data Capture in a few lines of code.

Other Use Cases

Beyond CDC, CDF can be useful in several use cases. For example, you can now update only the changes from your Silver table to Gold tables with substantially less processing cost. Another use case might be when you want to transmit data incrementally from Gold tables to external systems that can ingest change data output to reduce the processing overhead. Finally, for audit and compliance purposes, it might be necessary to keep a record of when, where, and how data has been changed. CDF, with its change log, helps to maintain the logs.

Universal Format

As enterprises move toward building the Lakehouse architectures, one of the decisions they need to make is which data format to choose. Ideally, they want to store data in an open-source format that provides data warehouse capabilities. There are three popular open-source formats available: Delta, Iceberg, and Hudi.

Now, if we go one level deeper in these formats, we see that all three are built on top of Parquet, with the differences being in the metadata layer. But these differences make these formats incompatible with being read by the same reader. The problem is further complicated when different departments within the organization try to use these different formats within the Lakehouse architecture (see Figure 15-7).

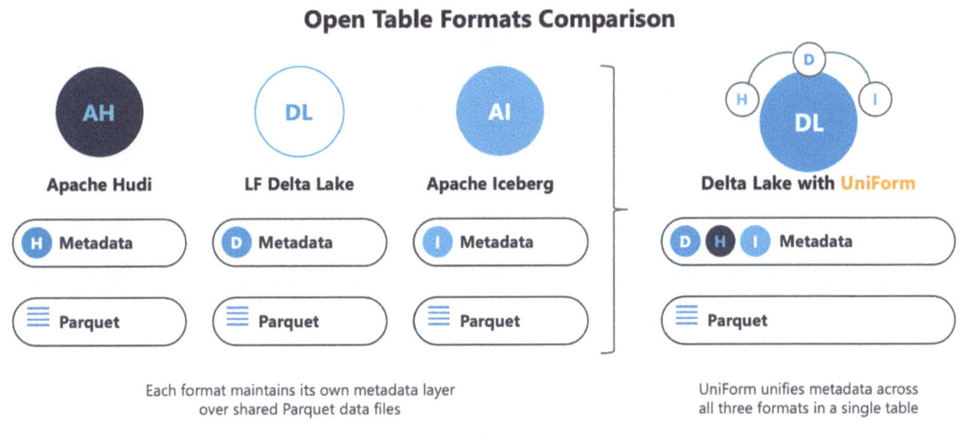

Figure 15-7. *UniForm vs. other open table formats*

To solve this problem, Databricks announced Databricks UniForm (Universal Format) with the Delta 3.0 release. As discussed earlier, all three formats are built on top of Parquet. UniForm leverages this fact to make Delta tables accessible to Iceberg or Hudi readers without data duplication, thereby reducing costs. When a table is created with UniForm activated, the metadata for the additional formats (e.g., Iceberg) is automatically instantiated and subsequently updated in response to any data mutation.

Note that prior to the release of Delta UniForm, switching between open table formats was either copy-based or conversion-based and only provided a point-in-time view of the data.

Let's see an example: there is a Delta reader and an Iceberg reader that are trying to read the Delta tables that are written by a Delta writer. UniForm in this scenario will generate Iceberg metadata asynchronously along with Delta metadata, thereby allowing both readers to read from the same Delta table. With Delta Lake 4.0, Unitform is supported by Unity Catalog and any Iceberg catalogs that support the Iceberg REST Catalog (see Figure 15-8).

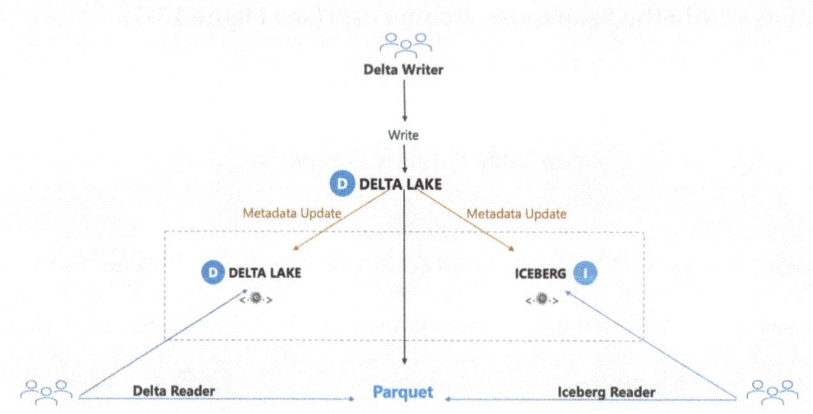

Figure 15-8. *Inner workings of Uniform*

You can enable Uniform on a new table by running the following command. Please note that Uniform is available only for UC-enabled tables, as shown in Listing 15-5.

Listing 15-5. Enabling Uniform in a Unity Catalog table

```
CREATE TABLE uniform.test.T (name STRING, age INT) TBLPROPERTIES(
 'delta.enableIcebergCompatV2' = 'true',
 'delta.universalFormat.enabledFormats' = 'iceberg');
```

If we investigate the table properties for the data, we can see something like
Figure 15-9.

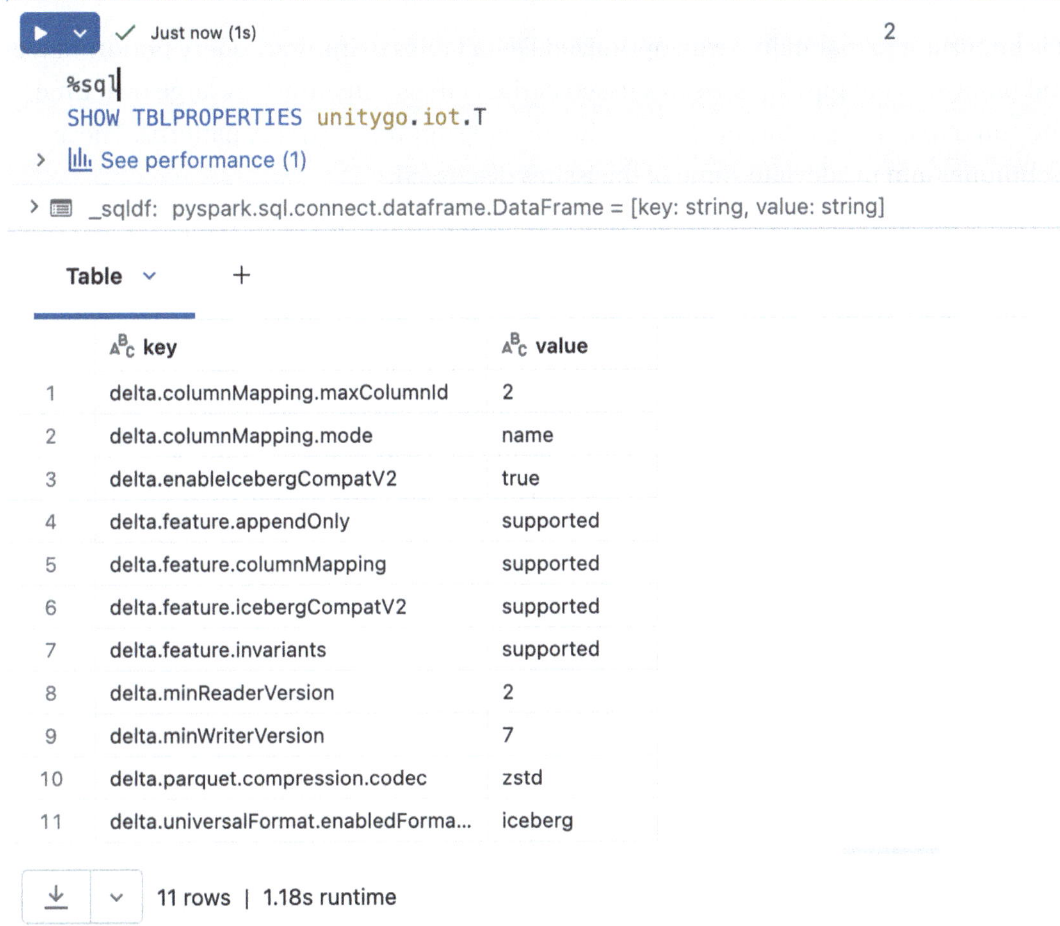

	key	value
1	delta.columnMapping.maxColumnId	2
2	delta.columnMapping.mode	name
3	delta.enableIcebergCompatV2	true
4	delta.feature.appendOnly	supported
5	delta.feature.columnMapping	supported
6	delta.feature.icebergCompatV2	supported
7	delta.feature.invariants	supported
8	delta.minReaderVersion	2
9	delta.minWriterVersion	7
10	delta.parquet.compression.codec	zstd
11	delta.universalFormat.enabledForma...	iceberg

Figure 15-9. *Properties of a Delta table with UniForm enabled*

Thus, with open table formats, organizations experience seamless data management,
ensuring data integrity and enabling smooth transactions across multiple users and
processing engines.

In the next part of the chapter, we will discuss some of the most common performance optimization techniques, such as VACUUM, OPTIMIZE, partitioning, Z-ORDER and Liquid Clustering. These techniques are not only optimization tools but also help slash storage costs, enhance parallelism, and reduce operating load on the infrastructure.

Delta Optimization

It is important to maintain clean, optimized Delta tables to improve query performance and build efficient pipelines. As discussed earlier, tables can grow very large over time and run into issues such as file layouts that do not support the query patterns. These techniques aim to alleviate some of the issues discussed.

- **Partitioning:** As the name suggests, partitioning organizes data files into separate directories based on the values of the partition key column. Partitioning data can significantly enhance query performance by allowing Spark to skip many unnecessary data partitions (i.e., subfolders) during scanning. Partitioning works best with low-cardinality columns, and one can choose columns that are commonly used in queries for partitioning, as shown in Listing 15-6.

Listing 15-6. Create or Alter table syntax

```
CREATE TABLE table_name
USING delta
PARTITIONED BY (column_name)

ALTER TABLE table_name ADD PARTITION (column_name = 'value')
```

> As a best practice, do not partition tables under 1TB, and do not partition data by a column if you expect each partition to be at least 1GB. Further, always choose a low-cardinality column—for example, year or date—as a partition key.

- **Optimize:** As discussed earlier, Delta folders may accumulate a very large number of small files (small-file problem), which affects query performance. Optimize compacts and packs these small files to a configurable size that optimizes performance for big data processing engines. Optimize keeps all data as is, but recalculates table statistics and cleans up metadata by removing unnecessary entries. The target file (1GB default) size of the new command can be changed by tweaking the attribute spark.databricks.delta.optimize.maxFileSize

You can also run Optimize on a Delta table by simply running the `OPTIMIZE` command

As a best practice, `OPTIMIZE` (with or without `ZORDER`) should be performed regularly, such as once a day or weekly, to maintain a good file layout and improve downstream query performance. Also, run `OPTIMIZE` on a separate job cluster equipped with compute-optimized VMs, as it is compute-intensive.

For managed tables, `OPTIMIZE` is run automatically.

- **Z-Order:** Z-ordering reorganizes data within Delta tables to improve query performance. It rearranges the data based on specified columns, allowing Delta Lake to skip irrelevant data during query execution. In short, the entire table is rewritten based on the columns specified in the z-order command.

  As a best practice, always choose high-cardinality columns (for example, `customer_id` in an `orders` table) for z-ordering. This is the opposite of partitioning, where low-cardinality columns are chosen. Further, choose the columns that are most frequently used in filter clauses or as join keys in the downstream queries. Finally, it is best to limit the number of columns to four or fewer, as more than that reduces the effectiveness of z-order.

- **Vacuum:** `VACUUM` deletes files that are no longer used in the Delta folders. By default, Delta retains older files up to 7 days and can be configured using the property `delta.deletedFileRetentionDuration`.

VACUUM is not reversible, so it should be used with caution. Furthermore, once it is on the table, your ability to use time travel is limited, but the vacuum reduces storage costs by deleting unnecessary files. Depending on the use case, you can decide whether to vacuum a particular table.

VACUUM is run automatically on managed tables.

After learning the fundamental optimization techniques, we will move on to two newer optimizations that Databricks recommends: liquid clustering and predictive I/O.

Liquid Clustering

Liquid clustering is a new feature introduced for the Delta table in Databricks Runtime 13.3 LTS and above. Let's examine how you can utilize this feature to enhance the performance of your Delta tables without much manual intervention.

As discussed, two of the most common techniques for optimizing your Delta tables for efficient storage and data retrieval are table partitioning and z-order.

When done right, these techniques help users increase the performance of their queries. But both require careful consideration. For example, use the right column to partition your data, and apply z-ordering each time new data is added to your table. Therefore, data engineers need to constantly work to keep the tables optimized.

Liquid clustering aims to replace both these features with much less manual intervention, thus reducing data management and tuning overhead. It's flexible and adaptive to changes in data patterns, scaling, and data skew.

With liquid clustering, keys (columns) can be chosen purely based on the query access pattern. You do not need to consider factors such as cardinality, key order, file size, potential data skew, and future changes in access patterns. Furthermore, the keys can be changed without rewriting the table's files; thus, as the query pattern changes over time, the data layout adapts accordingly.

As a best practice, enable liquid clustering for all new Delta tables. Some scenarios where liquid clustering is highly useful include tables with significant data skew, tables that are growing rapidly with new data, and queries that involve frequent filtering on high-cardinality columns.

Working with Liquid Clustering

Liquid clustering is enabled when creating a Delta table using the CLUSTER BY command and defining the clustering keys. Once enabled, run OPTIMIZE jobs to cluster data incrementally as shown in Listing 15-7.

Listing 15-7. End-to-end example running the OPTIMIZE command

```
CREATE TABLE table1(col0 int, col1 string) USING DELTA CLUSTER BY (col0);

CREATE EXTERNAL TABLE table2 CLUSTER BY (col0)
LOCATION 'table_location'
AS SELECT * FROM table1;

OPTIMIZE table2;
```

Another important aspect of liquid clustering is determining which clustering keys to use. To start, choose columns that are frequently used in query filters, regardless of their cardinality, since liquid clustering handles both high- and low-cardinality columns effectively. You can begin with one column and add up to four columns when needed. Finally, as the queries and workload evolve, use ALTER TABLE tbl CLUSTER BY to change the clustering keys as often as you want. Alternatively, we can use AUTO clustering to allow Databricks to choose for us. The best part is that there is no need to rebuild the table. The syntax is shown in Listing 15-8.

Listing 15-8. Other commands for liquid clustering

```
-- Using a LIKE statement to copy configurations
CREATE TABLE table3 LIKE table1;

-- Change the cluster key
ALTER TABLE table_name CLUSTER BY (new_column1, new_column2);

-- Cluster by AUTO
ALTER TABLE table_name CLUSTER BY AUTO;

-- Disable the cluster key
ALTER TABLE table_name CLUSTER BY NONE;
```

Current Limitations

According to the Databricks documentation, the following limitations exist:

- You can specify only columns for which statistics are collected for clustering keys. By default, the first 32 columns in a Delta table have statistics collected.

- You can specify up to four columns as clustering keys.

Predictive I/O

Predictive I/O is a collection of Databricks ML-powered optimizations that improve the performance of your data interactions. Its accelerated reads reduce the time to scan and read data, while accelerated updates reduce the amount of data that needs to be rewritten. Predictive I/O is enabled by default for serverless SQL and Pro SQL warehouses and clusters with a runtime of 14.0 or higher.

Here is how predictive I/O works with a simple analogy. Imagine that all data transactions are simply reads and writes. Think of the Windows defragmentation function, which has existed all the way back to Windows 95. File systems are often represented as data blocks, like containers or buckets, but over time, some room will remain in each block, regardless of its size. This created fragmentation, data scattered across noncontiguous blocks, rather than simply leftover space in blocks. Fragmentation increases read latency because the engine must jump around to assemble the full dataset; Delta's predictive I/O work to reduce that scatter so reads traverse fewer, larger, contiguous files (see Figure 15-10).

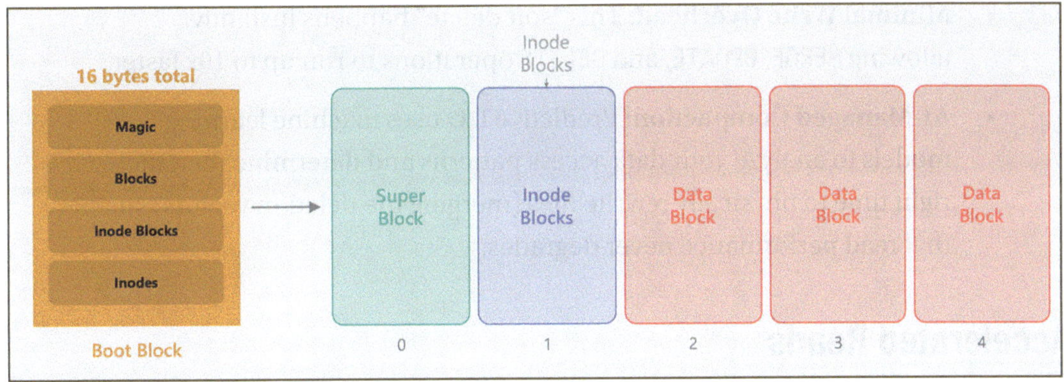

Figure 15-10. *A simple file system*

In file systems, defragmentation is the process of reorganizing files into contiguous blocks to optimize storage and read/write efficiency. You would agree that anything organized is easier to retrieve. The idea is simple. Recall in Figure 15-3, we presented the trade-off between Copy-on-Write and Merge-on-Read; it's confusing, but we have a solution.

ML/AI to the Rescue

By now, you may wonder: how do we tune these settings? What is an optimal file size? Do we need to choose between copy-on-write or merge-on-read?

Databricks Predictive I/O is an AI-powered optimization engine that improves how data is read and updated in Delta Lake tables. It essentially removes the trade-off between fast writes and fast reads by combining the speed of **Merge-on-Read** with the fast read efficiency of **Copy-on-Write**.

It works through two primary mechanisms: Accelerated Reads and Accelerated Updates.

Accelerated Updates (via Deletion Vectors)

Traditional updates require rewriting entire files even for a single row change (Copy-on-Write). Predictive I/O changes this by using Deletion Vectors:

- **Intelligent Deletions:** Instead of rewriting a full data file, Predictive I/O writes a tiny "deletion vector" (a compressed bitmap) that simply marks specific rows as deleted.

- **Minimal Write Overhead:** This "soft delete" happens instantly, allowing MERGE, UPDATE, and DELETE operations to run up to 10x faster.

- **AI-Managed Compaction:** Predictive I/O uses machine learning models to analyze your data access patterns and determine the exact right time to physically rewrite files (merging the deletion vectors) so that read performance never degrades.

Accelerated Reads

Predictive I/O for reads uses deep learning to act as an "intelligent index" without the manual overhead of creating or maintaining one:

- **Pattern Recognition:** It analyzes historical query patterns to anticipate which data will be needed next.

- **Data Skipping:** It calculates the probability that search criteria will match a row, allowing the engine to skip scanning entire portions of cloud storage that aren't relevant.

- **Optimized Scanning:** It determines the most efficient access pattern to read only the specific columns and rows required, reducing the time spent on I/O and decoding.

Figure 15-11 shows how Predictive I/O works alongside deletion vectors to optimize the MERGE performance.

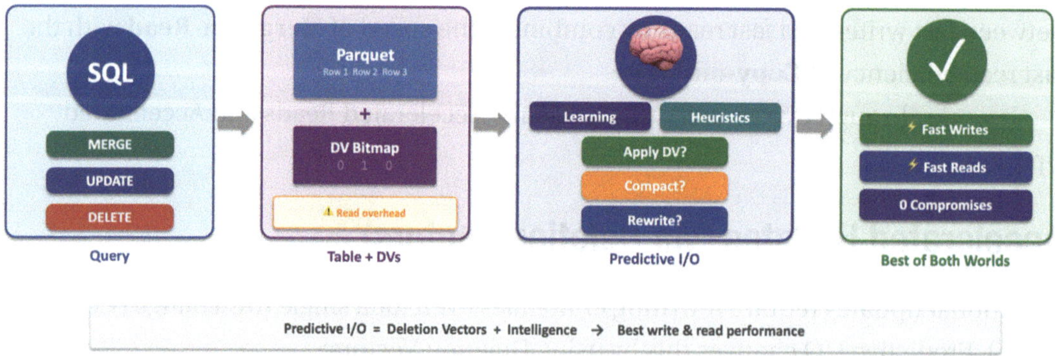

Figure 15-11. *Predictive I/O with deletion vector*

Delta Lake 4.0: The Future of Open Data Lakehouses

The release of **Delta Lake 4.0** marks a significant milestone in the project's evolution, with a focus on interoperability and on modernizing how users interact with data lakes. We will focus on some of the most important features around interoperability.

Delta Connect (Support for Spark Connect)

This is arguably the most significant architectural shift in version 4.0. It introduces support for **Spark Connect**, which decouples the client application from the Spark Driver.

Why It Matters: Previously, Delta Lake required a tight coupling with the Spark runtime. With Delta Connect, you can interact with Delta tables from lightweight environments (like local IDEs, notebooks, or edge devices) without needing a full Spark installation locally. This simplifies the development lifecycle and enables better integration with modern remote execution models.

Expanded UniForm (Universal Format)

UniForm allows Delta tables to be read as if they were Apache Iceberg or Apache Hudi tables without migrating or duplicating the data. In Delta 4.0, UniForm has seen significant improvements in metadata generation and compatibility.

Why It Matters: This addresses the "format wars" in the data lakehouse space. It allows organizations to standardize on Delta Lake for writes while still supporting downstream tools or engines that might only have native support for Iceberg or Hudi (no Delta support).

Delta Kernel

Delta Kernel is a set of libraries that centralizes the "Delta Protocol" logic. Instead of every engine having to re-implement how to read Delta logs, they all use the **Kernel** as the single source of truth, as shown in Figure 15-12. Because the kernel is also open source, developers can connect to this protocol freely, and the ecosystem will grow over time.

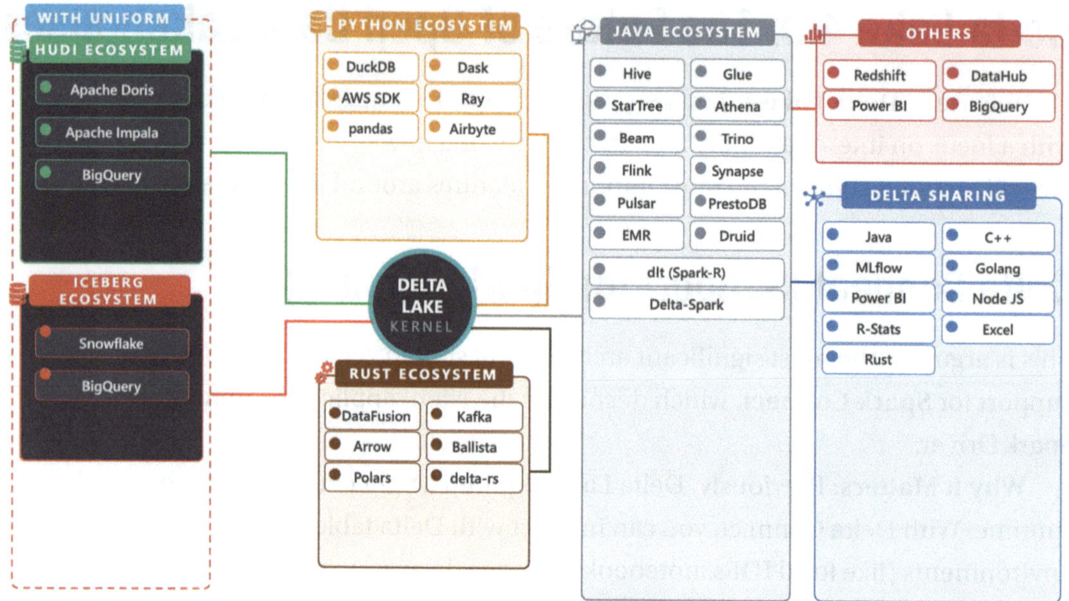

Figure 15-12. *The Delta Kernel ecosystem at launch*

Why It Matters: It effectively ends the "feature lag" between Spark and other engines. When a new feature like **Liquid Clustering** or **Deletion Vectors** is released, it becomes available across the entire ecosystem almost simultaneously through the Kernel. It creates a unified experience across all ecosystems.

Delta Sharing

The Delta format is one of the most important projects that Databricks developed and open-sourced to the community. Delta Sharing is the industry's first open protocol for secure, real-time data sharing across different computing platforms and clouds, which is also open source and donated to the Linux Foundation. Delta Sharing allows organizations to share live datasets, AI models, and notebooks without the need to replicate or move data. And being open-source means that there is no vendor lock-in and it is fully transparent.

While we are used to sharing files with tools like Google Drive and Dropbox, Delta Sharing is unique, not to mention it's fully open-sourced as well. Below are the key capabilities of Delta Sharing:

- **Zero-Copy Sharing:** Data is read directly from its source in cloud storage (e.g., S3, ADLS, GCS), ensuring recipients always have the freshest version without expensive egress or storage duplication.

- **Platform Independence:** Recipients can consume shared data using their tools of choice—such as Pandas, Apache Spark, Power BI, or Tableau—regardless of whether they use Databricks.

- **Broad Asset Support:** Beyond simple tables, it supports sharing streaming tables, SQL views, non-tabular volumes (PDFs, images), and AI models.

- **Governed Access:** Features built-in auditing and centralized governance via Databricks Unity Catalog, allowing providers to track who accessed which data and when.

There are two versions of Delta Sharing: Databricks-to-Databricks and Open Sharing. The sharing option is easily accessible from the Catalog tab in the menu by selecting a table, as shown in Figure 15-13.

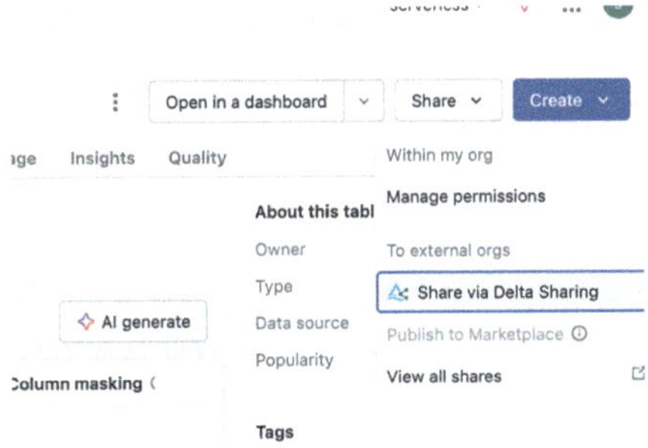

Figure 15-13. *Share via Delta Sharing*

Delta Sharing Use Cases

There are many internal and external use cases for Delta Sharing, but one of the most convenient products is Marketplace, where any company that produces data can publish it as a data product and charge consumers for their data usage.

Below are some use cases for external sharing:

- **Sharing Curated Datasets with Partners:** A bank, retailer, or SaaS company shares processed tables with vendors or partners. The partner can query live data without maintaining a copy.

- **Internal Collaboration:** Different teams (analytics, data science, BI) in the same organization access a single source of truth across different Clouds.

- **Data Monetization:** Companies can expose datasets to customers or subscribers for analytics or ML purposes.

Without the security and convenience, such a marketplace would not be possible. Figure 15-14 show datasets from S&P Global Market Intelligence.

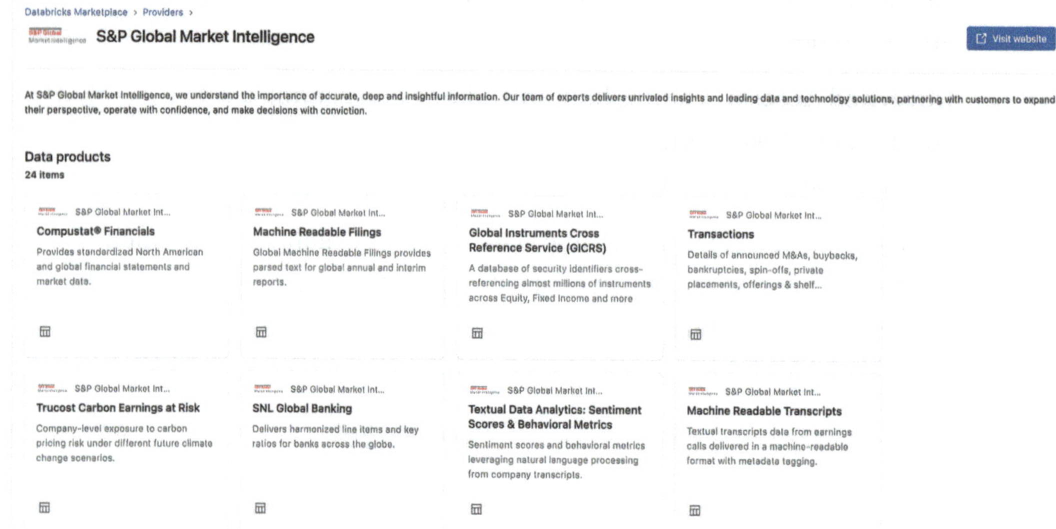

Figure 15-14. *S&P Global Market Intelligence from marketplace*

Conclusion

In this chapter, we investigated one of the building blocks of the Databricks lakehouse architecture: Delta Lake. This format provides both reliability and performance to your data. Delta Lake is the most critical part of your lakehouse because it provides all the warehouse-type capabilities for your data, such as ACID transactions, updates/deletes, merge functionality, schema enforcement and evolution, time travel, etc. We also learned about advanced features such as change data feed within the medallion architecture, UniForm, which allows multiformat readers (e.g., Iceberg reader) to read from the same Delta table, and Delta Sharing.

We studied optimization techniques such as optimize, z-order, and vacuum to improve the performance of your Delta tables. We also reviewed some of the Databricks recommended techniques, such as liquid clustering and predictive I/O.

Finally, we discussed how transformative Delta Lake 4.0 is as a lightweight driver to connect the entire data ecosystem. The Future of Open Data Lakehouses has arrived.

Lakeflow Declarative Pipelines: Managing Data and AI Workflows

It is no secret that well-curated, trustworthy data is the foundation of the lakehouse architecture. Organizations need clean, fresh, and reliable data to drive their analytics and data science projects, which in turn help them make decisions for key business initiatives.

However, most data engineers will agree that maintaining data quality and reliability at scale is quite complex and tedious. Apart from writing ETL transformations, they must spend much time on tasks such as handling table dependencies, recovery, backfilling, retries, and error handling. They must also manage the infrastructure, which turns simple ETL tasks into complex data pipelines.

In this chapter, we will introduce you to Lakeflow Declarative Pipelines, which enable data engineers to concentrate on writing the transformation logic (the "what"), while Databricks manages the rest (the "how"). We will start by understanding what Declarative Pipelines are and the concepts of declarative programming. Then we will look at key features of Lakeflow Declarative Pipelines, including change data capture (CDC), data quality and monitoring, enhanced autoscaling, and more.

What Are Declarative Pipelines?

Declarative Pipelines makes it easy to build and manage reliable batch and streaming data pipelines that deliver high-quality data on the Databricks lakehouse platform. Lakeflow Declarative Pipelines uses a simple declarative approach with SQL and Python to help data engineering teams simplify ETL development and management through pipeline development, automated data testing, and visibility for monitoring and recovery. Lakeflow Declarative Pipelines also automates infrastructure management by handling cluster sizing, error handling, performance tuning, and orchestration. Therefore, using Lakeflow Declarative Pipelines, data engineers can now spend less time managing tooling and more time on data transformations and deriving value from data.

So, what is the difference between Delta tables and Lakeflow Declarative Pipelines? Delta is a storage format, and the tables created on the underlying data are called *Delta tables*. Lakeflow Declarative Pipelines is a development framework that manages how data flows between Delta tables (see Figure 16-1).

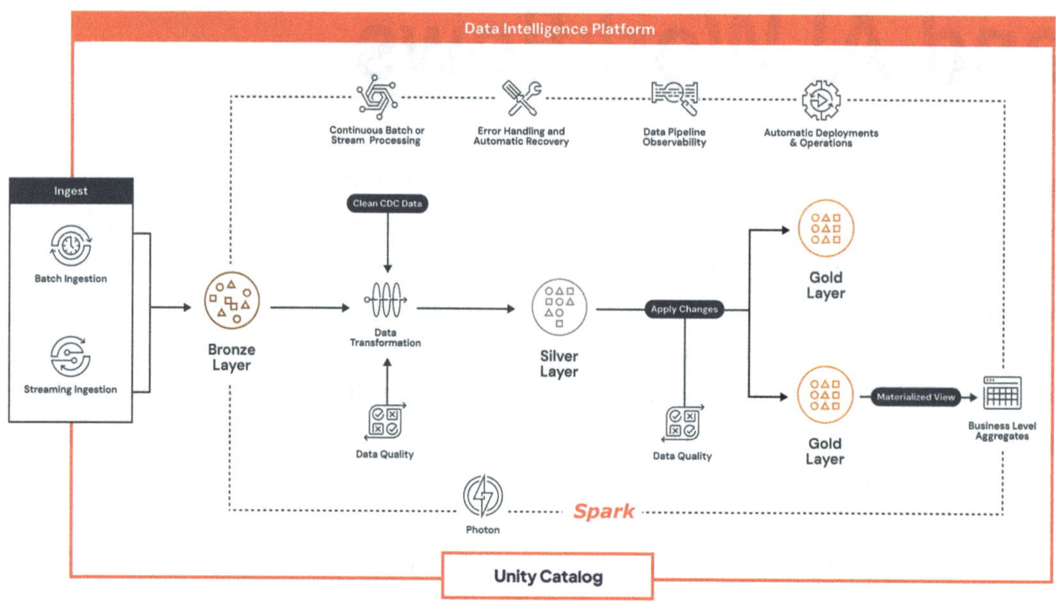

Figure 16-1. *Declarative pipelines overview*

Before we proceed, let's examine the different types of Lakeflow Declarative Pipelines datasets, namely, streaming tables, materialized views, and views.

- **Streaming Tables:** A streaming table is a Delta table that supports incremental data processing. It is most suitable for ingestion workloads and pipelines that require data freshness and low latency. It is designed to read append-only data sources like Kafka and Kinesis or file-based sources via Auto Loader.

- **Materialized Views:** A materialized view precomputes and stores query results and keeps them fresh over time. It is refreshed according to the pipeline's update schedule and, more importantly, incrementally, thus reducing processing costs. Each time the pipeline updates, query results are recalculated to reflect changes in upstream datasets.

It's critical to understand when to use materialized views. They are best used to reduce query latency in BI dashboards, serve curated metrics, and simplify complex ETL logic. While they provide consistent performance for real-time insights, they should be viewed as a specific optimization tool rather than a default storage pattern. Overusing them can lead to significant downsides, including increased costs and unnecessary maintenance burdens.

- **Views:** Views are temporary tables that should not be exposed outside of the Lakeflow Declarative Pipeline. They are used just like temporary tables in standard SQL processing. Views do not materialize or are not published to the target schema.

Next, we will build a simple declarative pipeline and explore some key features.

Data Ingestion Using Lakeflow Declarative Pipelines

The first step is to bring the requirements into the Lakeflow Declarative Pipelines. This could involve ingesting multiple raw files from a cloud storage folder or directly from a streaming source such as Kafka. It is important to note that data ingestion must be reliable and scale efficiently. Under the hood, Lakeflow Declarative Pipelines ingests data using Auto Loader. We discussed Auto Loader in detail in Chapter 13. To recap, Auto Loader incrementally processes new files as they land in the cloud storage. It can automatically infer schemas and evolve them as use cases require.

Listing 16-1 creates a Delta table called `loan_bronze`, and Listing 16-2 ingests JSON files from cloud storage into this table. Please note that Lakeflow Declarative Pipelines manages all Auto Loader configurations, like checkpointing, in the backend.

Listing 16-1. Creating a streaming table from a cloud storage source

```
CREATE OR REFRESH STREAMING TABLE loan_bronze
AS
SELECT * FROM cloud_files('/demos/dp/loans/raw_transactions', 'json',
map("cloudFiles.inferColumnTypes", "true"))
```

Listing 16-2. Data ingestion from Kafka

```
@dp.table
def sales():
  return (
    (spark.readStream
    .format("kafka")
    .option("subscribe", 'sales_trends')
    .option("kafka.bootstrap.servers", kafka_bootstrap_servers_tls)
    .option("kafka.security.protocol", "SSL")
    .option("startingOffsets", "earliest")
    .load()).select(col("key").cast("string").alias("eventId"),
    from_json(col("value").cast("string"), behavioral_input_schema).
    alias("json"))
  )
```

Before we build our silver layer, we will examine key concepts, including change data capture and expectations.

Real-Time Fraud Detection with Lakeflow

In the financial services sector, for platforms that manage high-volume transactions, latency is a critical factor. The ability to aggregate, orchestrate, and publish data at scale is essential for supporting business use cases such as fraud detection, payment risk evaluation, and real-time loan decisions.

The following section demonstrates how Lakeflow Declarative Pipelines enables the ingestion of streaming event data from Kafka to power real-time ML feature stores.

Streaming Event Deduplication

Internal services publish raw transaction events to Kafka topics. These streams often contain duplicate records due to "at-least-once" delivery guarantees and service retries. For accurate fraud modeling, downstream applications require a silver (cleansed and conformed) layer in which each unique transaction appears exactly once, ordered by its latest state, as shown in Figure 16-2.

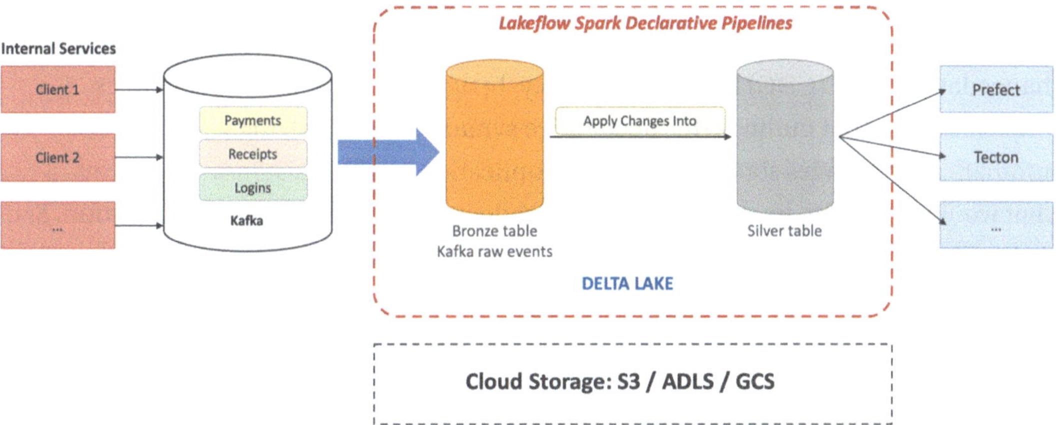

Figure 16-2. *Ingestion pipeline from Kafka to the downstream*

The first stage involves ingesting raw events into a Bronze table. Lakeflow seamlessly handles the connection to Kafka and the schema inference for the incoming JSON payloads, as shown in Listing 16-3.

Listing 16-3. Ingesting raw Kafka events

```
CREATE OR REFRESH STREAMING TABLE raw_payment_events
AS SELECT * FROM read_kafka(
  bootstrapServers => "<server:ip>",
  subscribe => "payments",
  startingOffsets => "earliest"
);
```

Once the raw data resides in the Bronze layer, the pipeline must deduplicate events and merge updates to create a reliable "single source of truth" for the risk models.

We can use this pattern to ensure that if two records share the same event_id, the pipeline preserves only the record with the most recent timestamp.

AUTO CDC API

The traditional approach works, but it often involves a lot of coding.

With standard Spark or SQL logic, engineers must manually implement stateful operations, such as tracking the latest version of each record, managing out-of-order events, handling deletes or updates, and ensuring proper deduplication. This usually requires complex joins, window functions, or custom state management code, which can become difficult to maintain as pipelines grow.

The AUTO CDC API simplifies this entire workflow by allowing you to define changedatacapture (CDC) logic declaratively. Instead of building the mechanics yourself. Figure 16-4 outlines this declarative syntax. The underlying system automatically manages state, ordering, deduplication, and consistency. This turns what would otherwise be dozens of lines of code into a single, readable declaration. See Listing 16-4 for examples of both Slowly Changing Dimension (SCD) 1 and 2.

Listing 16-4. Defining the feature engineering flow

```
-- First, define the target Silver table for the feature store:
CREATE OR REFRESH STREAMING TABLE payment_risk_features
COMMENT "Deduplicated, real-time payment features for fraud scoring";

-- SCD Type 1
-- Next, define the flow to manage deduplication and sequencing
CREATE FLOW compute_risk_features
AS AUTO CDC INTO payment_risk_features
FROM stream(raw_payment_events)
KEYS (event_id)
SEQUENCE BY event_timestamp
APPLY AS DELETE WHEN operation_type = "VOID"
COLUMNS * EXCEPT (_rescued_data, kafka_partition, kafka_offset);

-- SCD Type 2
CREATE FLOW compute_risk_features_scd2
AS AUTO CDC INTO payment_risk_features
FROM stream(raw_payment_events)
KEYS (event_id)
SEQUENCE BY event_timestamp
```

```
APPLY AS DELETE WHEN operation_type = "VOID"
COLUMNS * EXCEPT (_rescued_data, kafka_partition, kafka_offset)
STORED AS SCD TYPE 2;
```

This definition ensures that the payment_risk_features table remains strictly consistent with the latest state of each transaction, automatically handling out-of-order data and duplicates without custom state-management code.

Ensuring Data Quality in Real Time

One of the most important issues data engineers face when building data pipelines is ensuring data quality and establishing end users' trust in the data they use. Further, engineers often struggle to identify and resolve data quality issues once they discover them.

Lakeflow Declarative Pipelines provides a data quality management feature called Expectations that helps users define data quality and integrity constraints within their pipelines.

For example, a payment event must strictly possess a valid Event ID to be processable. Invalid records should be quarantined or dropped rather than propagated to the model. Listing 16-5 adds a quality constraint to the flow, ensuring only valid events enter the silver layer.

Listing 16-5. Applying data quality rules

```
CREATE FLOW compute_risk_features
AS AUTO CDC INTO payment_risk_features
FROM stream(raw_payment_events)
KEYS (event_id)
SEQUENCE BY event_timestamp
CONSTRAINT valid_event_id EXPECT (event_id IS NOT NULL) ON VIOLATION
DROP ROW;
```

In the previous query, we have defined a drop constraint on the Lakeflow Declarative Pipelines table. An expectation typically consists of three parts: the expectation name, the constraint to evaluate, and the action to take when the condition fails. An expectation name is a unique identifier and allows you to track the metrics for the particular constraint. A constraint returns a Boolean expression (True/False) based on the defined condition. Finally, action defines what to do if the condition fails.

There are three actions you can take for failed records.

1. **Warn:** In this action, the invalid records are written to the target tables, but failure is reported as a metric for the dataset.

2. **Drop:** The invalid records are dropped before data is written to the target table, and the number of records dropped is recorded.

3. **Fail:** The Lakeflow Declarative Pipeline is stopped, and the records have not been updated. Users need to check and update before manually restarting the pipeline.

Later in this chapter, we will see how you can view data quality metrics in the Lakeflow Declarative Pipelines monitoring UI (see Figure 16-3).

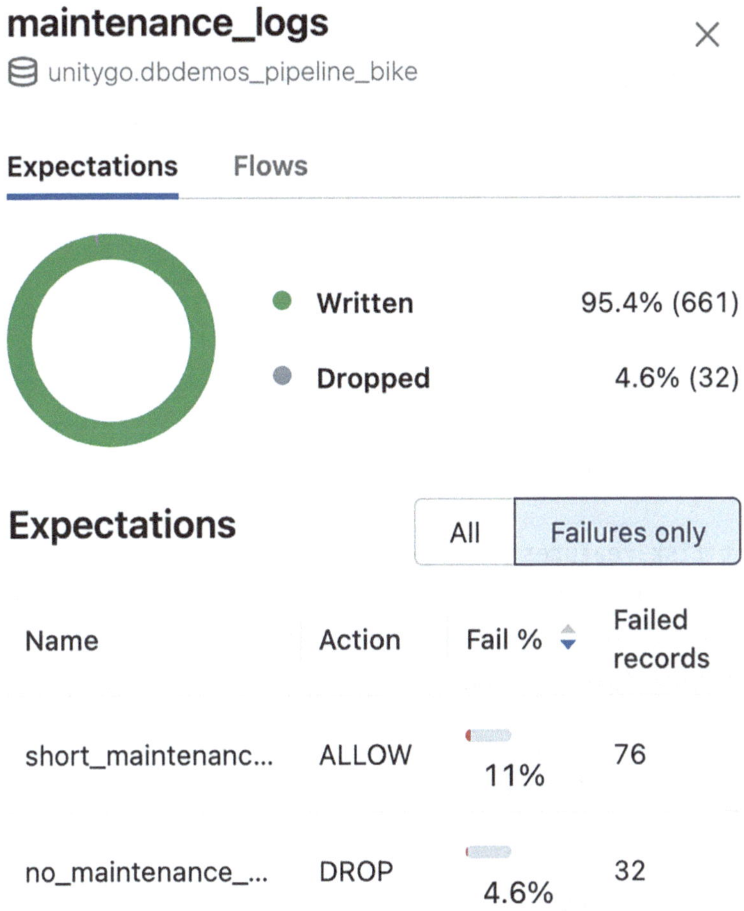

Figure 16-3. *Data quality metrics*

After creating the silver table, let's apply transformations and create a gold table (see Listing 16-6), which will be the final table in our medallion architecture.

Listing 16-6. Creating a gold materialized view

```
CREATE MATERIALIZED VIEW user_velocity_features
COMMENT "Real-time user features: Transaction counts and sums over 1-hour
windows"
AS SELECT
  user_id,
  window(event_timestamp, "1 hour") as time_window,
  count(transaction_id) as txn_count_1h,
  sum(amount) as total_spend_1h,
  avg(amount) as avg_spend_1h,
  max(amount) as max_spend_1h
FROM
  live.payment_risk_features
GROUP BY
  user_id,
  window(event_timestamp, "1 hour");
```

We have defined the logic for our Lakeflow Declarative Pipelines so far. Let's create our pipeline using the ETL designer.

Lakeflow Designer

After defining the logic for our bronze, silver, and gold tables, let's combine everything and create our first Lakeflow Declarative Pipeline. First, navigate to the +New icon at the top and click "ETL Pipeline," as shown in Figure 16-4. JSON and YAML modes are also available for quickly populating parameters, as shown in Figure 16-5.

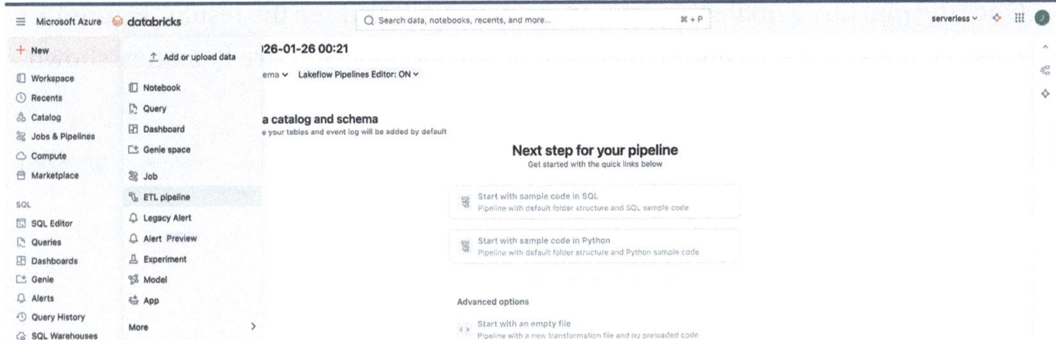

Figure 16-4. *Declarative Pipelines user interface*

Pipeline settings

```json
{
  "id": "2d9897bf-1fb6-4f57-8d83-449b47322cc9",
  "pipeline_type": "WORKSPACE",
  "name": "New Pipeline 2026-01-26 00:21",
  "libraries": [
    {
      "glob": {
        "include": "/Workspace/Users/jason.yip@tredence.com/New Pipeline 2026-01-26 00:21/transformations/**"
      }
    }
  ],
  "schema": "lichess",
  "continuous": false,
  "development": false,
  "photon": true,
  "channel": "CURRENT",
  "catalog": "unitygo",
  "serverless": true,
  "root_path": "/Workspace/Users/jason.yip@tredence.com/New Pipeline 2026-01-26 00:21"
}
```

Figure 16-5. *Sample Lakeflow Declarative Pipeline JSON file*

You can run the pipeline in dry run or production mode (run pipeline) to optimize pipeline execution. When the pipeline runs in dry run mode, you can check for problems in a pipeline's source code without waiting for tables to be created or updated. This feature is useful when developing or testing pipelines because it lets you quickly find and fix errors, such as incorrect table or column names.

Once the pipeline workflow is defined, we can run it and see the results. Figure 16-6 represents the high-fidelity lineage diagram for Lakeflow Declarative Pipelines, shown on the right panel.

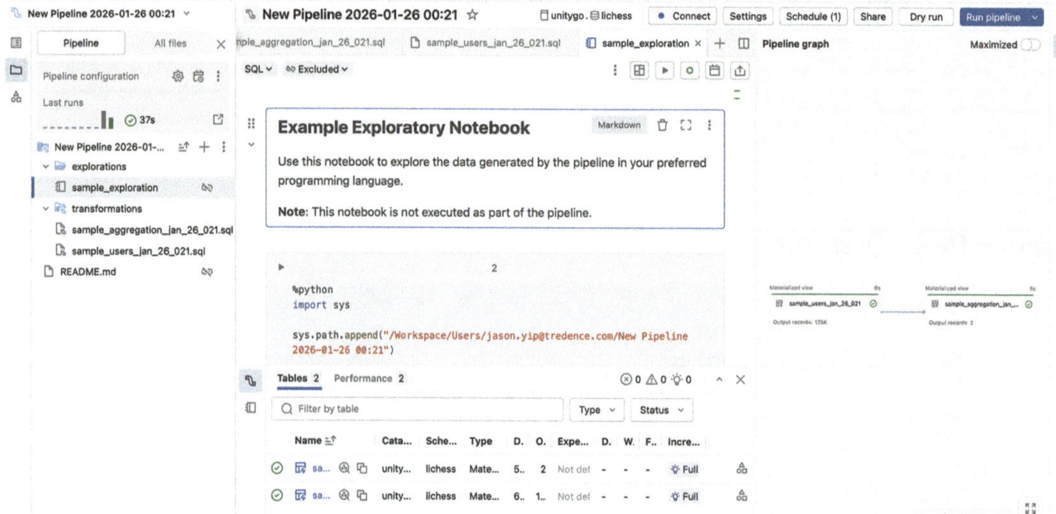

Figure 16-6. *User interface of Lakeflow Declarative Pipelines*

It is important to note that if a particular table in the pipeline is outdated or you want to refresh a specific table without rerunning the entire pipeline, select **Select table for refresh** and choose the tables you need to refresh or run again. The option appears in the lineage diagram shown in Figure 16-7.

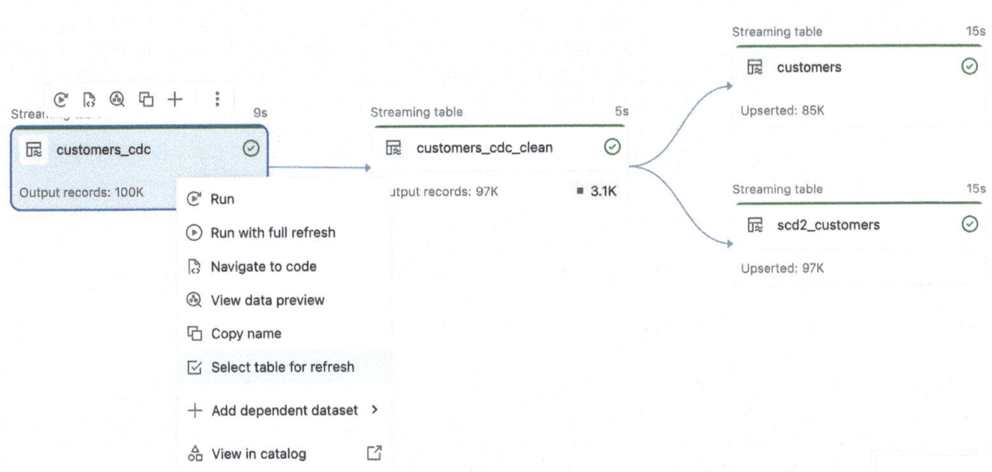

Figure 16-7. Select table for refresh option

Next, we will examine other aspects of Lakeflow Declarative Pipelines, including monitoring and logging, CI/CD, and enhanced autoscaling.

Pipeline Task Type

Once the pipeline is designed, we can either schedule it as is or make it part of our Lakeflow Jobs. Figure 16-8 shows a new task type called "Pipeline." With this new job type, we can chain up different pipelines together or connect them to an existing job.

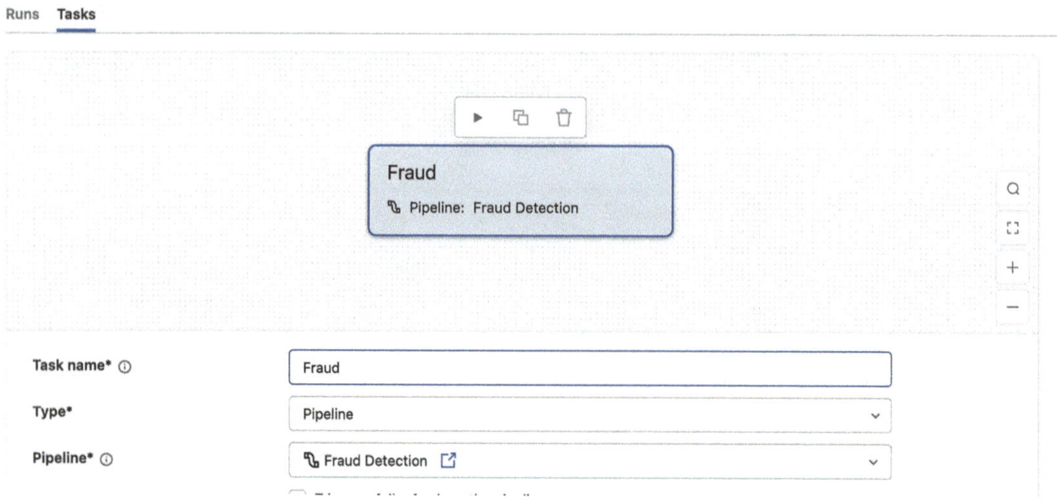

Figure 16-8. *Pipeline task type in Lakeflow Jobs*

Logging and Monitoring

Each Lakeflow Declarative Pipeline emits all event logs to a system table called `system.lakeflow.pipelines`. The system tables contain all information related to a pipeline, including audit logs, data quality checks, pipeline progress, and data lineage. You can find all Lakeflow-related system tables live in the `system.lakeflow` schema:

https://docs.databricks.com/aws/en/admin/system-tables/jobs#available-jobs-tableså

The logs are also visible in the Lakeflow Declarative Pipelines run UI page, where you can quickly investigate the errors. Lakeflow Declarative Pipelines provides a variety of error-handling capabilities, including retrying failed tasks, handling failed records, and detecting and fixing data quality issues, all presented through an intuitive graphical interface (see Figure 16-9).

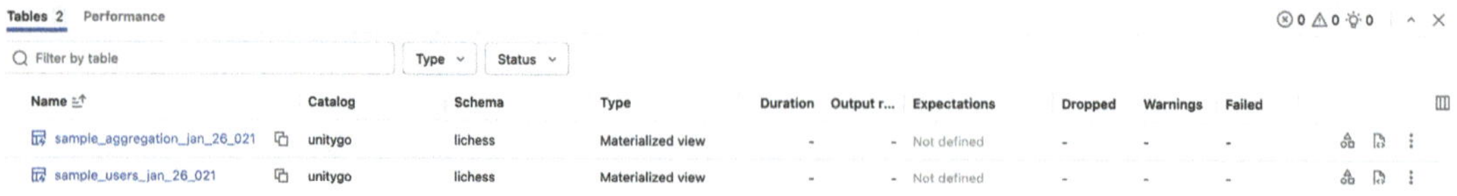

Name	Catalog	Schema	Type	Duration	Output r...	Expectations	Dropped	Warnings	Failed	
sample_aggregation_jan_26_021	unitygo	lichess	Materialized view	-	-	Not defined	-	-	-	
sample_users_jan_26_021	unitygo	lichess	Materialized view	-	-	Not defined	-	-	-	

Figure 16-9. *Lakeflow Declarative Pipelines task status*

These logs are exposed as Delta tables and used for monitoring, lineage, and data-quality reporting with the BI tool of your choice. Figure 16-10 shows a sample dashboard that can be built on AI/BI.

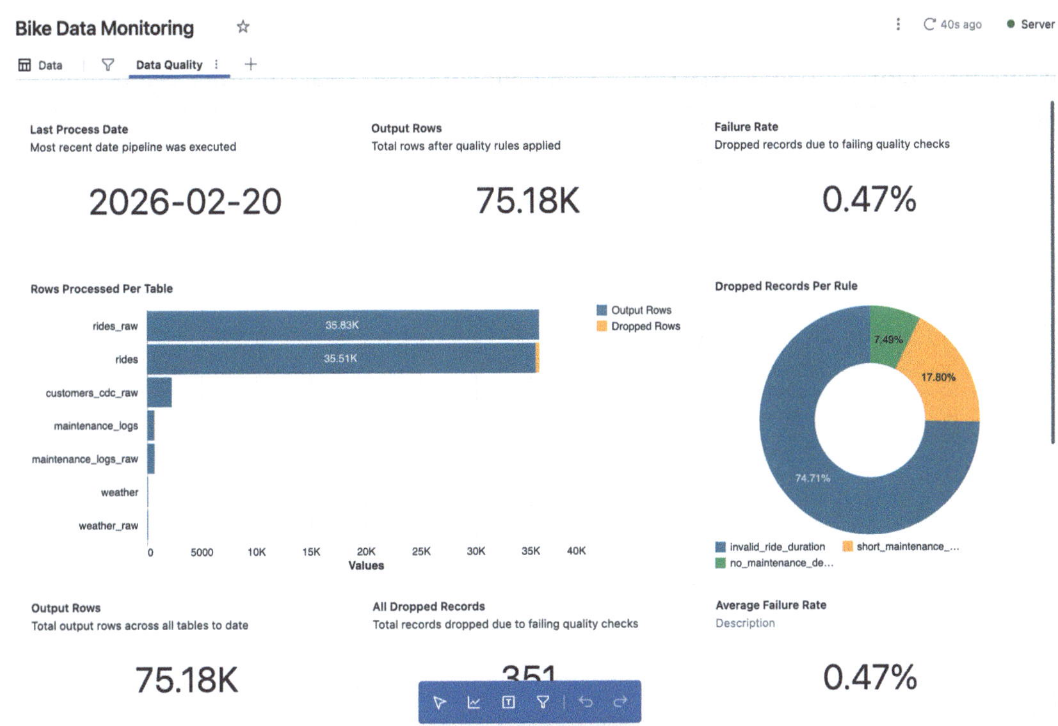

Figure 16-10. *Dashboard for monitoring Lakeflow Declarative Pipelines job statuses*

Conclusion

Data teams face ever-increasing demands to deliver reliable data quickly. However, with Lakeflow Declarative Pipelines, they can streamline reliable data pipelines and quickly find and manage enterprise data assets across various clouds and data platforms using Unity Catalog. Additionally, they can simplify enterprise-wide governance of both structured and unstructured data assets.

This chapter examined Lakeflow Declarative Pipelines, which provide a declarative framework for developing, managing, and deploying ETL pipelines. Lakeflow Declarative Pipelines automatically manage your infrastructure, ensure high data quality, and unify batch and streaming workloads. We built a Lakeflow Declarative Pipeline and explored key features such as change data capture, SCD Type 1 and 2 support, and Lakeflow Declarative Pipeline expectations, which help maintain data quality. We also discussed various performance optimizations that Lakeflow Declarative Pipelines uses via enhanced autoscaling. Finally, the runtime version is managed for you automatically by default. There is no need to worry about managing the latest runtime, as it is fully managed by the serverless platform.

Data Warehousing with Databricks SQL

If you're a data analyst who primarily uses SQL to write queries and reports and create comprehensive dashboards for analysis using your favorite business intelligence (BI) tools, Databricks SQL (DBSQL) provides a comprehensive environment for running ad hoc queries and creating dashboards on data stored in your data lakehouse.

Traditionally, SQL/BI use cases have most commonly been implemented by storing data in a data warehouse or a database, writing SQL queries in a SQL IDE, and, finally, using BI tools to build dashboards. However, with the lakehouse platform, you can handle all this without moving data to another storage system, such as a data warehouse or database.

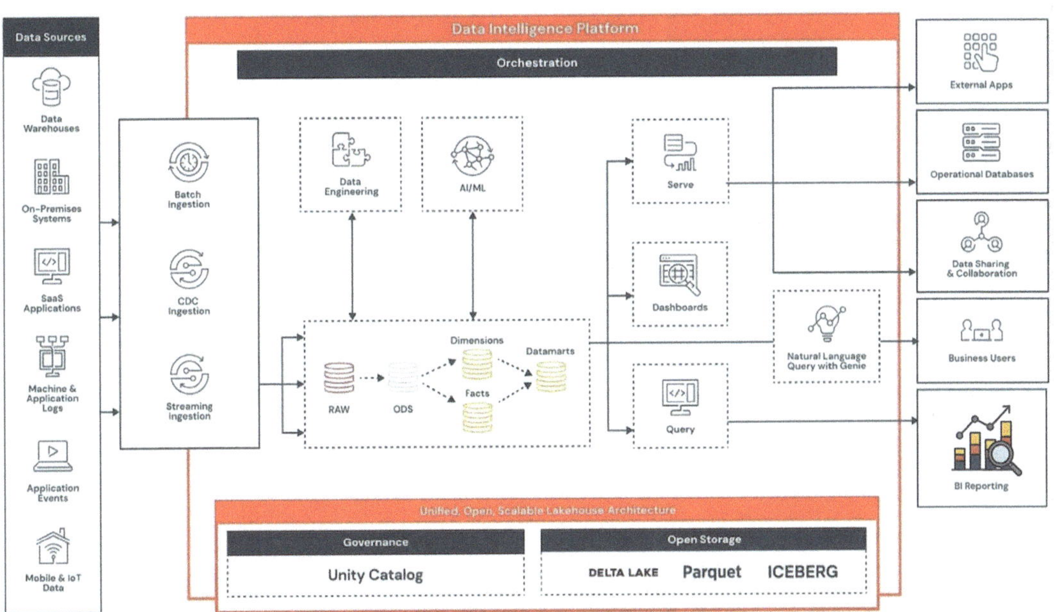

Figure 17-1. *Architecture diagram using Databricks SQL*

In this chapter, we will learn how the Databricks platform provides the most complete end-to-end data warehousing solution for all analytics use cases.

Next, we will move on to understand the components and key features of Databricks SQL (DBSQL).

What Is Databricks SQL?

Databricks SQL is a collection of services that bring data warehousing capabilities and performance to your existing data lake using open formats and standard ANSI SQL. The DBSQL platform provides not only a SQL editor but also powers the AI/BI dashboard that allows team members to collaborate with users directly in the Databricks workspace. Furthermore, Databricks SQL integrates with a variety of BI tools via connectors or JDBC/ODBC, allowing analysts to author queries and dashboards using their favorite BI tools without switching to a new platform. It is also powering the text-to-SQL engine called Genie (see Figure 17-2).

Figure 17-2. *SQL Persona section on Databricks sidebar*

In the following sections, we will investigate a few key services in DBSQL.

SQL Warehouses

SQL warehouses are compute resources within DBSQL that run your SQL queries on data objects. Simply put, SQL warehouses provide processing capabilities in DBSQL-like clusters in the data engineering part of the platform. There are three main types of SQL warehouses.

1. **Classic:** This offers limited Databricks SQL functionality and basic performance features. Only use a classic SQL warehouse to run interactive queries for data exploration with entry-level performance and Databricks SQL features.

2. **Pro:** This supports all the Databricks SQL functionality and delivers higher performance than Classic, including query federation, workflow integration, and data science and ML functions.

3. **Serverless:** This is the most powerful and cost-effective option. The serverless SQL warehouse offers the most advanced performance features and supports all Pro features, along with instant, fully managed compute. Serverless compute spins up almost instantaneously with best-in-class price/performance.

Figure 17-3 shows how to set up a SQL warehouse.

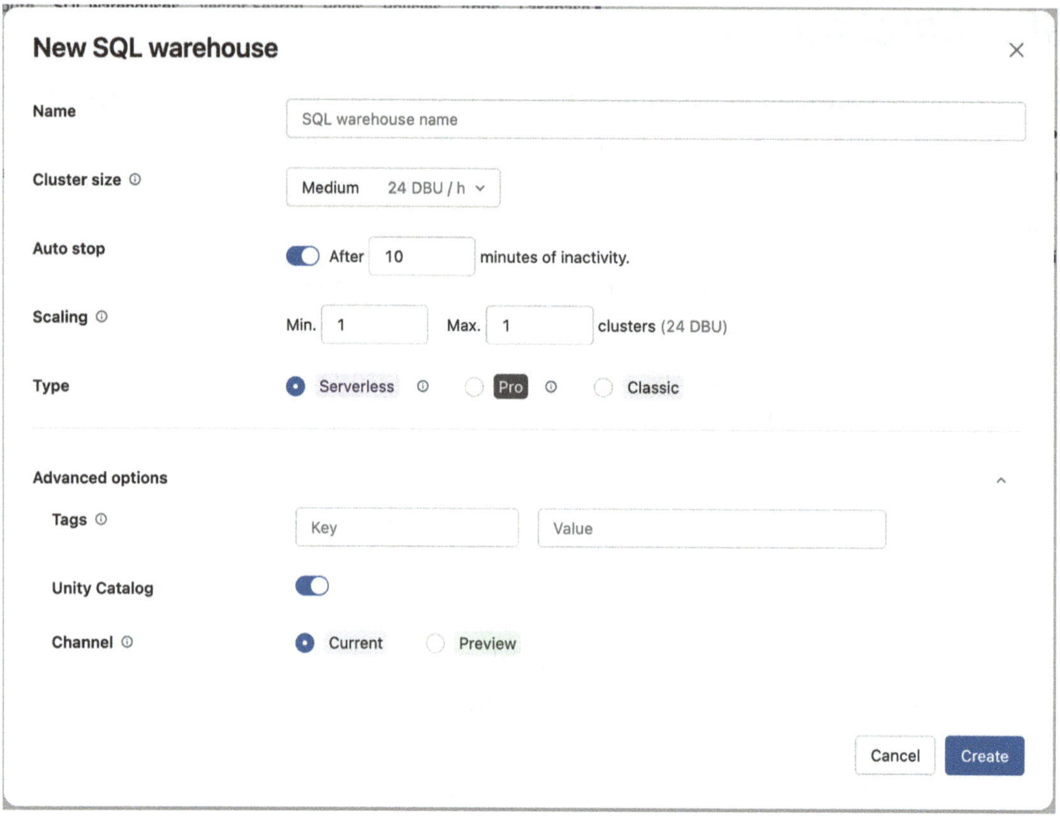

Figure 17-3. *Setting up a SQL warehouse in Databricks*

It is simple to spin up a SQL warehouse. Once you click Create Warehouse, the first parameter to enter is the warehouse name. Next, you can select the warehouse type: Classic, Pro, or Serverless.

Below are other important parameters to consider.

- **Cluster Size:** SQL warehouses come in T-shirt sizes from 2X-Small to 4X-Large. Choose the size based on your latency and throughput requirements. As a best practice, start with Medium and scale up or down as needed.

- **Scaling:** A traditional cluster comes with one driver and several workers. When you auto-scale a cluster, you are increasing the number of worker nodes. However, there is still only one driver node, and with high-frequency, low-latency workloads, it will become a bottleneck. With the SQL warehouse scaling feature, you determine the min. and max. number of clusters behind the endpoint, and it is these clusters (not workers) that can increase based on concurrency requirements.

Further, one of the key aspects of SQL warehouses that makes them performant is Photon. Let's understand what Photon is and how we can enable it.

Photon

Photon is the ANSI-compliant vectorized query engine developed by Databricks to support SQL and Spark workloads on Databricks. It comes with hundreds of built-in optimizations, providing the best performance for all tools, query types, and real-world applications. This includes the AI-powered predictive I/O that eliminates performance tuning like indexing by intelligently prefetching data.

It primarily accelerates SQL and DataFrame operations and is compatible with Apache Spark APIs, which means you don't have to rewrite your existing code (SQL, Python, R, Scala) to benefit from its advantages.

While Photon is an optional feature in interactive clusters, it is activated by default for SQL warehouses. You can also enable Photon for All-Purpose and Jobs clusters by toggling the switch on the Create Cluster page.

Figure 17-4 shows Photon's performance on the Databricks runtime (DBR), showing that the Photon-enabled runtime is almost three times faster than standard DBRs without Photon.

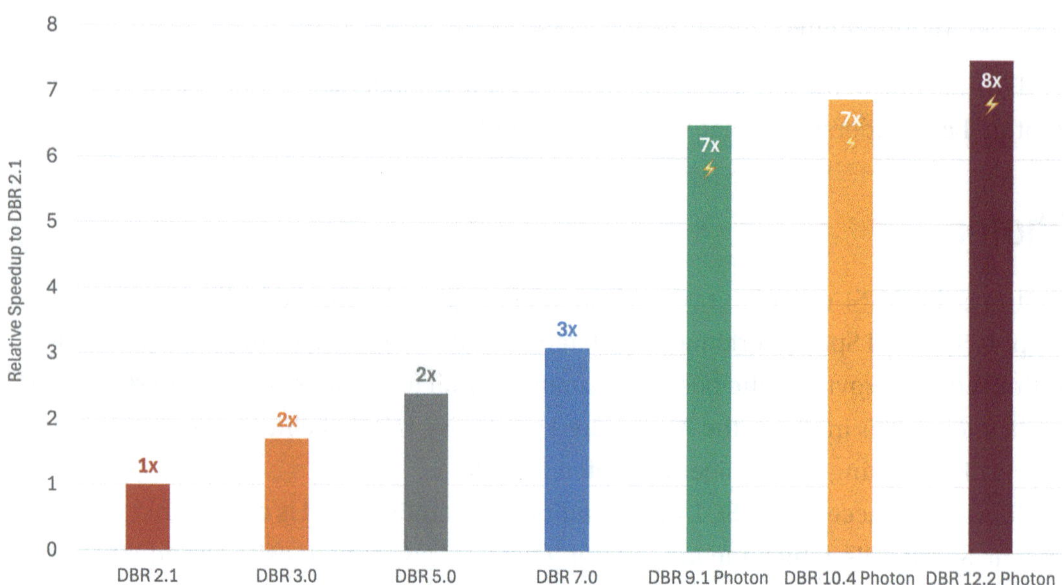

Figure 17-4. *TPC-DS 1TB performance by DBR version vs. Photon*

SQL Editor

The DBSQL UI provides a SQL editor (Figure 17-5) that you can use to author
SQL queries using a familiar ANSI SQL syntax, browse available data, and create
visualizations. You can also share your saved queries with other team members in the
workspace. SQL Editor also supports features such as autocomplete, autoformatting, and
autosave. Additionally, query updates can be scheduled to automatically refresh and
issue alerts when meaningful changes occur in the data.

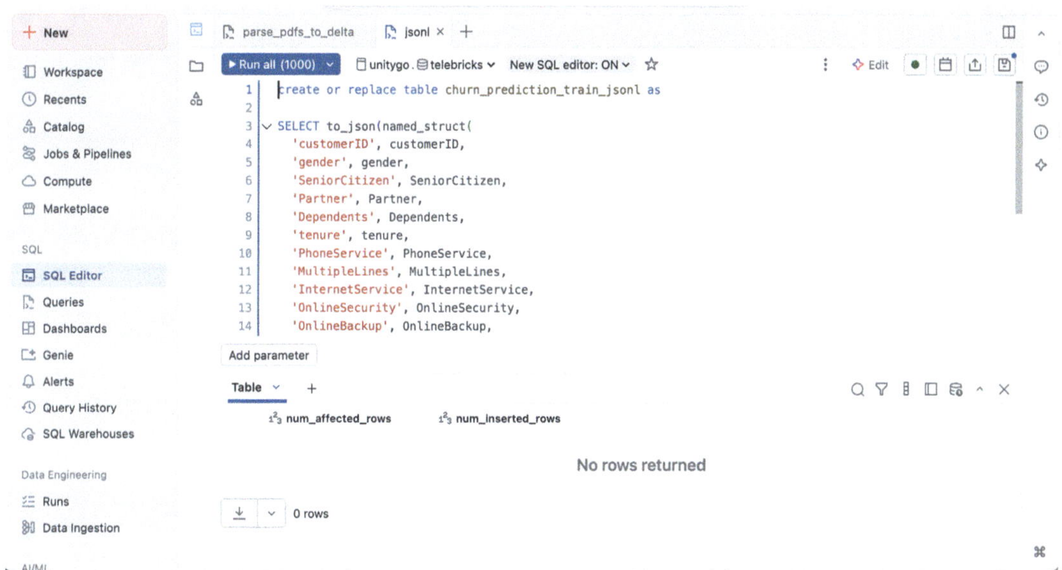

Figure 17-5. *SQL Editor in Databricks*

Introduction to AI/BI Dashboards

Databricks SQL enables data analysts to make sense of data through visualizations and drag-and-drop dashboards. Dashboards (now called *"legacy dashboards"*) within DBSQL allow users to combine visualizations (via the built-in SQL Editor) with text boxes to provide context for their data. Once built, dashboards can be easily shared with stakeholders, both within and outside the organization, via a web browser.

Databricks AI/BI dashboards (Figure 17-6) allow analysts to quickly build highly interactive dashboards using natural language. Furthermore, these dashboards are integrated with the Databricks platform, which ensures high performance at scale, while all security and governance policies are managed in Unity Catalog.

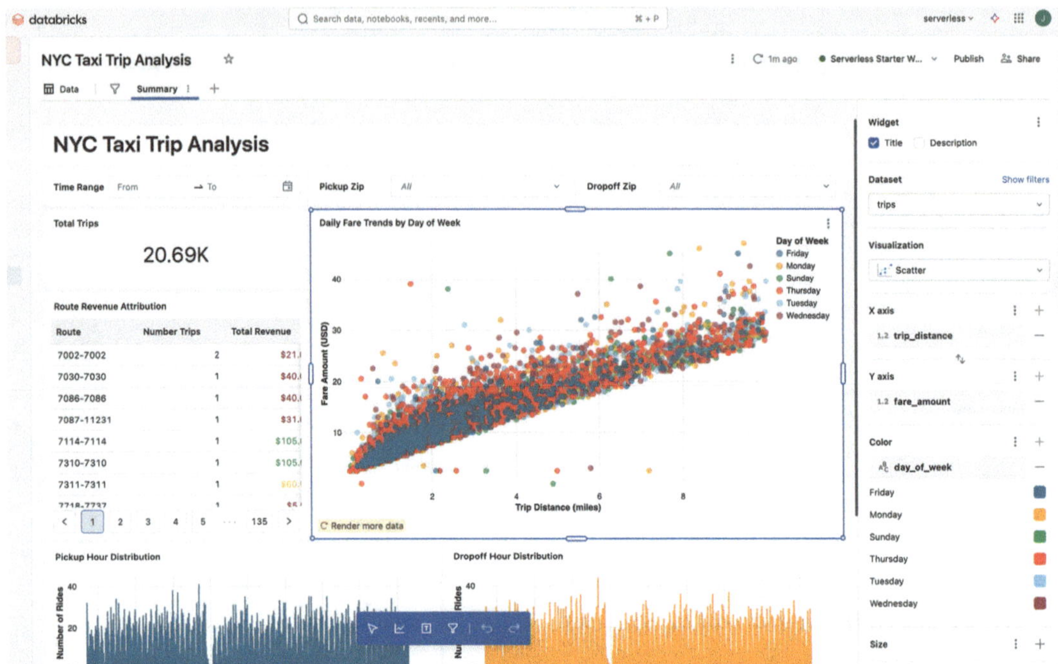

Figure 17-6. *AI/BI dashboards*

Let's check how you can quickly build AI/BI dashboards in DBSQL. The dashboard
has two tabs: a **Data tab** for searching for tables in the Unity Catalog or writing
queries that will serve as your dataset(s), and a **Canvas tab** where you create and
assemble visualizations, with the option to use natural language to generate tables
and visualizations. Some capabilities include sleek visualizations, cross-filtering, and
periodic PDF snapshots sent via email.

Finally, AI/BI dashboards allow users to publish their dashboards to the entire
organization. This means that any authenticated user in your identity provider (IdP)
can access the dashboard via a secure web link, even if they don't have Databricks
workspace access.

Alerts

DBSQL allows users to set up alerts that send notifications if a particular condition is not
met in the data. Take, for example, an inventory management table. One can set an alert
on the table if the quantity of a particular product or SKU falls below a certain threshold.
Notifications can be delivered via email or through other platforms such as Slack, Teams,
or webhooks. Figure 17-7 shows how we can set up an alert.

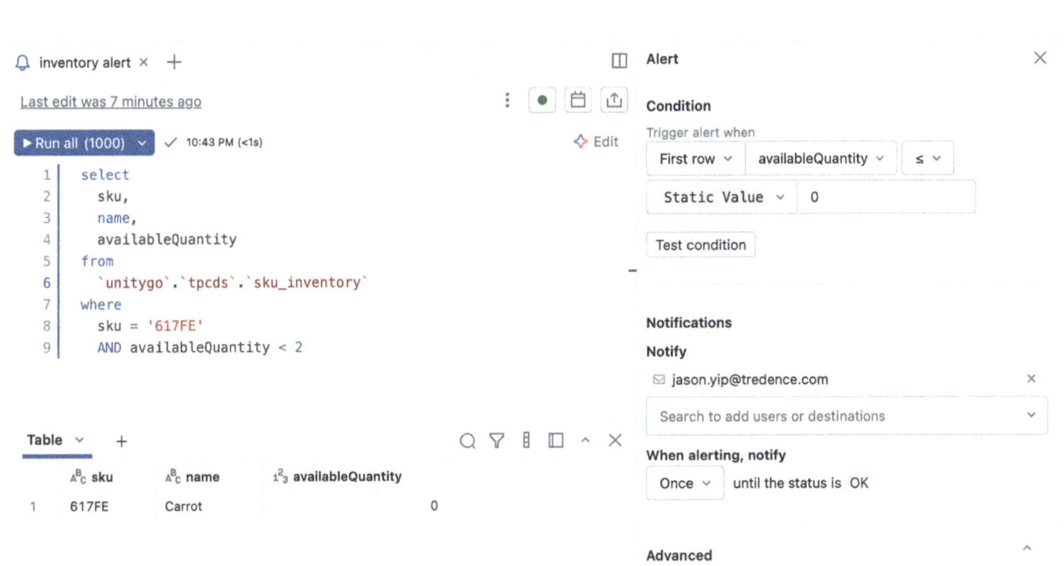

Figure 17-7. *Databricks alerts*

Note that while alerts can be set up in the Alert interface, destinations can only be configured by workspace admins and are available in "Settings" under "Notifications," as shown in Figure 17-8.

Figure 17-8. *Destination configuration under Settings*

Query History and Profile

Query history in DBSQL gives you full visibility and details of query execution for all the queries executed on the SQL warehouses for the last 30 days. With a unified view, you can not only see the number of queries executed at a particular time but also quickly zoom into specific queries and debug issues (see Figure 17-9).

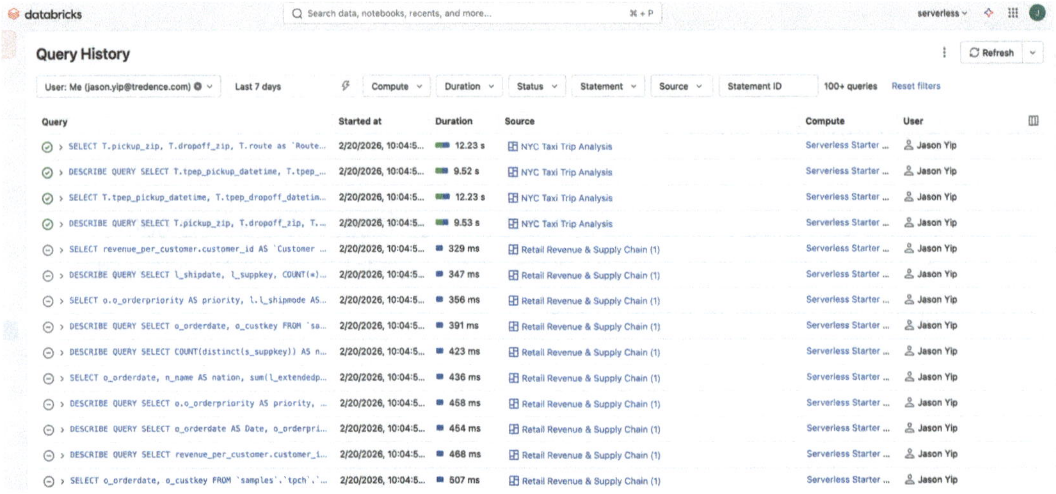

Figure 17-9. *DB SQL query history*

A query profile allows you to visualize the details of a query execution. The query profile helps you troubleshoot performance bottlenecks during the query's execution (see Figure 17-10). For example,

- Visualize each query task and its related metrics, such as the time spent, number of rows processed, and memory consumption.

- Identify the slowest part of a query execution at a glance and assess the impacts of modifications to the query.

- Discover and fix common mistakes in SQL statements, such as exploding joins or full table scans.

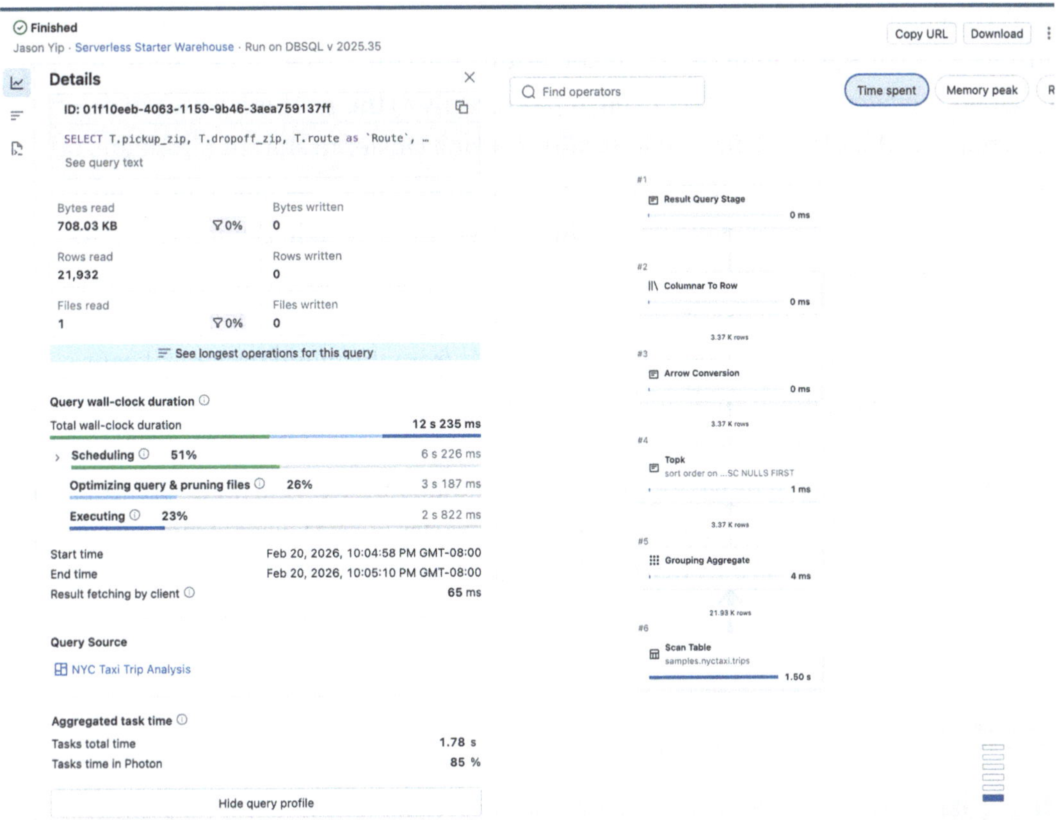

Figure 17-10. *Query profile*

The profiler is a significant improvement over the traditional Spark UI and is designed for SQL developers to gain insights quickly.

After the overview of DBSQL, we will dive into some important features.

Serverless Compute

Databricks Serverless represents a paradigm shift in Databricks compute. Serverless is a fully managed service that eliminates the burden of capacity management, patching, upgrading, and performance optimization for the cluster. Additionally, Serverless simplifies the billing. In other words, you need to pay Databricks only once for both Cloud Compute and Databricks costs. The paradigm has shifted from heavy maintenance and management to intelligent management, opening the door to auto-scaling and AI.

As discussed previously, with any type of compute, be it standard clusters, job clusters, or even SQL warehouse (non-serverless) the virtual machines are provided by the cloud provider, for which you need to pay directly to them. This has two main effects. First, not only does it take three to four minutes for a cluster to start or terminate, but users also have to manage these factors, including runtimes, machine types, and cluster sizes. Second, there are usually two-line items on your cloud bill: Databricks Costs ($DBU) and cloud VM costs.

With the introduction of Serverless, Databricks is basically "owning the compute." To put it simply, Databricks pre-purchases these VMs from the cloud provider, keeps some of them internally pre-warmed, and when you request a serverless compute resource, it attaches the specified number of VMs. Since this compute is fully managed by Databricks, it spins up or down in seconds rather than minutes. Furthermore, you only need to pay Databricks once, covering both Databricks and VM costs. Thus, serverless compute brings a truly elastic environment that's instantly available and scales with your needs.

Constraints in DBSQL

Many data analysts have experience with relational databases and building entity-relational models using primary key/foreign key relationships. After normalization, they usually build multidimensional data models (referred to as *star schemas*) so that it is easy to understand and analyze data across relational databases or data warehouses. Further, primary key/foreign key constraints help maintain data integrity and avoid errors during data processing and modification, thus helping maintain data quality.

Constraints on Databricks

Constraints in databases are rules that ensure data integrity and consistency by enforcing certain conditions or restrictions on the data stored in a table. Databricks supports standard SQL constraint management clauses, which can be divided into two categories:

1. **Enforced Constraints:** These are enforced on the tables/columns to ensure data quality and integrity: NOT NULL and CHECK.

2. **Informational Constraints:** These constraints are not enforced but explain the relationships between fields in the tables: primary keys and foreign keys.

Enforced Constraints

Enforced column constraints are rules that apply to a single column in a table. Delta tables support the following column constraints:

- NOT NULL: This constraint ensures that a column must have a value for each row and cannot be null. This ensures data completeness and consistency.

- CHECK: This constraint validates that a column's value meets a specific condition or a range of conditions, such as ensuring that a particular column is within a certain range or a number is greater than a specific value. It helps ensure data accuracy and consistency. If the expression evaluates to false, an error is raised, and the statement is rolled back.

Below is an example of how to define these constraints on a table. Constraints can be set either when creating a new table or on an existing table. You can add constraints in a new table as shown in Listing 17-1.

Listing 17-1. Create a table statement with a CHECK constraint

```
CREATE TABLE T1 (
  id INT NOT NULL,
  quantity INT,
  date DATE,
  CONSTRAINT chk_quantity CHECK (quantity > 0)
);
```

To add a constraint to an existing table, you can use ALTER TABLE ADD CONSTRAINT, and to drop such a constraint, you can use ALTER TABLE DROP CONSTRAINT, for example, in Listing 17-2.

Listing 17-2. Alter table statement with CHECK constraint

```
ALTER TABLE T1 ADD CONSTRAINT dateWithinRange CHECK (Date > '1900-01-01');
```

Informational Constraints: Primary Key and Foreign Key

A Databricks Lakehouse with Unity Catalog gives users the ability to build entity relationships that are simple to maintain and evolve. Note that primary key and foreign key constraints are currently informational only and are not enforced. To leverage primary keys/foreign keys (PKs/FKs), your workspace should be UC-enabled with Databricks Runtime (DBR) version 11.1 and above.

Let's see how we can implement a primary key/foreign key relationship with an example. We can create a traditional "star schema" with three dimention tables and one fact table. They will be called fact_sales, dim_product, dim_store, and dim_customer, respectively, as shown in Listing 17-3.

Listing 17-3. Creating primary keys and foreign keys in Delta tables

```
--STORE DIMENSION
CREATE OR REPLACE  TABLE dim_store(
  store_id BIGINT GENERATED ALWAYS AS IDENTITY PRIMARY KEY,
  store_name STRING,
  address STRING
);

--PRODUCT DIMENSION
CREATE OR REPLACE  TABLE dim_product(
  product_id BIGINT GENERATED ALWAYS AS IDENTITY PRIMARY KEY,
  sku STRING,
  description STRING,
  category STRING
);
```

```
--CUSTOMER DIMENSION
CREATE OR REPLACE  TABLE dim_customer(
  customer_id BIGINT GENERATED ALWAYS AS IDENTITY (START WITH 0 INCREMENT
  BY 10) PRIMARY KEY,
  customer_name STRING,
  customer_profile STRING,
  address STRING
);

CREATE OR REPLACE TABLE fact_sales(
  sales_id BIGINT GENERATED ALWAYS AS IDENTITY PRIMARY KEY,
  product_id BIGINT NOT NULL CONSTRAINT dim_product_fk FOREIGN KEY
REFERENCES dim_product,
  store_id BIGINT NOT NULL CONSTRAINT dim_store_fk FOREIGN KEY REFERENCES
dim_store,
  customer_id BIGINT NOT NULL CONSTRAINT dim_customer_fk FOREIGN KEY
REFERENCES dim_customer,
  price_sold DOUBLE,
  units_sold INT,
  dollar_cost DOUBLE
);
```

The "View relationships" button (Figure 17-11) in the Overview or Schema tab conveniently shows the relationship between tables (Figure 17-12).

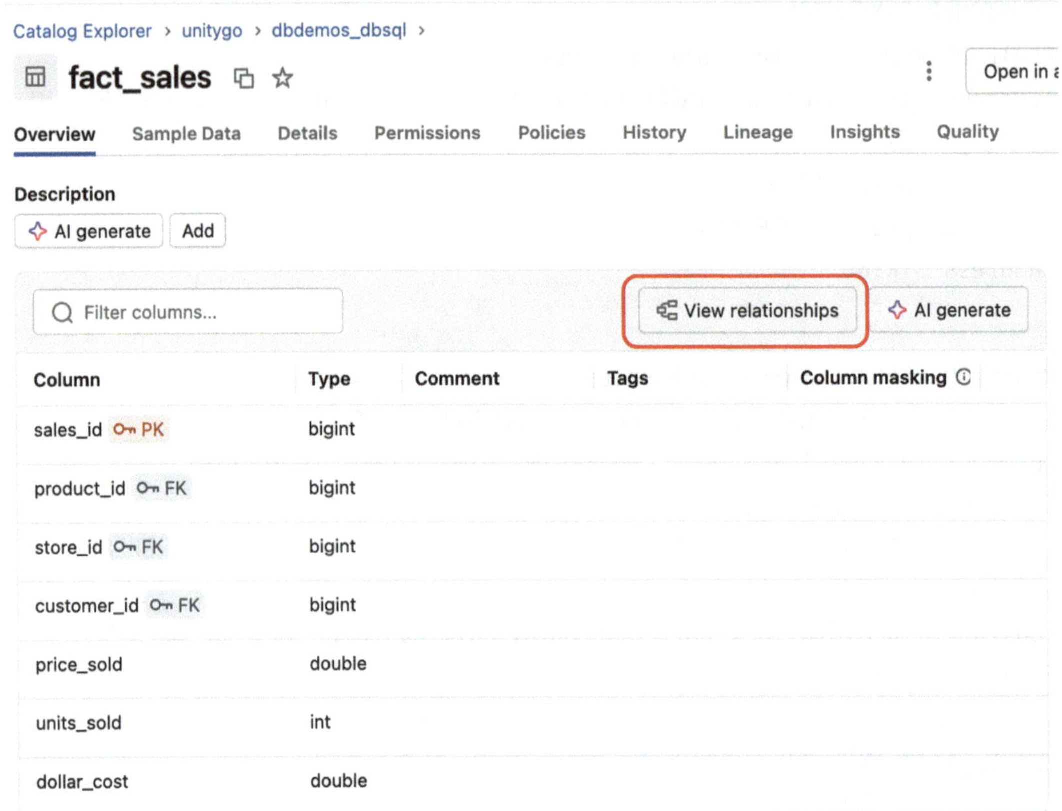

Figure 17-11. *"View relationships" button*

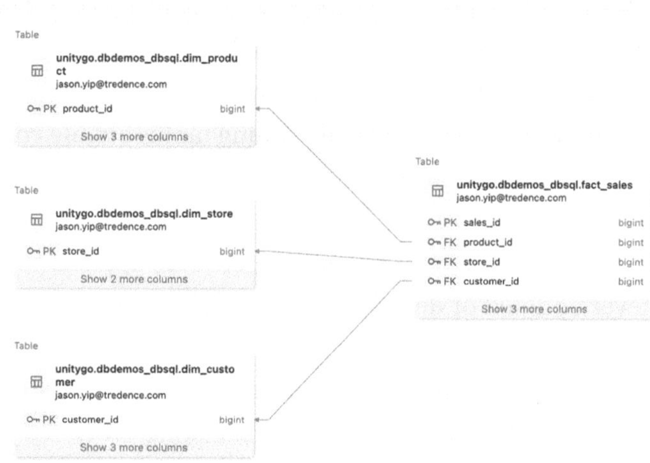

Figure 17-12. *ER diagram in Databricks*

Next, we move into streaming tables and materialized views. We touched on these two briefly in Chapter 9, but we will do a deeper dive here.

Streaming Tables and Materialized Views

Some common challenges data analysts face when working in data warehouses include being unable to self-service ingest and fix data issues, not having the most recent data for BI dashboards, and slow BI dashboards due to the large volume of underlying data.

Streaming tables and materialized views in DBSQL will allow SQL analysts to perform data engineering tasks, enabling real-time data capabilities within their existing workflows. It is important to note that both streaming tables and materialized views require Unity Catalog and serverless enabled in your workspace. In the next section, we will discuss these two features in detail.

Streaming Tables

A streaming table is a special type of table that enables ingestion in DBSQL. It is managed by Unity Catalog and supports append-only, incremental, and streaming data processing from various data sources. In the context of medallion architecture, streaming tables are ideal for bringing data into the Bronze layer and for incrementally processing data in subsequent layers. Streaming tables enable continuous, scalable ingestion from any data source, including cloud storage, message buses (Event Hubs, Kafka), and more.

Streaming from a source requires the source to be append-only, meaning existing records are never updated or deleted. To configure the streaming table to ingest your source, you must specify the STREAM keyword. Say, for example, you have an S3/ADLS container, and a lot of new files are continuously arriving. You can create a streaming table by using the following syntax as shown in Listing 17-4.

Listing 17-4. Creating a streaming table

```
CREATE OR REFRESH STREAMING TABLE mystream
  AS SELECT * FROM STREAM read_files('s3://<bucket>/<path>/<folder>')
```

By default, read_files processes all the files in the folder. To avoid this, you can set the property includeExistingFiles option to false as shown in Listing 17-5.

Listing 17-5. Streaming table with includeExistingFiles flag

```
CREATE OR REFRESH STREAMING TABLE mystream
AS
SELECT * FROM STREAM read_files('s3://<bucket>/<path>/<folder>',,
includeExistingFiles => false
)
```

Once the previous command is executed, a Lakeflow Declarative Pipeline is created under the hood for each streaming table. You can keep these tables updated and refreshed.

To load data from a system like Kafka, use the following command in Listing 17-6.

Listing 17-6. Reading from a streaming source

```
SELECT * FROM STREAM read_kafka(
  bootstrapServers => '<server:port>',
  subscribe => '<topic>',
  startingOffsets => 'latest'
);
```

Materialized Views

A materialized view is a database object that stores a query's results. Unlike regular virtual database views, which derive their data from the underlying tables any time they are run, materialized views contain precomputed data that is incrementally updated on a schedule or on demand. This precomputation of data allows for faster query response times and improved performance in certain scenarios.

A materialized view is a special type of view that precomputes and stores the results of a SQL query and keeps them up-to-date automatically through incremental maintenance. Unlike regular views, which compute results on-the-fly, Databricks materialized views track changes to the underlying data and apply only the incremental updates, enabling fast query performance even when the source data changes frequently. This makes them well-suited for dashboards and repeated queries over complex aggregations without manual pipeline code.

Create a Materialized View

Databricks SQL materialized view CREATE operations use a Databricks SQL warehouse to create and load data in the materialized view. Because creating a materialized view is a synchronous operation in the Databricks SQL warehouse, the CREATE MATERIALIZED VIEW command blocks until the materialized view is created and the initial data load finishes. A Lakeflow Declarative Pipeline is automatically created for every Databricks SQL materialized view. When the materialized view is refreshed, the Lakeflow Declarative Pipeline is started to process the refresh as shown in Listing 17-7.

Listing 17-7. Creating a materialized view

```
CREATE MATERIALIZED VIEW mv1
AS SELECT
  date, sum(sales) AS sum_of_sales
FROM
  table1
GROUP BY
  date;
```

Refresh a Materialized View

In Databricks SQL, you can set up automatic refresh for a materialized view on a predefined schedule. This schedule can be configured when creating the materialized view using the SCHEDULE clause or added later with the ALTER VIEW statement. Once a schedule is established, a Databricks job is automatically created to handle the updates. Listing 17-8 shows the syntax for scheduling and manual refresh.

Listing 17-8. Scheduling and manual refresh of a materialized view

```
CREATE OR REPLACE MATERIALIZED VIEW catalog.schema.hourly_metrics
  SCHEDULE EVERY 1 HOUR
AS SELECT
    date_trunc('hour', event_time) AS hour,
    count(*) AS events
FROM catalog.schema.raw_events
GROUP BY 1;

REFRESH MATERIALIZED VIEW mv1;
```

You can also trigger the refresh using TRIGGER ON UPDATE as shown in Listing 17-9.

Listing 17-9. Triggering updates when data source changes

```
CREATE OR REPLACE MATERIALIZED VIEW catalog.schema.hourly_metrics
  TRIGGER ON UPDATE
AS SELECT
    date_trunc('hour', event_time) AS hour,
```

```
    count(*) AS events
FROM catalog.schema.raw_events
GROUP BY 1;
```

Next, we move on to another important feature: Lakehouse Federation, which lets you query data stored in data sources without moving the data.

Lakehouse Federation

Lakehouse Federation provides query federation capabilities for the Databricks platform. Query federation enables users and systems to run queries across multiple data sources without migrating all data to a single central location.

Most organizations have valuable data distributed across multiple data sources— databases, data warehouses, object storage systems, etc. This siloed data leads to incomplete insights, hindering the ability to make informed decisions based on the full available data.

To query data across multiple data sources, users typically need to move or migrate their data to a central location, which usually takes time and effort. Lakehouse Federation addresses these critical pain points and makes it simple for organizations to expose, query, and govern siloed data systems as an extension of their lakehouse. Databricks supports federated queries across systems such as MySQL, PostgreSQL, Amazon Redshift, Snowflake, Azure SQL Database, Azure Synapse, BigQuery, and others. With Lakehouse Federation, you can query these external sources directly without moving or duplicating data, providing a unified, streamlined experience for accessing distributed datasets.

Further, Unity Catalog's advanced security features, such as row- and column-level access controls, discovery features like tags, and data lineage, are available in Unity Catalog across these federated data sources, ensuring consistent governance.

To make a dataset available for read-only querying using Lakehouse Federation, you create the following (Figure 17-13):

- **A connection** that specifies a path and credentials for accessing an external database system

- A **foreign catalog** that mirrors a database in an external data system, enabling you to perform read-only queries on that data system in your Databricks workspace, managing access using Unity Catalog

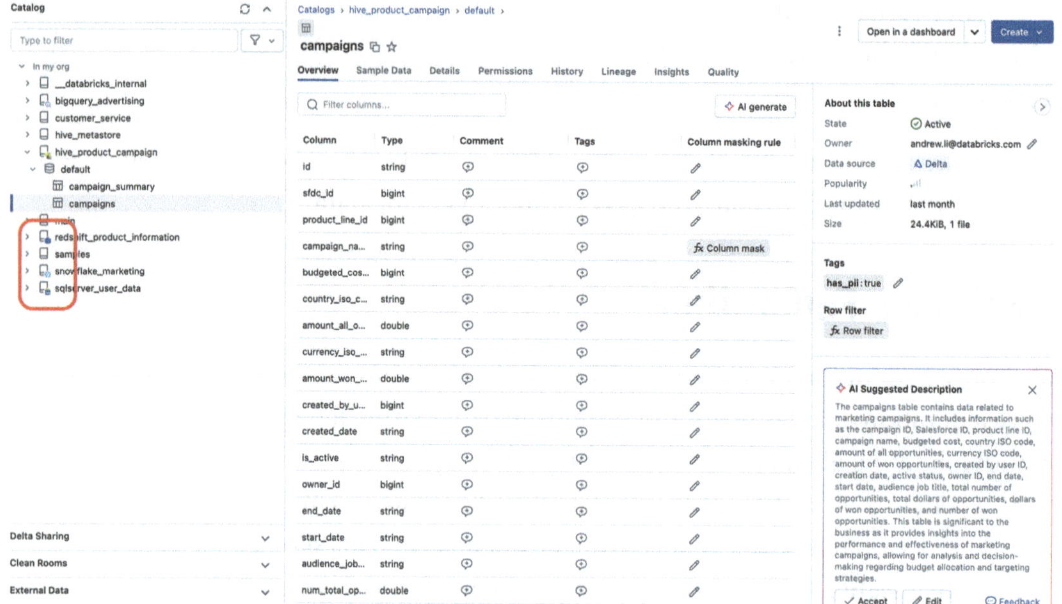

Figure 17-13. *Lakehouse Federation in action*

You can start to write queries against these tables in Databricks SQL and create visualizations to view the data.

Lakehouse Query Federation is ideal when you want to query live data directly from external sources without copying it into your lakehouse, i.e., when you need fresh, live data. However, as a best practice, Lakehouse Federation is read-only and adds network latency for remote queries, so it should not be used for real-time data processing, where latency is paramount, or complex data transformations, where vast amounts of data need to be ingested and processed.

AI Functions in DBSQL

In the age of AI, there is an urgent need to combine AI output into a BI report so management can take action based on the results. However, this inference pipeline not only creates another layer of complexity but also requires seasoned data scientists and an ML/LLM Ops team to maintain, which can become costly.

Consume LLM Models in DBSQL

Now, there are multiple ways to consume these large language models within Databricks. The traditional way is to leverage the code provided on Huggingface. Though this approach is flexible, it will require integrating the sample code into an existing pipeline, which requires development work.

The next approach is to use the Model Serving API. Databricks has curated popular models and made them part of the platform. These are then exposed as the Foundation Model API. With the Foundation Model API, developers can access these carefully curated models out of the box without going through the deployment process and getting enhanced performance.

The third approach we will learn in detail here is AI functions in serverless SQL.

AI functions enable analysts to integrate LLMs in SQL to enrich data and extract actionable insights.

There are two types of AI functions provided by Databricks:

1. **Built-in functions** backed by the Foundation Model APIs

2. **Custom functions** backed by a serverless model serving endpoint

Built-in functions invoke a state-of-the-art generative AI model to perform tasks such as sentiment analysis, classification, and translation. We will examine some common built-in functions.

- **ai_analyze_sentiment:** Given text, output the sentiment of the text, like positive, negative, neutral, or mixed.

- **ai_classify:** Ask the LLM to do classification. A good use case is to ask an LLM to determine if the text contains PII, which is to ask it if the text is `["contains PII" and "no PII"]`.

- **ai_extract:** Ask the LLM to extract any entities. Like regex patterns, but you no longer need to write a regex. You only need to tell the function what you want to extract. For example, "Place" will allow you to extract a place name.

- **ai_gen:** Prompting at scale. Given a list of questions, ask the LLM to output a list of answers, given in table format.

For a list of AI functions, please visit the Databricks website: `https://docs.databricks.com/aws/en/large-language-models/ai-functions`.

Next, we will put the AI functions in action. Consider this Kaggle Amazon review dataset.

We can download it to Databricks and create a Delta table. Using DB SQL's built-in AI functions, we can extract the sentiment from the text, successfully connecting AI with BI (see Figure 17-14).

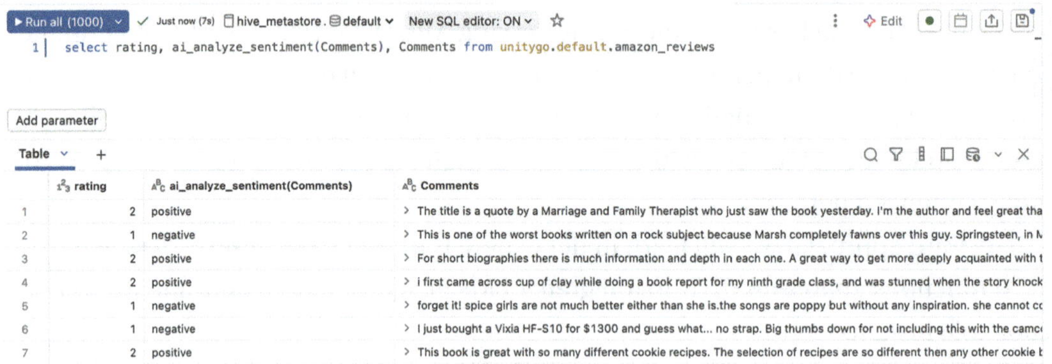

Figure 17-14. *Sentiment analysis with DB SQL*

Finally, after persisting the results in Unity Catalog, we can publish the enriched dataset to Databricks AI/BI Dashboard or any other BI tool of choice.

Custom Functions Backed by a Serverless Serving Endpoint

The `ai_query()` function allows you to query your ML and LLM models served by Databricks Model Serving endpoints using SQL. To do so, this function invokes an existing Databricks Model Serving endpoint and parses and returns its response. You can use `ai_query()` to query endpoints that serve custom models hosted by a model-serving endpoint, foundation models made available using Foundation Model APIs, and external models, which are third-party models hosted outside of Databricks.

Below is an example that queries the model behind the sentiment-analysis endpoint with the text dataset and specifies the request's return type as in Listing 17-10.

Listing 17-10. Specifying a return type in ai_query

```
SELECT text, ai_query(
    "sentiment-analysis",
    text,
    returnType => "STRUCT<label:STRING, score:DOUBLE>"
  ) AS predict
FROM
  catalog.schema.customer_reviews
```

In the next part of the chapter, we will examine how you can connect your BI tools through DBSQL.

Integrate BI Tools with Databricks

Organizations typically leverage BI tools such as Power BI, Tableau, and Looker for enterprise-wide dashboards and reporting. Moreover, many data analysts have been proficiently using these tools for quite some time. Databricks provides validated integrations with your BI tool of choice, allowing you to connect to your data using SQL warehouses or clusters. As a recommended practice, analysts get the best experience when they connect their BI tools to optimized gold tables via SQL warehouses.

In this section, we will focus on connecting Power BI to Azure Databricks. There are two main ways of doing this. The first is to publish to Power BI Online from Databricks. The second popular method is to connect Power BI Desktop to Databricks. Let's explore both these methods.

Publish to Power BI Online from Databricks

This allows users to publish tables from Databricks Catalog Explorer UI directly to Power BI workspaces. In short, this is a one-click publish of UC datasets to Power BI workspaces. This method supports both DirectQuery and Import modes. Moreover, you can publish entire schemas with table relationships (PK/FK). Some requirements are that the data must be registered in Unity Catalog, the compute must be UC-enabled, users must have a premium Power BI license, and users must enable "Users can edit data models in Power BI service (preview)" under workspace settings and data model settings. In Figure 17-15, we can go to the Catalog tab and select either the full schema or a particular table. Next, in the drop-down, select "Use with BI tools" and then "Publish to Power BI workspace."

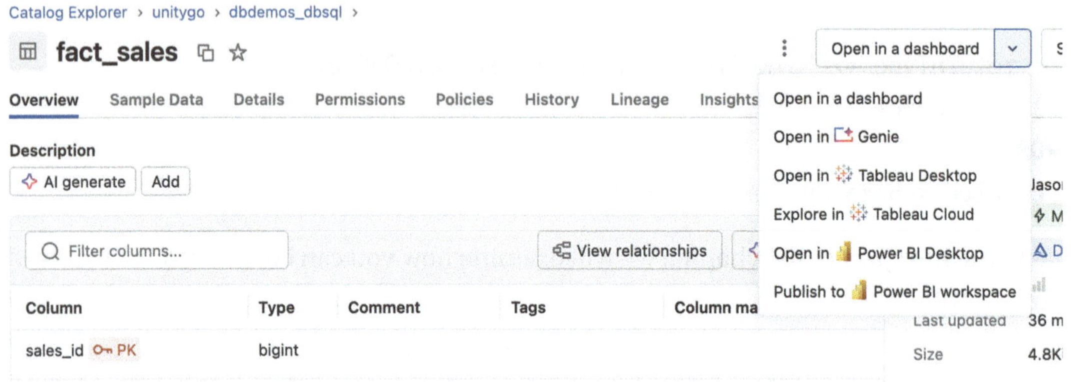

Figure 17-15. *Publishing to Power BI online*

You can select Publish to Power BI. This will ask you to authenticate with your Microsoft account via Microsoft Entra ID. Once the authentication is completed, the user can select the Power BI workspace and dataset mode (Direct Query or Import Mode). Thereafter, click Publish to Power BI, and you can start to query this dataset in your Power BI workspace.

Connect Power BI Desktop to Databricks

Users can also connect their Power BI Desktop to Delta Lake tables from Unity Catalog via the SQL warehouse for a full modeling experience in Power BI. Furthermore, Power BI offers three main storage modes for tables. The first is Import mode, in which all data is preloaded into Power BI's in-memory cache. Second is DirectQuery mode, in which the data remains in the source system, and the metadata is stored in Power BI. Finally, a newer feature is Hybrid Tables, which combine Import and DirectQuery modes by using partitions. The user can select the mode they want to use depending on the use case.

Let's now look at how to connect a Databricks SQL warehouse with Power BI. In Partner Connect (under menu Marketplace > Partner Connect integrations), when you click Power BI and select the Databricks Compute resource you want to connect to, the connection file is downloaded. You can open that file with Power BI Desktop, and your connection will be automatically configured. After selecting the connectivity mode, you can start querying the tables from Power BI as shown in Figure 17-16.

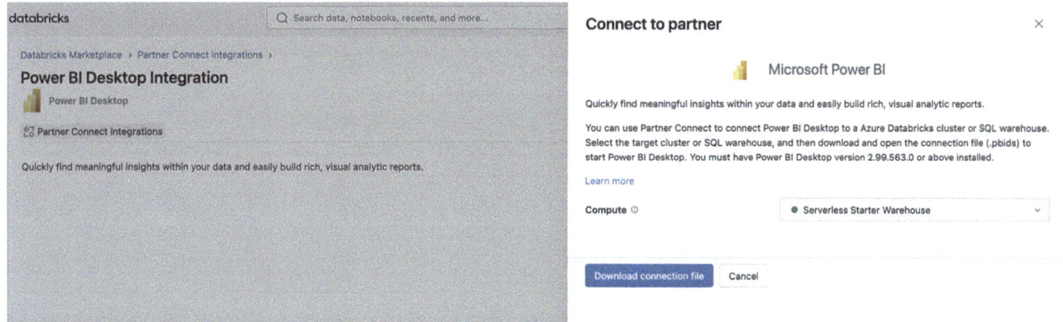

Figure 17-16. *Power BI connection in Partner Connect*

Conclusion

Databricks SQL provides comprehensive warehousing capabilities on the lakehouse platform and offers features for data analysts across various BI use cases. SQL warehouses, especially serverless ones, offer enhanced compute to process SQL queries and provide connections to various BI tools.

Some of the key features discussed in the chapter include Lakehouse Federation, AI functions, materialized views, streaming tables, and constraints, which include primary key/foreign key constraints. Finally, we saw how easily you can integrate the BI tool of your choice with the Databricks platform, along with Power BI as a case study.

Databricks Pricing and Observability Using System Tables

In this chapter, we will investigate how pricing for running workloads on Databricks works. It is important to understand how Databricks calculates the cost of your workloads. We will see what factors determine the pricing model and recommend which compute SKU should be used for running your specific workloads.

Then we will look at the concept of observability and how you can do observability on the Databricks platform using system tables.

Costs Associated with the Databricks Platform

Almost all Databricks costs are tied to compute resources used, not to the number of seats, unlike many SaaS vendors that charge per user. In Databricks, regardless of how many people access the platform, costs are driven primarily by the compute you use, which is especially attractive in the agentic era. Since Databricks decouples storage and compute, storage costs (which are provisioned in your cloud account) are billed directly to the cloud provider. Further, for compute, which is again provisioned in your cloud account, the costs can be divided into two parts—Databricks costs and cloud compute costs.

Cloud compute costs refer to the underlying hardware, such as virtual machines, disks, etc., within the customer's cloud account. The cloud compute is billed separately from the Databricks costs and is again paid directly to the cloud provider. It is important to note that the pricing model changes slightly with serverless compute, which we will discuss later in this chapter.

© Jason Yip, Nikhil Gupta and Marcin Wojtyczka 2026

J. Yip et al., *Databricks Data Intelligence Platform*, https://doi.org/10.1007/979-8-8688-2524-8_18

For this chapter, we will mostly look at the Databricks cost for running the compute resources in the lakehouse platform. But before we do that, let's look at some of the cloud cost components involved in running workloads on Databricks.

Cloud Infrastructure Costs

First, we will break down the costs associated with the lakehouse platform:

- **Storage Costs:** Within the lakehouse platform, data is stored in cloud storage (e.g., S3 on AWS, ADLS on Azure). The storage costs are paid directly to the respective cloud provider. The charges for storage are normally usage-based, i.e., they depend upon the amount of data being stored and the number of operations performed on it.

- **Compute Costs:** The cloud compute costs are the costs of using the cloud compute infrastructure (e.g., EC2 on AWS, VMs on Azure). Cloud infrastructure costs include fees for virtual machines (VMs), disks, and other resources, which are paid directly to the cloud provider. Since Databricks clusters are ephemeral, the cloud provider charges for the duration that the VMs are running in the Databricks cluster.

- **Networking Costs:** There are networking costs associated with deploying/running workloads within workspaces. Some of these are the costs of NAT gateways, load balancers, and private links (if enabled). Furthermore, if the data and workspace are in different regions, egress costs apply as well. All of these costs are again paid directly to the cloud provider.

Next, we will learn how to calculate Databricks' DBU costs and infrastructure costs.

Databricks Pricing

What Are Databricks Units?

A *Databricks unit* (DBU) is a normalized unit of processing power. Databricks consumption occurs through clusters (job or all-purpose compute), SQL warehouses, and serverless compute, all of which are priced in terms of DBUs. DBUs are the underlying unit of consumption within the platform. However, the billing is based on per-second usage.

Next, we will examine the factors that determine DBU consumption for Databricks compute. These are the three key factors that influence the cluster price:

- **Compute Size and Type:** This refers to the size and type of VMs one chooses, both for the worker and master nodes in the cluster. Depending on the VM size, the number of DBUs also changes. Further, the number of DBUs consumed depends on whether Photon is enabled on the cluster. The compute size and type determine the number of DBUs that are consumed. This is greatly simplified when using serverless compute.

- **Product SKU:** The Product SKU determines the charge per DBU-hr. For example, the per second price is DBU-hr / 3600. There are several different SKUs for the compute resources. This includes all-purpose cluster, jobs cluster, Lakeflow Declarative Pipelines, SQL warehouse (Classic and Pro), serverless compute, etc. Generative AI uses a slightly different method to calculate cost, but it is still based on DBUs. Depending on the SKU used, the amount charged for DBU-hr.

- **Account Tier:** This is the Databricks account pricing tier in which the workspace runs. One can select Standard, Premium, or Enterprise for Azure Premium or Enterprise for AWS and GCP. Depending on the tier, the number of DBUs charged varies.

After reviewing some of the levers that determine DBU consumption, let's move on to an example that shows how to calculate the dollar value of a cluster's pricing. In the following example, we assume a nine-node cluster, together with the driver node, a total of 10 VMs powers the cluster, as shown in Figure 18-1.

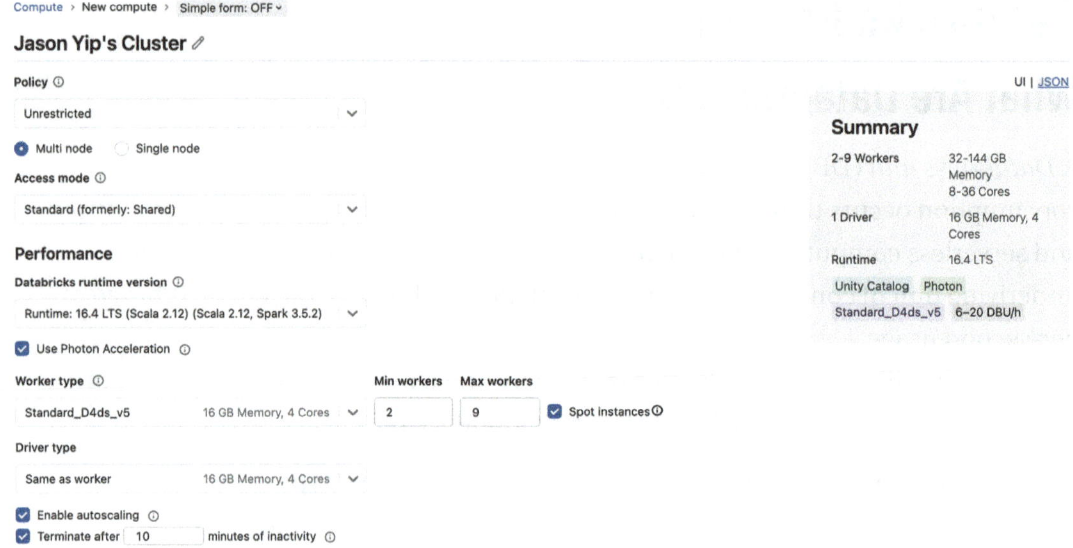

Figure 18-1. *Databricks cluster configuration*

In the **Summary** box in Figure 18-1, you can see that the cluster of this configuration would consume 15 DBUs/hr. Next, we can see how much that will cost in dollars. Figure 18-2 shows the Azure Databricks pricing page. (Please note that this pricing is as of the writing of this book. For actual and most current pricing, visit: `https://www.databricks.com/product/pricing`.)

Workload	Standard Tier Prices	Premium Tier Prices
Interactive Serverless Compute***	-	$1/DBU-hour
All-Purpose Compute**	$0.40/DBU-hour	$0.55/DBU-hour
Automated Serverless Compute***	-	$0.47/DBU-hour
Jobs Compute**	$0.15/DBU-hour	$0.30/DBU-hour
Jobs Light Compute	$0.07/DBU-hour	$0.22/DBU-hour
SQL Compute	-	$0.22/DBU-hour
SQL Pro Compute	-	$0.55/DBU-hour
Serverless SQL***	-	$0.70/DBU-hour

Figure 18-2. *Azure Databricks pricing information*

The cluster shown in Figure 18-1 is an all-purpose compute and is created in a Premium Databricks account. Therefore, as shown in Figure 18-3, this cluster is priced at $0.55/DBU-hr. Since the cluster is consuming 15 DBU/hr, we can easily calculate that the price for running this cluster would be as follows:

Total DBU cost: 15* $0.55 = $8.25/hr.

You can also estimate the Databricks-related costs using the Databricks Pricing Calculator:

```
https://www.databricks.com/product/pricing/product-pricing/instance-types
```

Further, we will calculate the price of the VMs that are being used for the cluster. Since this is Azure Databricks and the VMs are Standard_DS3_V2, we can go to the VM pricing page and find the cost to run 10 VMs for 1 hr.

Instance	vCPU(s)	RAM	DBU Count	DBU Price	Pay As You Go Total Price	1 Year Savings Plan (% Savings) Total Price	3 Year Savings Plan (% Savings) Total Price	Spot (% Savings) Total Price
DS3 v2	4	14.00 GiB	0.75	$0.413/hour	$0.692/hour	$0.606/hour ~13% savings	$0.544/hour ~21% savings	$0.465/hour ~33% savings

Figure 18-3. *Pricing of DS3 v2*

Total VM cost: 10 * $0.2930 = $2.93/hr

Therefore, the user will need to pay $11.18/hr ($8.25 + $2.93). Further, as noted earlier, the Databricks DBU cost would be paid to Databricks, and the VM cost of $2.93 would be paid to the cloud provider.

To check the price of different VMs in all three clouds, we can refer to the following websites:

- Azure: `https://azure.microsoft.com/en-us/pricing/details/virtual-machines/linux/`

- AWS: `https://aws.amazon.com/ec2/instance-types/`

- GCP: `https://cloud.google.com/compute/docs/general-purpose-machines`

In the previous example, we have enabled Photon acceleration. Please note that the number of DBUs required for the Photon engine is 2x higher. Therefore, if we disable Photon, the cost of the DBU would be $4.125, and the total cost will become $4.125 + $2.93 = $7.055.

A logical question we would ask is, should we enable Photon to pay a premium price? It is important to note that although Photon appears to be twice as expensive, its performance will be roughly 3x higher than without it. Therefore, for most workloads, Photon offers a better price-to-performance ratio than workloads running without Photon enabled. Figure 18-4 illustrates that when running a sample NYC Taxi query, the performance is three times faster, and in our experience, the performance guarantee is quite consistent.

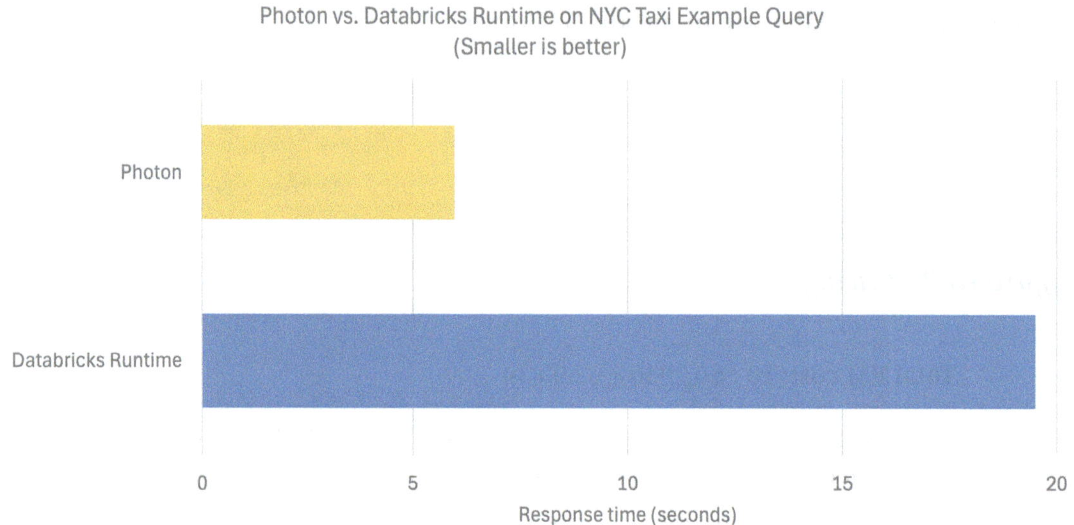

Figure 18-4. *Photon vs. Databricks Runtime on the NYC taxi example query*

Databricks SQL comes with Photon free of charge. For more information, please refer to Chapter 17.

Before we move on, we will discuss another important aspect of job compute pricing. Referring to the previous example, let's assume we had spun up a jobs cluster instead of the all-purpose compute. In Figure 18-3, the job's compute is $0.3 DBU/hr, which is almost 50% less than the all-purpose compute. The cost of running the same cluster would be 15 * $0.3 = $4.5, and the cloud compute cost of $2.93 would remain the same. Therefore, it is strongly recommended that all automated jobs always utilize job clusters.

Another important aspect that will help you reduce the Databricks bill is the committed-use contracts. By default, your Databricks account will run using a pay-as-you-go (PAYG) model, where you are billed only for the compute you actually use, offering maximum flexibility but potentially higher costs for steady or growing workloads. A commitment deal involves agreeing to a fixed amount of usage over a period at a discounted rate. By committing, you can further reduce your costs compared with pay-as-you-go while ensuring a more predictable budget for your workloads.

SQL Warehouse Pricing

In this section, we will learn about the pricing of Databricks' SQL warehouse compute and how pricing differs from other computes in Databricks.

There are three types of SQL warehouse computes available in Databricks: Classic, Pro, and Serverless. Classic and Pro are similar to all-purpose computers, where the DBU cost and underlying infra cost are paid separately, but Photon is included in the price. Second, Serverless is one all-inclusive price.

As discussed earlier, Databricks fully manages the underlying cloud compute instances for Serverless. Therefore, rather than having two separate charges (i.e., the DBU compute cost and the underlying cloud compute cost), as it is the case in Classic and Pro Warehouses, the user pays only a single charge to Databricks for both. The pricing calculation for Classic and Pro SQL warehouses remains the same as discussed earlier. Next, we will investigate how we calculate the pricing of serverless SQL warehouse compute.

In Figure 18-5, we have an X-Large cluster size that will consume 80 DBU/hr.

Figure 18-5. *Creating a new SQL warehouse*

According to the Databricks pricing page as of writing this book, Classic SQL is $0.22/hr, Pro is $0.55/hr, and Serverless is $0.7/hr. Therefore, the total cost for the example X-Large serverless SQL warehouse for one hour would be $0.70 × 80 = $56. This is the total cost, including the underlying infrastructure. This applies only to Serverless, as the other cluster types will require users to pay for the underlying infrastructure as well.

Databricks SQL does not allow you to choose the infrastructure, unlike all-purpose and job clusters. However, Databricks has carefully selected the most suitable VMs for each cloud and tuned performance to deliver the best price/performance for their analytical SQL workloads. To understand what Databricks chose for the underlying infra, please refer to the following:

- Azure: `https://learn.microsoft.com/en-us/azure/databricks/compute/sql-warehouse/warehouse-behavior`

- AWS: `https://docs.databricks.com/en/compute/sql-warehouse/warehouse-behavior.html`

- GCP: `https://docs.gcp.databricks.com/en/compute/sql-warehouse/warehouse-behavior`

Databricks Cost Management Best Practices

In this section, we will explore best practices for cost management on the Databricks Platform.

1. **Cluster Policies**

   Cluster policies allow users and groups to follow pre-defined rules when configuring or spinning up clusters. With cluster policies, admins can limit cluster size and define the types of VMs that can be used when creating clusters. Furthermore, admins can set the maximum DBUs per hour for the clusters. Therefore, by enforcing cluster policies, admins can manage compute costs on the platform.

2. **Cluster Access Controls**

   Cluster access controls allow admins to control which users can create clusters. Cluster-level permissions control whether a user can attach to, manage, or restart a cluster. As a best practice, cluster creation abilities should be given to admins who can manage and govern access to the clusters.

3. **Cluster Autoscaling and Cluster Termination**

   Databricks cluster autoscaling automatically adds and removes worker nodes in response to changing workloads to optimize resource usage. Autoscaling makes it easier to achieve high cluster utilization because you do not need to provision the exact number of nodes to match the workloads. This cannot only enable workloads to run faster than on an under-provisioned cluster but also reduce overall costs compared to a statically sized cluster by improving resource utilization.

   Cluster auto-termination terminates a cluster after a specified inactivity period and starts whenever needed. As a best practice, always enable auto-termination for all-purpose clusters to prevent these clusters running overnight or over weekends.

4. **Spot Instances**

    When you are requesting certain types of clusters, you are
    requesting new virtual machines from the cloud provider. Cloud
    providers also need to maximize resource utilization. Spot
    instances let you use idle compute or spare capacity at a discount
    in the cloud provider's region, up to the price you can optionally
    specify in advance. One thing to note about spot instances is that
    the cloud provider can recall them at any time, even while your job
    is running, so this is not suitable for critical workloads with strict
    SLAs. Spot instances can be deployed in some long-running jobs
    where, if the cluster restarts midway, the job continues from where
    it left. Further, you can also configure it to fall back to on-demand
    instances.

5. **Instance Pool**

    Instance pools apply at the workspace level, where administrators
    can pre-allocate popular virtual machine instances, either on-
    demand or via spot. Then, when clusters are starting, there is
    no need to acquire them from the cloud provider, which speeds
    up cluster startup. However, administrators must be careful
    not to pre-allocate too many virtual machines, because once
    provisioned, the cloud provider will charge the account regardless
    of whether they are actively used.

6. **Cluster Tags**

    Cluster tags can be associated with Databricks clusters to attribute
    cost for chargeback purposes. For example, if an organization has
    multiple BUs, each BU can tag its clusters with specific tags.

For an in-depth analysis of different cluster strategies, please refer to the Databricks
website:

`https://docs.databricks.com/en/compute/cluster-config-best-practices.html.`

Databricks Observability: System Tables

Observability is a key aspect of modern cloud data platforms. In simplistic terms, observability is how well one can understand the IT system from its generated outputs, such as logs, metrics, and traces. Therefore, observability gives admins a way to optimize and control their platforms based on the data they generate.

Some of the typical use cases platform admins might be interested in include the following:

- Monitoring costs

- Monitoring security and audit

- Monitoring platform usage/pipeline states

- Data observability and optimization

- Performance/resource utilization

Databricks' system tables, integrated within Unity Catalog, provide curated datasets that enable users to query and answer these use cases.

Introduction to System Tables

Let's explore how to use system tables to improve observability of the Databricks platform.

System tables are a Databricks-hosted analytical store for operational and usage data. They are fully integrated with Unity Catalog.

System tables can be used to monitor and analyze the performance, usage, and behavior of Databricks platform components. By querying these tables, users can gain insights into how their jobs, notebooks, users, clusters, ML endpoints, and SQL warehouses function and change over time. This historical data can be used to optimize performance, troubleshoot issues, track usage patterns, and make data-driven decisions.

System tables enhance observability and provide valuable insights into the operational aspects of Databricks usage, enabling users to better understand and manage their workflows and resources. Based on the schemas/tables available as of writing the book, you can address the following use cases:

- Cost and usage analytics

- Efficiency analytics

- Audit analytics

- GDPR regulation

- Service-level objective analytics

- Data quality analytics

System tables are available to customers who have **Unity Catalog** activated in at least one workspace in their account. The data in these tables is collected across all workspaces in the account, regardless of whether Unity Catalog is enabled on those workspaces. However, system tables would be visible and queryable only in the workspace with Unity Catalog enabled.

Since system tables are governed by Unity Catalog, you need at least one Unity Catalog–governed workspace in your account to enable system tables. That way, you can map your system tables to the Unity Catalog metastore. System tables must be enabled by an account admin. You can enable system tables in your account either by using the Databricks CLI or by calling the Unity Catalog API in a notebook.

The system tables are organized in a catalog named `system`, a fundamental component of every Unity Catalog metastore. Inside this catalog, you'll find schemas such as access and billing that house the system tables. These tables offer a comprehensive view of your Databricks environment, enabling you to make informed decisions and optimize resource allocation (see Figure 18-6). Note that the billing schema is enabled by default; the others must be enabled manually.

```
∨ ⬓ system
    > ⊜ __internal_logging
    > ⊜ access
    > ⊜ ai
    > ⊜ ai_gateway
    > ⊜ billing
    > ⊜ compute
    > ⊜ information_schema
    > ⊜ lakeflow
    > ⊜ mlflow
    > ⊜ query
    > ⊜ serving
    > ⊜ storage
```

Figure 18-6. *Databricks system table catalogs*

For details of the system table schema, please refer to the Databricks documentation:

`https://docs.databricks.com/en/admin/system-tables/index.html`

System table access is governed by Unity Catalog. By default, no users have access to system tables. To grant access, a metastore admin or other privileged user must grant USE SCHEMA and SELECT permissions on the system schemas.

Common Schemas/Tables Available with System Tables

These are the most popular schemas/tables available with the system tables:

- **Audit Logs:** Includes records for all audit events across your Azure Databricks account.

- **Billing Usage:** Includes records for all billable usage across your account. Each usage record is an hourly aggregate of a resource's billable usage.

- **Table Lineage:** Includes a record for each read or write event on a Unity Catalog table or path.

- **Workflow:** Allows you to view records related to job activity in your account. Further, you can join job system tables with billing tables to monitor the cost of jobs across your account.

These system tables provide valuable insights into the activities, resource utilization, and data lineage within your Databricks account and can be used for historical KPI tracking, monitoring and alerting, and forecasting expected usage for a data intelligence platform. There are many more schemas, such as Pricing, Cluster, SQL Warehouse, etc., that users can analyze to ascertain the operational health of the Databricks platform.

To give readers an actionable example of how system tables can be used to monitor the performance of SQL Warehouses, the following query can be used to identify SQL Warehouses with large query queues, which may indicate a need for autoscaling. SQL alerts can also be created on this query to notify administrators automatically when scaling is required, as shown in Listing 18-1.

Listing 18-1. Example query to monitor a warehouse performance

```
SELECT compute.warehouse_id, SUM(waiting_for_compute_duration_ms) AS
queueing_ms
FROM system.query.history
WHERE start_time BETWEEN '2026-02-01' AND '2026-04-01'
GROUP BY ALL HAVING queueing_ms > 0
ORDER BY 2 DESC;
```

System Table: Billing Usage Example

In the data and AI era, wherever there is data, there is an opportunity for AI. The granularity of the billing tables is detailed enough to use as an input for a time-series forecast model. Databricks has built a demo, and the notebooks are available here:

`https://notebooks.databricks.com/demos/uc-04-system-tables/index.html`

Figure 18-7 illustrates that we can use cluster SKU and workspace ID, along with the historical cost trend, as training data to predict the future cost and feed into a dashboard for monitoring purposes.

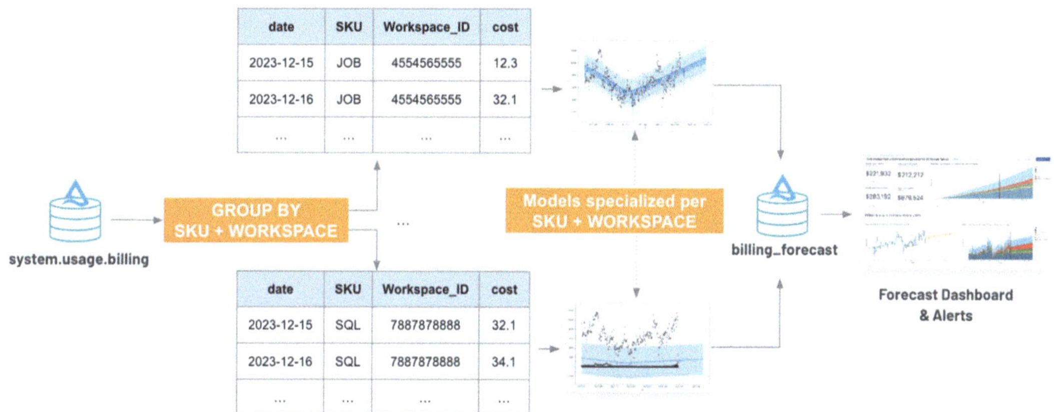

Figure 18-7. *Predictive analysis of utilization and pricing*

You can build your custom solutions by leveraging the monitoring tables for predictive analysis and achieve greater savings in terms of cluster pooling, termination time, and beyond.

Conclusion

In this chapter, we looked at how Databricks cost is calculated. There are two types of costs associated with Databricks compute: cloud compute costs for VMs, paid directly to the cloud provider (non-serverless clusters only), and DBU costs, paid to Databricks. We looked into how to calculate costs for various compute SKUs like all-purpose clusters, job clusters, and serverless SQL warehouses.

Then we moved into observability on the Databricks platform using system tables.

System tables in the Unity Catalog provide great insights to administrators who want to dig deeper into the platform, such as audit logs, pricing, and lineage. We have also demonstrated that, beyond standard operational reports, teams can create predictive analytics with the data, making it great for the finance team to do budgeting.

From Ideation to Creation: Building Intelligent Products with GenAI

In the last edition, we created a chatbot for understanding diabetes; you are encouraged to check out the chapter. Despite the chatbot now being achievable with Agent Bricks' Knowledge Assistant, it's still critical to understand the end-to-end workflow. In this second edition, we will walk through a practical healthcare case study in which we will attempt to fine-tune a vision language model (VLM), a.k.a. Qwen3-VL-2B-Instruct, to watch real surgery videos and generate post-surgical reports to train new surgeons. It's an ambitious and advanced attempt that will demonstrate the full capabilities of the Databricks Data Intelligence Platform.

Multimodal language models are evolving rapidly, but enterprise tooling is still catching up. That's why most of the use cases we see today are around OCR or documents. No doubt, our society relies on many documents, but being able to describe and detect objects from a prompt without a custom neural network is a huge step forward.

To simplify the process, we will fine-tune a model with only one GPU; this is meant to show what's possible and not for production and to ensure anyone without access to premium resources can also run the code. We will then generate a report using Databricks Genie.

The Problem Statement

Throughout this book, we will be using the `CholecSeg8k` dataset from Hugging Face. The full dataset name is called `minwoosun/CholecSeg8k`. `CholecSeg8k` is a rare public semantic segmentation dataset for laparoscopic cholecystectomy (gallbladder removal) surgeries. Semantic segmentation is a pixel-level annotation where each pixel is labeled according to the object class it belongs to (e.g., car, road, pedestrian). A common use case is autonomous driving, where it uses both bounding boxes (for object detection) and semantic segmentation (for scene understanding). For example: detecting a car (bounding box) and understanding the road/pedestrian area (segmentation). Figure 19-1 illustrates this concept from the perspective of an autonomous vehicle. An example can be found on Tesla's website, which provides detailed videos highlighting this process: `https://www.tesla.com/AI`

Figure 19-1. *Sample object detection in an autonomous vehicle (image by Jason Yip)*

The Cholec80 dataset contains 80 videos of cholecystectomy surgeries performed by 13 surgeons, without annotations. `CholecSeg8k` is based on the original Cholec80 dataset. The team at National Taiwan University extracted 8,080 laparoscopic cholecystectomy image frames from 17 video clips in Cholec80 and annotated them. The goal was to enhance the accuracy of computer-assisted surgeries. However, in this chapter, we will explain the annotations to new surgeons, as annotations alone, without an English description, can be challenging at times. Figure 19-2 shows one of the annotations found in the `CholecSeg8k` dataset.

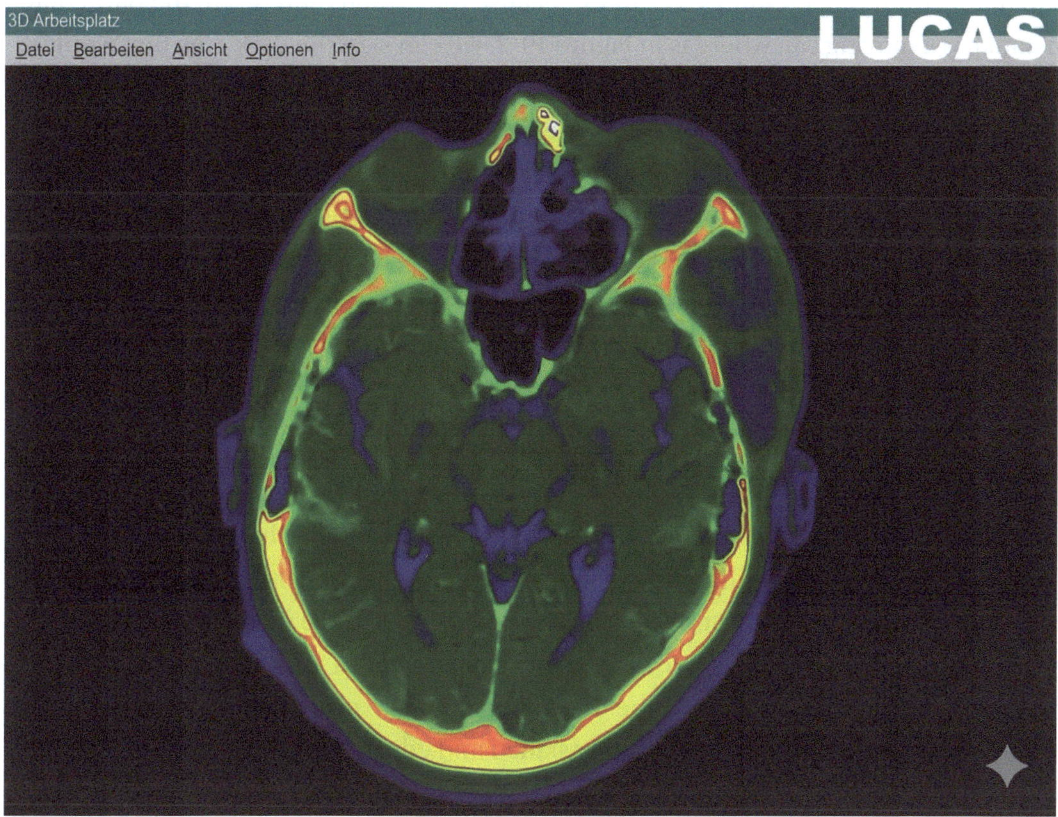

Figure 19-2. *Semantic segmentation example in a surgical setting
(Rüdiger Marmulla, CC BY-SA 3.0 <https://creativecommons.org/licenses/
by-sa/3.0>, via Wikimedia Commons) (This image is not included in the book's
Creative Commons license.)*

While pixel-level annotation can provide very precise information, as shown in Figure 19-1, semantic segmentation is often used alongside bounding boxes because the computational cost of drawing a bounding box is much lower than classifying every pixel, especially in real-time scenarios. Furthermore, by localizing the detection area, we can reduce the annotation area. In the case of autonomous driving in Figure 19-1, we can skip the surrounding buildings and trees because the primary interest in the particular instance is a moving car, whereas in Figure 19-2, we can skip the dark area that would otherwise be meaningless in a surgery.

At the same time, we are leveraging fine-tuning a pre-trained model rather than training a custom deep learning model from scratch, because base models in general have been trained on many images and their patterns, regardless of whether they were trained on surgery videos. The images required to teach a VLM something new are much less than what's required for training from scratch. That makes 8,080 images a perfect fit for this task and gives us a chance to achieve meaningful results with one GPU.

End-to-End Architecture

The ideation process of this project is an attempt to go beyond a chatbot and text-based finetuning, as well as a continuation of the last edition's healthcare and life science use case. We want to kick off some discussions on a difficult use case. Rather than a mock end-to-end use case, we are interested in leaving room for future exploration and development. It's not hard to imagine that elite healthcare professionals usually translate to surgeons because they are both highly skilled and educated. If we can open a door in this high specialty, we can potentially save lives using Databricks.

Figure 19-3 shows the architecture of the fine-tuning portion of this project. Since Genie Deep Research is just a click of a button away, we will focus on generating high-quality inference outputs. With these outputs, we will feed into Genie for its research agent to make sense out of them.

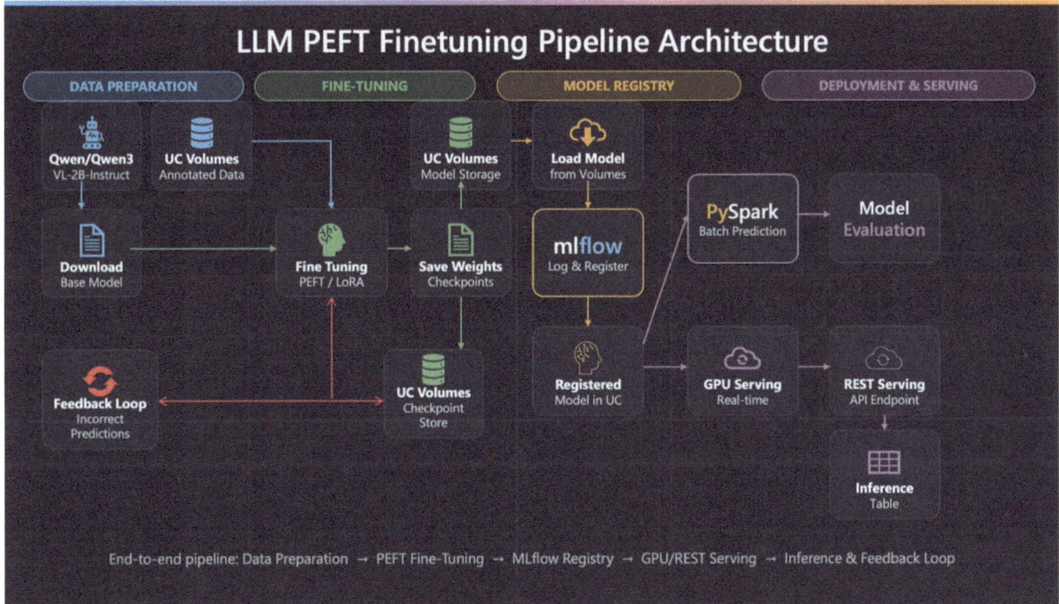

Figure 19-3. *End-to-end architecture of the fine-tuning process*

This architecture is designed to take a general-purpose base LLM, specialize it using specific data, deploy it in different ways, and continuously improve it through a feedback loop. Therefore, it's not specific to the Qwen model we chose. Below is the high-level end-to-end architecture:

1. **Infrastructure and Inputs**

   The entire pipeline operates within a Databricks Workspace sitting on top of the serverless GPU architecture. It starts with two key inputs:

   - **Base Model:** A pretrained Large Language Model, specifically Qwen/Qwen3-VL-2B-Instruct, is selected and downloaded from Hugging Face.

   - **Training Data:** Specific annotated data (labeled data suited for the target task) is loaded from Hugging Face into UC (Unity Catalog) Volumes, which serves as the central storage layer.

2. **The Fine-Tuning Process**

   The goal of this process is to combine the base model and the annotated data to converge to a model that understands specific annotation requirements.

   Using Parameter-Efficient Fine-Tuning (PEFT) techniques, the model is fine-tuned on the annotated data. PEFT ensures that instead of retraining the entire base model, only a small set of extra parameters (weights) are updated, making the process faster and more computationally efficient.

3. **Saving Artifacts**

   Once fine-tuning is complete, the essential outputs are preserved. The new model weights (the PEFT adapters), training checkpoints, and the associated tokenizer are saved back into UC Volumes for persistent storage. These weights can be retrieved later for inference without having to retrain the model.

4. **Model Registration via MLflow**

   MLflow is used to log experiment details and formally register the model in UC. This creates a versioned, "Registered model" object that acts as the single source of truth for deployment.

5. **Deployment Pathways**

   The Registered Model can then be used for batch inferencing or served via Databricks Model serving.

   - **Batch Prediction:** The model can be registered as a PyFunc model for high-volume, offline processing. The output of these batch jobs can be used for post-surgical analysis, and performance metrics can be monitored in Databricks Lakehouse Monitoring.

   - **Real-Time Serving:** The model is deployed using Mosaic AI GPU Model Serving infrastructure to expose a REST Serving endpoint. This allows external applications to send real-time requests and get immediate predictions, which will assist surgeons in pinpointing exact locations to operate. All inputs and outputs from this endpoint can be logged into an inference table.

6. **The Feedback Loop (Continuous Improvement)**

   Finally, the architecture includes a mechanism for iteration using the Databricks Review App. Experts can review the output and give feedback and correct the labels if required. These corrections can then be fed back into the fine-tuning pipeline to enhance accuracy. They become part of the updated **annotated data**, allowing the model to be retrained and improved in future cycles.

Data Ingestion

In an enterprise setting, Unity Catalog is the de facto standard for governing both structured and unstructured data. Please refer to Chapter 13 if you are not familiar with Databricks' ingestion methods. In our case, we are in the development phase, which is why we are pulling directly from Hugging Face. If you are not familiar with Hugging Face, it is one of the most popular repositories for open-source models and datasets, with a footprint that includes all the top open-source large language models.

Loading Multimedia Files

Normally, loading files from Hugging Face is very straightforward with one line of code, `load_dataset`, from the `datasets` library. However, if we try to load the `minwoosun/CholecSeg8k` dataset as shown in Listing 19-1, we will get an error saying "Dataset scripts are no longer supported," as shown in Figure 19-4. That's because Databricks has moved away from allowing the execution of arbitrary scripts within datasets for security reasons.

Listing 19-1. Loading the CholecSeg8k dataset

```
from datasets import load_dataset
dataset = load_dataset("minwoosun/CholecSeg8k")
```

```
1  from datasets import load_dataset
2  dataset = load_dataset("minwoosun/CholecSeg8k",
3                         cache_dir='/Volumes/unitygo/cholecseg8k/images')
```

/databricks/python_shell/lib/dbruntime/huggingface_patches/datasets.py:24: UserWarning: D
ess bar widgets that can cause performance issues for your notebook or browser. To avoid
ar()` to turn off the progress bars.
  warnings.warn(

🔴 > RuntimeError: Dataset scripts are no longer supported, but found CholecSeg8k.py

Figure 19-4. *Data loading error for the CholecSeg8k dataset*

Unlike text data, multimedia files often require additional processing to be meaningful because they do not contain a human-readable description. Without a human-readable description, machine interactions might not be very meaningful. That's why, in the case of CholecSeg8k dataset, it contains a script to extract different images in addition to the original one, known as masks, as shown in Table 19-1.

Table 19-1. *Mask type and its purpose*

Mask Type	Purpose
Color Mask	**Visualization:** Designed for human experts to easily identify different classes. Each class is painted in a distinct color.
Watershed Mask	**Programming/Processing:** Created for easier computational use. Each annotated pixel has the same class ID value across all three color channels, making it simpler to read programmatically.
Annotation Mask	**Tooling:** This is the original hand-drawn mask used by the PixelAnnotationTool to generate the other two masks (color and watershed).

An example of the original image and the three masks are available on the Hugging Face website: https://huggingface.co/datasets/minwoosun/CholecSeg8k.

Not everything is visible to the naked eye, but we can extract information from these images, and that's what we will do in the next section. However, in this section, we need a way to resolve the script error because it is no longer safe to execute a script remotely in a dataset, as it can easily cause security problems.

One idea is to download the zip file to Unity Catalog, extract the files, and access the masks. This is probably the most straightforward way to process the data. However, it will take a very long time to download them, and the approach is not portable. Fortunately, Hugging Face provides a way to access its file system directly via HfFileSystem. We can then directly access the zip file stored in the remote location and implement the script in our environment. This is a critical step that will save us hours of time trying to access the files. Listing 19-2 shows that we can read the zip file as if it were located on our local system.

Listing 19-2. Opening a zip file from Hugging Face

```
fs = HfFileSystem()
hf_zip_path = "datasets/minwoosun/CholecSeg8k/data/CholecSeg8k.zip"
zip_file = fs.open(hf_zip_path, 'rb')
```

Bounding Box

As discussed above, the color mask is designed for human eyes, whereas the watershed mask is designed for computers to process, with each pixel containing a class number. In the context of this dataset, there are 13 classes as shown in Table 19-2.

Table 19-2. *Classes identified in the photos*

Class ID	Class Name
Class 0	Black Background
Class 1	Abdominal Wall
Class 2	Liver
Class 3	Gastrointestinal Tract
Class 4	Fat
Class 5	Grasper
Class 6	Connective Tissue
Class 7	Blood
Class 8	Cystic Duct
Class 9	L-hook Electrocautery
Class 10	Gallbladder
Class 11	Hepatic Vein
Class 12	Liver Ligament

The watershed mask contains one of the 13 classes, which is shaped exactly like the color mask. As a result, we can extract the coordinates and compute a bounding box for these classes to narrow the search for semantic segmentation. We will use this text output as instructions for our fine-tuning. Listing 19-3 illustrates how to do it in Python.

Listing 19-3. Extracting bounding boxes and detecting tool-organ interactions

```python
import numpy as np
from scipy.ndimage import binary_dilation

def generate_detection_text(mask_image) -> str:
    """Convert segmentation mask to bounding box text format."""
    try:
        # Handle dict format from Hugging Face datasets
        if isinstance(mask_image, dict):
            if "bytes" in mask_image and mask_image["bytes"]:
                mask_image = Image.open(BytesIO(mask_image["bytes"]))
            elif "path" in mask_image and mask_image["path"]:
                mask_image = Image.open(mask_image["path"])

        mask_array = np.array(mask_image.convert("L"))
        unique_ids = np.unique(mask_array)
        height, width = mask_array.shape

        detected_objects = []
        masks_by_id = {uid: (mask_array == uid) for uid in unique_ids if
        uid in CHOLEC_CLASSES}

        for uid in unique_ids:
            if uid == 0 or uid not in CHOLEC_CLASSES:
                continue

            current_mask = masks_by_id[uid]
            rows, cols = np.where(current_mask)

            if len(rows) == 0:
                continue

            class_name = CHOLEC_CLASSES[uid]
```

```python
    # Compute bounding box scaled to [0, 1000] range
    y1 = int((rows.min() / height) * 1000)
    y2 = int((rows.max() / height) * 1000)
    x1 = int((cols.min() / width) * 1000)
    x2 = int((cols.max() / width) * 1000)

    # Detect tool-organ interactions
    interaction = ""
    if uid in TOOLS:
        structure = np.ones((21, 21), dtype=bool)
        dilated = binary_dilation(current_mask,
        structure=structure)

        for organ_id in ORGANS:
            if organ_id in masks_by_id and np.any(dilated & masks_
            by_id[organ_id]):
                interaction = f" interacting with {CHOLEC_
                CLASSES[organ_id]}"
                break

    obj_str = f"{class_name} <loc{y1:04d}><loc{x1:04d}><loc{y2:04d}
    ><loc{x2:04d}>{interaction}"
    detected_objects.append(obj_str)

return "\n".join(detected_objects) if detected_objects else "No
objects detected."

except Exception as e:
    logger.error(f"Error generating detection text: {e}")
    return "Error processing mask."
```

Core Logic Breakdown

Below, we will break down the logic of the function.

1. **Input Handling and Preprocessing**

   The function first ensures the input is a valid image by
   normalizing it. It converts the image to grayscale ("L") and gets
   the unique pixel values (object IDs) present in the image.

2. **Bounding Box Generation**

For every unique object ID found in the mask, it performs the following operations:

- **Coordinates:** It finds all pixels belonging to that object using `np.where(current_mask)`.

- **Extraction:** It calculates the bounding box (y_{min}, `x_{min}`, `y_{max}`, `x_{max}$`).

- **Scaling:** It scales coordinates to a range of **0 to 1000**. This is a common practice for vision language models so that coordinates are resolution-independent.

3. **Interaction Detection**

Because we have tools during the surgery, we need to check if a tool is touching an organ.

- **Dilation:** It takes the mask of a tool and "expands" its borders using `binary_dilation` with a 21 x 21 pixel brush.

- **Overlap Check:** It then checks if this expanded tool area overlaps with any organ mask using a bitwise AND (&).

- **Result:** If they overlap, it appends a string like `"interacting with gallbladder"`.

4. **Output Formatting**

Finally, it assembles the data into a specific string format:

```
[ClassName] <locY_min><locX_min><locY_max><locX_max> [Interaction]
Example Output:
Grasper <loc0250><loc0400><loc0450><loc0600> interacting with
Gallbladder
```

These are normalized coordinates with values between 0 and 1000. The below website discusses the concept of bounding boxes more in detail:

`https://developers.google.com/ml-kit/vision/object-detection`

Fine-Tuning Loop

Databricks offers many fine-tuning products. From the most automated product, which is Agent Bricks' Information Extraction and Custom LLM, to the fine-tuning experiments found in the Experiment tab, to the fully customized serverless GPU offering. In this case, we will be leveraging Databricks serverless GPUs because we are fine-tuning multimedia using a vision language model. Rest assured, serverless GPU can perform fine-tuning very efficiently. In this section, we will briefly introduce the concept, but a deeper dive can be found at Databricks' website:

```
https://www.databricks.com/blog/efficient-fine-tuning-lora-guide-llms
```

Most large language models today still share a common architecture called the Transformer. These models consist of many repeated transformer layers. Previously, fine-tuning required updating all parameters of a pre-trained model. For example, Qwen3-VL-2B contains 2 billion parameters, which requires significant computational resources to update.

LoRA (Low-Rank Adaptation of Large Language Models) is an enhanced fine-tuning technique that departs from the conventional approach of fully fine-tuning all weights in the pre-trained large language model's weight matrix. Instead, LoRA introduces two smaller low-rank matrices whose product approximates the weight update to the larger matrix. These smaller matrices constitute the LoRA adapter. Once fine-tuned, this adapter is incorporated into the pre-trained model for inference purposes.

QLoRA represents a refinement (by using quantization) of LoRA, prioritizing memory efficiency. It achieves this by quantizing the pre-trained base model weights to 4-bit precision (NF4), significantly reducing the memory footprint while keeping the LoRA adapter weights in higher precision for training. This reduction significantly diminishes the memory footprint and storage demands. In the QLoRA framework, the pre-trained model is loaded into GPU memory with quantized 4-bit weights, rather than the higher precision (e.g., 16-bit) typically used in standard LoRA. Figure 19-5 illustrates this process for a single layer.

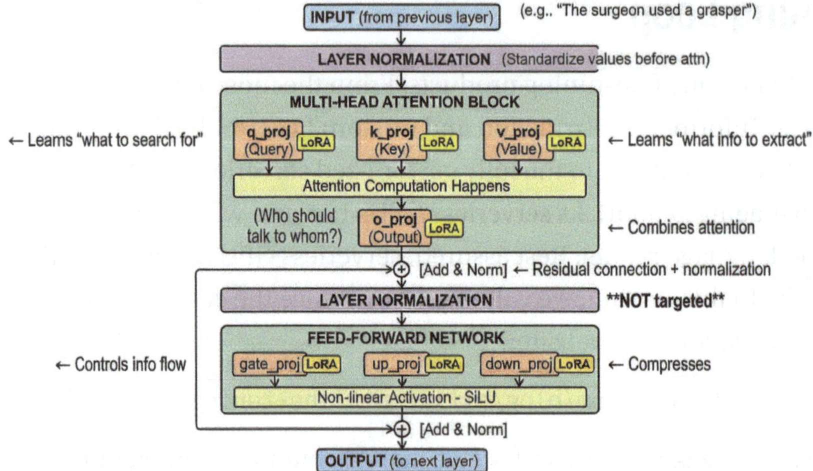

Figure 19-5. *Applying LoRA to a single transformer layer*

The Qwen3-VL-2B model has 32 transformer layers. As shown in Figure 19-5, each layer gets LoRA on 7 projections, including 4 in attention (q, k, v, o) and 3 in feed-forward (gate, up, down). That makes the total LoRA adapters equal to 32 layers × 7 projections = 224 adapters. Through this fine-tuning process, the 224 adapters can adapt to surgical terminology, recognize surgical instruments, understand medical context, and generate accurate bounding box coordinates, all while preserving the base model's general knowledge. Figure 19-6 illustrates the end-to-end fine-tuning process, combining a vision encoder with text tokens for vision language fine-tuning.

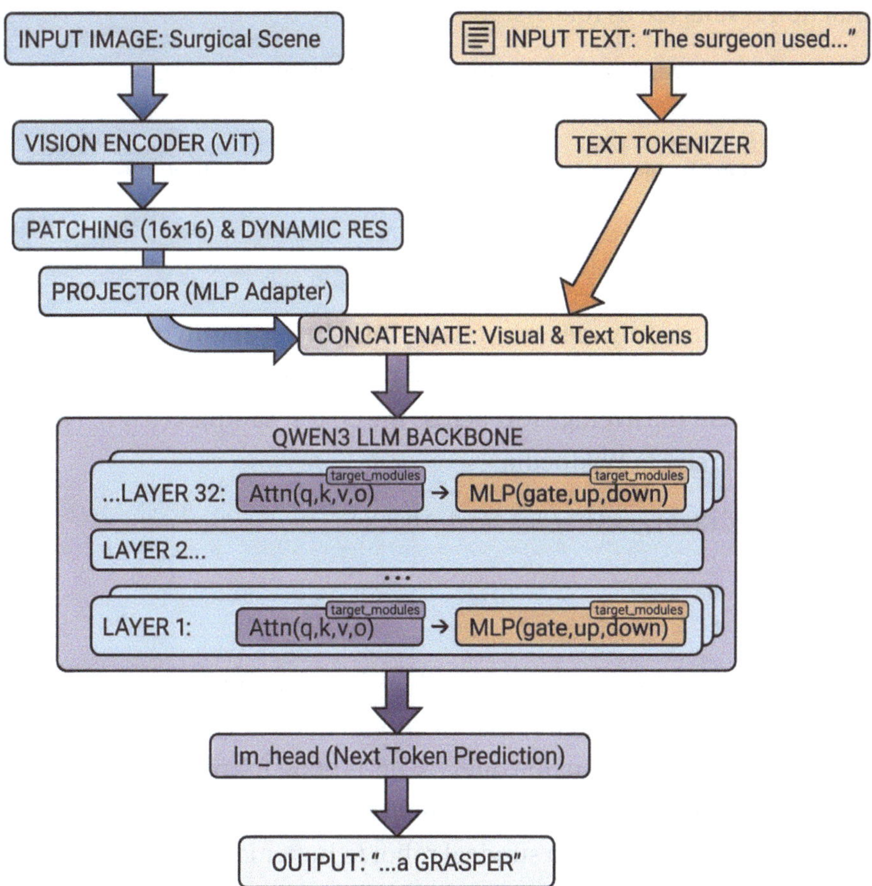

Figure 19-6. *End-to-end vision language fine-tuning process*

For development with one GPU, we opted to select 500 images for one label (e.g., liver ligament) as training data, as well as down-sample the image due to memory constraints. Despite that, we can still get some basic results that align with the human labeling. For example, with 50 test samples (450 train, 50 test), we can achieve the below metric in Table 19-3.

Table 19-3. *Eval metric for 50 test samples*

Precision	Recall	F1	Mean IOU
0.60	0.47	0.53	0.86

To prediction results are as follows:

- **Prediction**: Liver Ligament: <loc0660><loc0099><loc0999> <loc0899>

- **Ground Truth**: Liver Ligament <loc0622><loc0108><loc0995> <loc0893>

- **IoU**: 0.874

The accuracy is commonly measured using **IoU (Intersection over Union)** or the Jaccard Index. In this example, we achieved 0.874, which is close to 90% accuracy and slightly higher than the average IoU across the 50 test samples (0.86). For further explanations, we can refer to Wikipedia:

```
https://en.wikipedia.org/wiki/Jaccard_index
```

Deployment

As shown in Figure 19-4, deployment can be done in two different ways. The first is for batch inference, where we can register the model into Unity Catalog and infer the bounding box of a set of images in one query. This is good for post-surgical analysis and report writing. Another way to do it is to do it in real time as the surgery is ongoing; we can annotate the organ to help a surgical robot to better isolate the target organ for surgery, similar to the autonomous driving example in Figure 19-1.

PyFunc: The Generic Flavor for Custom Models

This final deployment process follows the pattern of a machine learning workflow. We can wrap the generic PyFunc model implementation in a class as shown in Listing 19-4. We can then use this class as a reference to log our model to Unity Catalog. The full class is not provided here for brevity, but the predict function is basically calling the ChatCompletion API, which can also be found on this book's GitHub repo.

Listing 19-4. Partial code for SurgicalVLMWrapper

```python
class SurgicalVLMWrapper(mlflow.pyfunc.PythonModel):
    """MLflow PyFunc wrapper for Surgical VLM"""

    def __init__(self, model=None, processor=None, model_path=None):
        self.model = model
```

```python
        self.processor = processor
        self.model_path = model_path
        self.device = None

    def load_context(self, context):
        from transformers import Qwen3VLForConditionalGeneration,
        AutoProcessor
        from peft import PeftModel
        import json

        model_path = context.artifacts.get("model_path")

        if torch.cuda.is_available():
            self.device = "cuda"
        else:
            self.device = "cpu"

        self.processor = AutoProcessor.from_pretrained(model_path, trust_
        remote_code=True)

        adapter_config_path = Path(model_path) / "adapter_config.json"

        if adapter_config_path.exists():
            with open(adapter_config_path) as f:
                adapter_config = json.load(f)
            base_model_name = adapter_config.get("base_model_name_or_path",
            "Qwen/Qwen3-VL-2B-Instruct")

            base_model = Qwen3VLForConditionalGeneration.from_pretrained(
          base_model_name,
          device_map="auto" if self.device == "cuda" else None,
          trust_remote_code=True,
          torch_dtype=torch.float16 if self.device != "cpu" else torch.
          float32,
            )
            self.model = PeftModel.from_pretrained(base_model, model_path)
        else:
            self.model = Qwen3VLForConditionalGeneration.from_pretrained(
```

```python
            model_path,
            device_map="auto" if self.device == "cuda" else None,
            trust_remote_code=True,
            torch_dtype=torch.float16 if self.device != "cpu" else torch.
            float32,
                )

        self.model.eval()

    def predict(self, context, model_input):
        import re
        import base64

        # System prompt MUST match training prompt
        system_prompt = """Detect surgical objects. Output format:
<ClassName> <loc####><loc####><loc####><loc####>
Classes: Liver, Gallbladder, Grasper, L-hook Electrocautery, Fat, Cystic
Duct, Blood, Connective Tissue, Abdominal Wall, Gastrointestinal Tract,
Hepatic Vein, Liver Ligament
Coords: 4-digit (0000-1000) as y1,x1,y2,x2. Tools add: interacting with
<Organ>"""
        user_prompt = "Detect all surgical instruments and anatomical
        structures. Provide precise bounding box coordinates."

        ... omitted for brevity ...

        return pd.DataFrame(results)
```

Once the class is defined, we can use it to define the model logging and registration mechanism as shown in Listing 19-5. This process should be nothing new to machine learning engineers who are used to deploying ML models in Databricks. But if you are new to this process, the MLflow documentation explains this PyFunc model flavor in detail:

```
https://mlflow.org/docs/latest/model/#pyfunc-model-flavor
```

Listing 19-5. Logging and registering the PyFunc model to UC

```
# Create wrapper
wrapper = SurgicalVLMWrapper(model=model, processor=processor, model_
path=final_dir)

# Define signature
input_schema = Schema([ColSpec("string", "image_path")])
output_schema = Schema([ColSpec("string", "raw_text"), ColSpec("string",
"detections")])
signature = ModelSignature(inputs=input_schema, outputs=output_schema)

# Example input
input_example = pd.DataFrame({"image_path": ["/path/to/surgical_
image.png"]})

# Log model as PyFunc
mlflow.pyfunc.log_model(
    artifact_path="surgical-vlm",
    python_model=wrapper,
    artifacts={"model_path": final_dir},
    pip_requirements=[
        "torch>=2.0.0",
        "transformers>=4.37.0",
        "accelerate>=0.26.0",
        "peft>=0.7.0",
        "qwen-vl-utils",
        "pillow>=9.0.0",
        "pandas>=1.5.0",
    ],
    input_example=input_example,
    signature=signature
)

run_id = mlflow.active_run().info.run_id
model_uri = f"runs:/{run_id}/surgical-vlm"

mlflow.set_registry_uri("databricks-uc")
registered_model = mlflow.register_model(model_uri, REGISTERED_MODEL_NAME)
```

After successfully registering the model, we can find it in Unity Catalog, as shown in Figure 19-7.

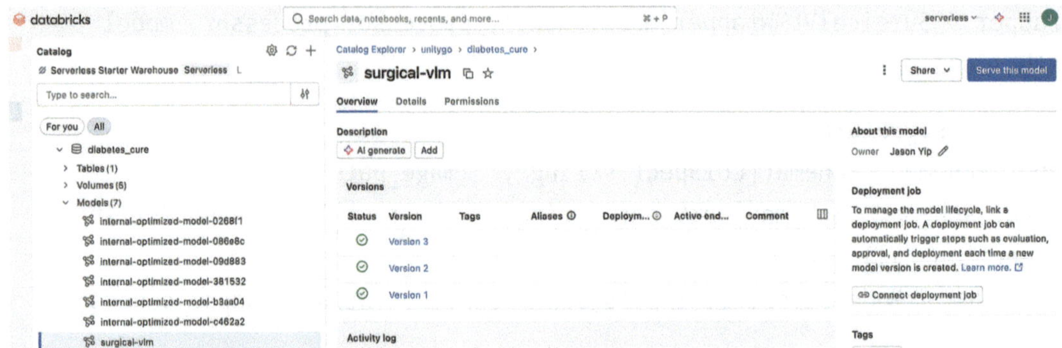

Figure 19-7. *Model registration found in Unity Catalog*

The model is now ready for batch inference! Listing 19-6 shows how to use the batch_decode function for batch inference. Later in this chapter, we will use the inference results and feed them into Genie to write a report as educational material for new surgeons.

Listing 19-6. Batch inference function

```
def batch_decode(model, processor, images, batch_size=8, max_new_
tokens=512):
    """

    Perform batch inference on multiple images
    """

    # System prompt (must match training)
    system_prompt = """Detect surgical objects. Output format:
<ClassName> <loc####><loc####><loc####><loc####>
Classes: Liver, Gallbladder, Grasper, L-hook Electrocautery, Fat, Cystic
Duct, Blood, Connective Tissue, Abdominal Wall, Gastrointestinal Tract,
Hepatic Vein, Liver Ligament
Coords: 4-digit (0000-1000) as y1,x1,y2,x2. Tools add: interacting with
<Organ>"""

    user_prompt = "Detect all surgical instruments and anatomical
    structures. Provide precise bounding box coordinates."
```

```python
# Prepare all messages at once (vectorized)
messages_batch = [
    [
        {"role": "system", "content": system_prompt},
        {"role": "user", "content": [
            {"type": "image", "image": img},
            {"type": "text", "text": user_prompt}
        ]}
    ]
    for img in images
]

# Apply chat template to all messages at once
texts = [
    processor.apply_chat_template(msgs, tokenize=False, add_generation_
    prompt=True)
    for msgs in messages_batch
]

# Process all images and texts in one batch
inputs = processor(
    text=texts,
    images=images,
    return_tensors="pt",
    padding=True
).to(model.device)

# Generate predictions for entire batch in parallel
with torch.no_grad():
    output_ids = model.generate(
        **inputs,
        max_new_tokens=max_new_tokens,
        do_sample=False
    )
```

```
    # Decode all predictions at once
    predictions = processor.batch_decode(
        output_ids[:, inputs.input_ids.shape[1]:],
        skip_special_tokens=True
    )

    return predictions

logger.info("Batch inference function defined")
```

Real-Time Serving

After we register the model, we can serve it using the blue button in the top-right corner of Figure 19-7. This button will create a serving infrastructure powered by Mosaic AI Model Serving. We can choose from CPUs to GPUs for serving. While the inference compute requirement is much lower than for model training, we still need to scale up compute if we want to do a lot of annotation at the same time. Just imagine that it might take a finite set of GPUs to train a large language model. When it is deployed to the public, demand can surge beyond what's required for model training. The configuration is shown in Figure 19-8.

Serving endpoints ›

Create serving endpoint

Experimenting with LLMs? Try pay-per-token Foundation Model APIs!

General

Name
Endpoint name cannot be changed after creation.

URL preview: https://adb-8333330282859393.13.azuredatabricks.net/serving-endpoints/

Served entities

Entity details   ×

Entity   **Version**   **Traffic (%)**

unitygo.diabetes_cure.surgical-vlm   3 ⌄   100

Validate this model's payload and dependencies. See how here ☐.

Compute type

CPU   ⌄

CPU	✓
GPU Small	T4
GPU Large	A100
GPU Large x2	2xA100
GPU Large x4	4xA100

Scale to zero
Not recommended for production use. Expect higher latency on the first request as the
endpoint scales up.

Figure 19-8. *Model serving user interface*

In Databricks, real-time inference can be achieved through two primary serving
options. The first option is to create a feature serving endpoint, which allows external
applications to fetch low-latency features via a REST API. This is ideal for cases where
the model logic resides outside of Databricks. Alternatively, we can use a model
serving endpoint to deploy a model directly within Databricks. In this setup, the model
automatically performs online feature lookups from Unity Catalog, eliminating the need
for manual feature-joining logic during inference.

Deep Research Agent

This is the final step of the project: write a post-surgical report using a research agent.
In Chapter 5, we learned that a deep research agent is basically a planning agent and a
ReAct agent wrapped around a control flow like LangGraph. If there are specific parts

443

of the research that we want to have tighter control over, this will be a very good tool when we understand what's in the data and how best to organize it. On the other hand, Databricks Genie is designed to experiment with the data using intelligent queries.

Without domain expert knowledge, we are just going to design a bare-basic schema and rely on the Databricks Genie research agent to decide how to make sense out of the data. Table 19-4 is the schema of our input table, which is basically the unprocessed data from our inference output.

Table 19-4. *Schema of the Genie input table*

Column	Description
frame	Represents the specific frame extracted from the video. This column contains a reference to the actual image.
coordinates	Contains the spatial coordinates related to the objects or features identified within the frame, essential for understanding their locations.
box_count	Indicates the total number of bounding boxes detected in the frame, providing insight into the density of objects present.

Earlier in this chapter, we discussed that the inference output is just coordinates like `<loc####><loc####><loc####><loc####>`. I'd argue that even with an experienced surgeon, it's very difficult to write something meaningful by looking at an array of coordinates without watching the video itself. We will attempt it anyway. Listing 19-7 is what we'd ask Genie.

Listing 19-7. Context provided to Databricks Genie for research

```
You are given tokenized 3D coordinates used in a specialized medical or
surgical system to precisely locate and track anatomical structures like
the Hepatic Vein and surgical instruments like the Grasper within a defined
spatial volume during laparoscopic cholecystectomy. Use the data provided
to determine what are the key points in the procedure to train next
generation surgeons to perform similar procedures.
```

Research Focus

For untrained non-medical professionals like the authors, it'd be extremely difficult to come up with a research focus given the coordinates. That's where the research agent gives value. Figure 19-9 shows that the agent decided to research the spatial movement, which naturally makes complete sense because we only provide the agent with coordinates. The hard part is to tie the movement to the phase of the surgery.

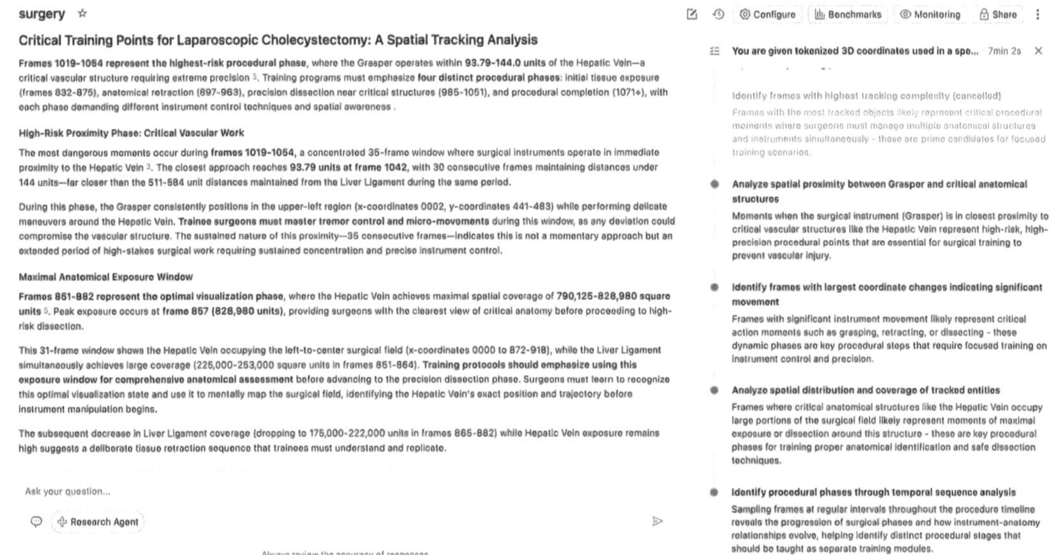

Figure 19-9. *Sample output of the Genie research agent*

Frame Analysis Results

Recall that our input schema simply lists the file name and the frame's inference output, without providing any context. Genie is smart enough to parse out the string, e.g., loc999, because we can't calculate anything with the extra "loc" term as a prefix and perform statistical analysis. Listing 19-8 is an excerpt from the agent output showing that the agent developed a four-phase procedural architecture.

Listing 19-8. Partial output from Genie research agent

Four-Phase Procedural Architecture

```
The complete procedure follows a clear four-phase temporal progression,
each requiring distinct surgical skills:
```

Phase 1: Initial Exposure and Positioning (frames 832-875) The Grasper operates in the mid-to-lower surgical field (y-coordinates 543-661) while establishing maximal Hepatic Vein exposure (x-coordinates extending to 714-902). This phase combines the high-movement repositioning identified in movement analysis with the maximal anatomical exposure window. **Trainees must learn aggressive but controlled tissue manipulation** to achieve optimal visualization before proceeding.

Phase 2: Anatomical Retraction (frames 897-963) The Grasper transitions upward (y-coordinates 603-635) while the Hepatic Vein progressively contracts (x-coordinates reducing from 756 to 614). This 66-frame sequence represents **deliberate tissue retraction to create working space** around critical structures. Surgeons must maintain steady retraction pressure while monitoring for vascular compromise.

Phase 3: Precision Dissection (frames 985-1051) The Grasper descends significantly (y-coordinates dropping from 549 to 462), moving into the high-proximity zone identified earlier where distances to the Hepatic Vein reach minimum values of 93.79-144.0 units. The Hepatic Vein re-expands (x-coordinates increasing from 660 to 764) as dissection proceeds. **This 66-frame phase demands absolute precision and represents the highest-skill portion of the procedure.**

Phase 4: Completion (frame 1071+) The Grasper returns to upper position (y-coordinate 640) while Hepatic Vein exposure reduces (x-coordinate 529), suggesting final verification steps or transition to closure. Instrument movements become less frequent as the critical work concludes.

Next Steps

With Genie Deep Research, we have successfully unlocked the spatial insights from the surgery. It's a good starting point. Any hospital would have the patient's guidelines and additional vitals during surgery, which are not visible in the video. We can easily connect Genie to additional data from the hospital's internal datasets governed by Unity Catalog. Optionally, we can connect the agent with a knowledge assistant for internal/external medical publications, which can unlock additional insights not known at the time of the surgery. The possibilities are endless.

Conclusion: From Pixels to Saving Lives

The Databricks Data Intelligence Platform not only provides the intelligence required to power next-generation applications, but it is also built on the Lakehouse architecture, providing a strong foundation for workloads of any size and complex data transformations. Throughout this chapter, we have successfully demonstrated the ability to create a sophisticated application from ideation to creation, all within the Databricks ecosystem.

- Fine-tune advanced vision language models like Qwen3-VL-2B-Instruct using efficient techniques such as QLoRA on serverless GPU infrastructure.

- Govern and manage both structured and unstructured data seamlessly through Unity Catalog.

- Deploy models flexibly for either high-volume batch inference or real-time REST serving powered by Mosaic AI.

- Unlock deep insights from specialized data using Databricks Genie, which can translate raw coordinates into meaningful procedural phases for training and analysis.

We moved beyond standard text-based chatbots to tackle a complex, multimodal challenge: fine-tuning a vision language model (Qwen3-VL-2B-Instruct) to interpret real-world surgery videos. By leveraging efficient techniques like QLoRA and serverless GPU infrastructure, we showed that even advanced fine-tuning is accessible without requiring massive, premium computing resources.

This workflow also highlighted the large language model lifecycle, which is similar to the traditional machine learning lifecycle. We utilized Unity Catalog to govern unstructured data, MLflow to register and manage model versions, and Model Serving to deploy these insights for both batch and real-time use. Finally, by feeding our inference results into Databricks Genie, we transformed raw, abstract coordinate data into actionable, critical training points to educate new surgeons.

This end-to-end journey illustrates that the potential of GenAI extends far beyond simple automation; it can enhance highly specialized fields, bridging the gap between raw data and expert human understanding. However, it remains challenging to achieve full automation, especially in healthcare settings. That's why Databricks has offered comprehensive toolsets, including GenAI evaluations, LLM judges, prompt registry, prompt optimization, and traces. Databricks' tight integration allows data teams and domain experts to stay on the same platform, collaborate closely, and iterate through feedback loops seamlessly. With Databricks, we can power this agentic era.

Index

A

Access control list (ACL), 308
Access control mechanisms, 318
Agent bricks, 12–13, 23, 61, 62, 80, 153
    Bespoke orchestration to integrated workflows, 58
    business value, 59
    categories, 39–41
    components, 157
    consumers, 157
    information extraction, 41–45
    optimization and production
        batch inference results, 47
        customer-agent interaction, 45
        evaluation, 46, 47
        performance, 46
        research and engineering, 48
        visual comparison graphs, 46
    review session, 161
    single trace endpoint, 157
    unstructured files, 50, 51
    use cases, 40
    user-facing perspective, 39
Agentic AI workflow, 87
Agent learning from human feedback (ALHF), 82, 162, 163, 165
AI/BI dashboards, 383, 384
AI/BI Genie
    agentic reasoning, 141
    architecture, 141, 142
    chat interface, 144
    definition, 141
    general instructions, 143
    instructions, 145, 146
    metadata, 143
    metadata editing, 145
    POS spaces, 142, 143
    Q/A with space, 147
    research, 147, 148
    space configurations, 144
    SQL statements, 143
    steps, 144
AI models' images, 17
AI-powered governance
    ai_classify SQL function, 138
    ai_mask function, 138
    comments, 135, 136
    data classification, 137, 138
    elements, 135
    judges, 139
    lineage, 136
    query in serverless SQL, 137
    security framework, 140
    transparency, 136
ai_query SQL function, 47
AI runtime, 208
ALHF, *see* Agent learning from human feedback (ALHF)
Anthropic's think tool, 96
Apache Spark, 210
Application performance monitoring (APM), 204
Approximate nearest neighbors (ANN), 67

Attribute-based access
     control (ABAC), 197
Auto loader, 274
Automated schema evolution, 300, 301
Auto-tracing, 237–239

B

batch_decode function, 440
Batches, 263
Batch mode, 282
Behavior drift, 158
BI, *see* Business intelligence (BI)
Big data, 1
Billing issues, 43
BootstrapFewShot, 257, 258
Bronze layer, 337
Built-in functions, 399
Business intelligence (BI), 377, 401, 402
Byte-pair encoding (BPE), 33

C

California Consumer Privacy
     Act (CCPA), 137
CDC, *see* Change data capture (CDC)
CDF, *see* Change data feed (CDF)
Chain, 63
Chain-of-thought, 155, 256
Change data capture (CDC), 299, 300,
     304, 344
Change data feed (CDF), 343, 344
ChatCompletion API, 157
ChatGPT, 17
Checkpointing, 270
Cholec80 dataset, 422
CholecSeg8k dataset, 422, 427, 428
Chunking

agentic, 76
definition, 72
document-based, 75
fixed-size, 73, 74
goals, 73
methods, 76
process, 73
recursive, 74
semantic, 74, 75
Chunks, 72
clone command, 342
CloudFiles, 274, 294
Cloud ingestion, 288–293, 303, 304
Cloud object storage layer, 333
Cluster access controls, 413
Cluster autoscaling, 413
Cluster auto-termination, 413
Cluster policies, 413
Cluster tags, 414
Cognitive architectures for language
     agents, 180
Cognitive memory, 179–181
Column-level masking, 197
Column-level security (CLS), 322
Column masks, 322, 323
Compute costs, 406
Concept drift, 158
Constraints
    categories, 388
    enforced, 389
    informational, 390, 391, 393
    primary key/foreign key, 388
    view relationships button, 392
Content moderation, 139
Context windows, 72
Continuous feedback
     mechanism, 154
Continuous mode, 177

T

U

V